Wild Plants for a Sustainable Future

110 multipurpose species

To Maite García-Urtiaga
for her vision, enthusiasm
and continuous support.

Wild Plants for a Sustainable Future

110 multipurpose species

Edited by Tiziana Ulian, César Flores, Rafael Lira, Avhatakali Mamatsharaga, Kebadire K. Mogotsi, Patrick Muthoka(†), Samodimo Ngwako, Desterio O. Nyamongo, William Omondi, Abdoul K. Sanogo, Sidi Sanogo and Efisio Mattana

Kew Publishing
Royal Botanic Gardens, Kew

First published in 2019 by the Royal Botanic Gardens,
Kew, Richmond, Surrey, TW9 3AB, UK
www.kew.org

ISBN 978 1 84246 673 5
e-ISBN 978 1 84246 674 2

Distributed on behalf of the Royal Botanic Gardens, Kew in North America by the University of Chicago Press, 1427 East 60th St, Chicago, IL 60637, USA.

British Library Cataloguing in Publication Data
A catalogue record for this book is available from the British Library.

Copy-editing and indexing: Sharon Whitehead
Scientific editing and proofreading: Steve Davis
Design and page layout: Nicola Thompson, Culver Design
Production management: Georgina Hills

For information or to purchase all Kew titles please visit shop.kew.org/kewbooksonline or email publishing@kew.org

Kew's mission is to be the global resource in plant and fungal knowledge, and the world's leading botanic garden.

Kew receives approximately one third of its running costs from Government through the Department for Environment, Food and Rural Affairs (Defra). All other funding needed to support Kew's vital work comes from members, foundations, donors and commercial activities, including book sales.

Printed and bound in Italy by L.E.G.O S.p.A.

Contents

IER / CRRA SIKASSO
IER
Project MGU/Useful Plants Project (UPP)
Conservation et gestion durable
des plantes utiles aux communautés rurales
ARBORETUM DE PLANTES MEDECINALES, ARTISANALES,
FOURRAGERES ET ALIMENTAIRES
DATE D'INSTALATION: JUILLET 2009
VILLAGE PARTENAIRE: COMMUNAUTE VILLAGEOISE DE IFOLA
SITE: IFOLA/COMMUNE RURALE DE FARAKALA

Editors

Tiziana Ulian[6], César Flores[4], Rafael Lira[4], Avhatakali Mamatsharaga[5], Kebadire K. Mogotsi[1], Patrick Muthoka(†)[2], Samodimo Ngwako[1], Desterio O. Nyamongo[2], William Omondi[2], Abdoul K. Sanogo[3], Sidi Sanogo[3] and Efisio Mattana[6]

Scientific editor

Steve Davis

Authors

Cecilia Amosso, Steve Davis, César Flores, Daniel Franco-Estrada, Willem Froneman, Ester Gaya, Pablo Gómez-Barreiro, Paulina Hechenleitner, Alex J. Hudson, Bokary A. Kelly, Martin López Carrera, Vusi Lukhele, Jabulani Mahlangu, Gloria N. Mashungwa, Albertinah T. Matsika, Efisio Mattana, Myrna Mendoza Cruz, David Meroka, Kebadire K. Mogotsi, Amanda Moreno Rodríguez, Patrick Muthoka(†), Veronicah Ngumbau, Samodimo Ngwako, Desterio O. Nyamongo, William Omondi, Victor Otieno, Fernando Peralta–Romero, Stéphane Rivière, Moctar Sacande, Lirio J. Sánchez Hernández, Abdoul K. Sanogo, Rokia Sanogo, Sidi Sanogo, Lucy Shai, Bamphitlhi Tiroesele, Seoleseng O. Tshwenyane, Tiziana Ulian, Karin van der Walt, Paul Wilkin and Guadalupe Zavala

Photographers

C. Amosso, P. Bakewell-Stone, P. Ballings, H. Beentje, R. v. Blittersdorff, P. Brena-Bustamante, J. Burrows, H. Cervantes-Maya, R. Dackouo(†), D. Franco-Estrada, W. Froneman, L. García, H. Gaya, P. Gómez-Barreiro, A. Heath, R. Heath, A. Hudson, R. Kesseler, L.-N. Le Roux, M. Legwaila, E. Mattana, A. McRobb, M. Meso, H. Migiro, K. K. Mogotsi, L. Mogotsi-Yimbo, O. Morebodi, N. Muema, T. Mukwevho, S. Ngwako, J. Nichols, A. Notten, D. O. Nyamongo, V. Otieno, F. Peralta-Romero, H. Pickering, A. K. Sanogo, S. Sanogo, J. Stevens, W. Stuppy, O. Téllez-Valdés, T. Ulian, C. Varet, K. van der Walt and B. Wursten

Contributing institutions

1. Botswana: Botswana University of Agriculture and Natural Resources (formerly Botswana College of Agriculture).
2. Kenya: Kenya Forestry Research Institute, Kenya Agricultural and Livestock Research Organization (formerly Kenya Agriculture Research Institute), National Museums of Kenya.
3. Mali: Institut d'Economie Rurale, Centre Régional de Recherche Agronomique de Sikasso.
4. Mexico: Facultad de Estudios Superiores Iztacala de Universidad Nacional Autónoma de México.
5. South Africa: Lowveld National Botanical Garden, South African National Biodiversity Institute.
6. UK: Royal Botanic Gardens, Kew.

Arboretum of medicinal plants in the village of Ifola, Mali. (Photo: S. Sanogo, IER.)

Suggested citation

How to cite this book:

Ulian, T., Flores, C., Lira, R., Mamatsharaga, A., Mogotsi, K. K., Muthoka, P., Ngwako, S., Nyamongo, D.O., Omondi, W., Sanogo, A. K., Sanogo, S., Mattana, E. (eds) (2019). *Wild Plants for a Sustainable Future*: *110 Multipurpose Species.* Royal Botanic Gardens, Kew.

How to cite a single species profile (example):

Ngwako, S., Mogotsi, K. K, Sacande, M., Ulian, T., Davis, S., Mattana, E. (2019). *Adansonia digitata* L. In: Ulian, T., Flores, C., Lira, R., Mamatsharaga, A., Mogotsi, K. K., Muthoka, P., Ngwako, S., Nyamongo, D.O., Omondi, W., Sanogo, A. K., Sanogo, S., Mattana, E. (eds) (2019). *Wild Plants for a Sustainable Future*: *110 Multipurpose Species.* Royal Botanic Gardens, Kew. pp. 6–9.

Acknowledgements

This book has been funded by Maite García-Urtiaga (MGU), a philanthropist based in Spain, as part of the 'MGU — Useful Plants Project' (UPP) managed by the Royal Botanic Gardens, Kew (RBG Kew). The editors and authors are deeply grateful to Maite García-Urtiaga, her three children and her niece Adriana, who made the compilation and publication of this book possible.

The valuable support of the local communities and their authorities as well as the project teams and RBG Kew's Millennium Seed Bank Partnership teams in Botswana, Kenya, Mali, South Africa, and Mexico is gratefully acknowledged, particularly the help of Patricia Davila (FESI-UNAM) from Mexico and Erich van Wyk (SANBI) from South Africa.

Many people at RBG Kew have supported this publication: Beth Ambrose, Emily Ambrose, Steven Bachman, Gabriele Bradamante, Richard Deverell, Mark Nesbitt, Tim Pearce, Hugh W. Pritchard, Martina Rogledi, Monique Simmonds, Michiel van Slageren, Wolfgang Stuppy, Robert Turner, Michael Way, Paul Wilkin and Katherine J. Willis. In particular, the editors would like to thank Moctar Sacande, Paul Smith and Roger Smith for their invaluable guidance throughout the project and for their contribution to the early development stage of this book. Finally, we would like to thank many colleagues who have provided technical and scientific support to the project, including staff of RBG Kew whose work has made the Millennium Seed Bank's collections possible, as well as the Kew Foundation, in particular Andrea Diez de Sollano, who helped with donor's liaison.

(†) Sadly, Patrick Muthoka, from Kenya, an editor and author of this work, passed away before the final publishing of this book. We would like to thank Patrick for his valuable contribution from the very first stages of this project and dedicate this work to his family and colleagues at the National Museums of Kenya. We are also deeply saddened by the premature departure of Réné Dackouo, from Mali, who helped with the identification of plants and kindly provided some of the photographs for the Mali chapter.

Disclaimer

Forewords

Maite García-Urtiaga

In 1939, my family and I arrived in Mexico as refugees escaping the Spanish Civil War. I was only a child then, and Mexico's colourful rich and exotic culture had a huge impact on me. My father, a man fascinated by nature, who throughout his life planted thousands of trees wherever he went, played a key role in raising our awareness.

I was especially surprised by the relevance that the indigenous population gave to plants. Through local people, we were introduced to all sorts of remedies made from medicinal plants: infusions, pastes and sometimes leaves that they would place on their brows. This extraordinary knowledge was not written in text books at the time, it was simply passed on from generation to generation.

Years later, while attending boarding school in Canada, I read a book that not only fascinated me but connected the dots with my curiosity for Mexican plants. The story described a young woman lost in the desert, who survived by eating the plants and roots she found.

Late in my life, and after much research, I came across the Royal Botanic Gardens, Kew (RBG Kew). I found in this wonderful institution, the vessel to carry out my lifelong dream. I wanted to donate the necessary funding to help guarantee the preservation of and access to the planet's medicinal plants for future generations.

My project was received with enthusiasm and the MGU — Useful Plants Project was born. I am proud to say that RBG Kew's Millennium Seed Bank Partnership has done a superb job and this fills me with pride. This publication will allow you to further understand our project and its intentions.

With my appreciation for all that has been accomplished by the MGU — Useful Plants Project and the hope that we all go on taking care of our beautiful planet.

Maite García-Urtiaga

Roger Smith

Globally, a very wide range of undomesticated plants sustains the well-being of humans who depend on low-cash livelihoods. However, people who are not reliant upon them often underestimate the value of these plants. These valuations vary within and between communities and through time. An extreme example, though pertinent here, would be the greatly increased use of native plants for many purposes recorded during the two-and-a-half year siege of Madrid during the Spanish Civil War. When times are difficult, Europeans too fall back on a much wider range of plant diversity for their livelihoods than might be thought.

Although necessary and laudable, conserving the genetic resources of these valued species in seed banks for future generations does little to sustain and improve the livelihoods of the present generation who use them regularly. Yet their stewardship will determine whether the use of such species is sustainable or leads to their loss through over use. The need to better balance the interests of present-day stewards with concerns for future generations had become apparent in the early days of the Kew seed bank when collecting trips overseas became more usual. The problem was how to achieve this balance without maintaining any long-term presence in-country.

The funding of RBG Kew's Millennium Seed Bank Project (now Partnership, MSBP) with its wide-ranging network of partners, many of whom were from countries where a large proportion of the population live in low-cash situations, inadvertently created the first element to address this problem. The MSBP now had, through its partnership, the possibility to undertake long-term projects in-country that built on seed conservation skills to meet other aspirations of the UN Convention on Biological Diversity.

Maite García-Urtiaga knowingly created the second element. Through her experiences, so elegantly described in her foreword, she has learned to value these species and desires that their usefulness is recorded for posterity. Through her research, she saw the possibility that the MSBP could conserve the genetic resources of these species for future generations while also recording information about their use and value. Through her vision, Maite encouraged the MSBP to step outside its usual practice and work directly with the present-day stewards of these useful species. Her hope was that the use of these species would become more sustainable if any problems encountered in their propagation could be overcome using the knowledge and skills developed through banking seed. Through her generosity, she provided the encouragement and financial resources that would enable this to happen.

The third element came from RBG Kew partners in Botswana, Kenya, Mali, Mexico and South Africa. These partners stepped forward to meet Maite's challenge based either on their pre-existing experience of working with communities in low-cash economies or their willingness to develop such skills quickly.

The fourth element reflects the considered way in which low-cash communities steward the useful species they value highly. After considerable consultation and discussion, these communities recognised the value to them of what was being proposed and committed time and energy — their most prized assets — to 'The MGU — Useful Plants Project'. What was learned together is recorded here and presented in a way that is aimed at practitioners not scholars. It is for those who wish to act rather than anguish.

I share with Maite great pride in how much this project has achieved since our first discussions about what might be possible. My thanks and admiration are due to all the former colleagues and partners who have, directly or indirectly, contributed to this success. However, special thanks must be offered to Tiziana Ulian, who throughout the project has skilfully shepherded the many different personalities involved in one direction, despite their initial inclination to tred other paths!

Roger Smith

Former Head of the Seed Conservation Department

Royal Botanic Gardens, Kew

Paul Wilkin

The concept of Natural Capital has recently started to provide a new rationale for the conservation of biodiversity. The wealth of life on Earth and the genetic diversity that underpins it form a set of natural assets that provide ecosystem services (human benefits) to people. For example, the trees in and around cities help reduce air pollution, provide clean water, stabilise soil, regulate local and global climate via carbon capture and enhance health and well-being. The value of these services is becoming increasingly apparent and can often be quantified. In rural contexts and especially in less-developed economies, people are much more extensively and directly reliant on wild to managed plants and fungi (and other natural assets) for resource provision.

As a result, the department that I lead at RBG Kew, Natural Capital and Plant Health, uses collections-based science to identify and evaluate the roles that plants and fungi play in providing services to humankind. In particular, it seeks to address overarching issues that include ecosystem service provision, the effects of climate change, resilience and resource security. It does so through a portfolio of different international projects such as 'The MGU — Useful Plants Project' (UPP). Thanks to the generosity of Maite García-Urtiaga, RBG Kew was able to develop, in collaboration with partner institutions and communities in five different countries, a project developing the understanding of plant use in conjunction with the conservation of resource- and ecosystem-service-providing species. The creation of an inventory of natural capital assets, including this book, is a key step towards enhancing the lives and livelihoods of the inhabitants of UPP project communities and beyond.

Paul Wilkin

Head of Natural Capital and Plant Health Department

Royal Botanic Gardens, Kew

Richard Deverell

Many rural communities are dependent upon natural vegetation for their everyday needs such as food, medicine, fuel and building materials. However, these plants, so critical to the daily well-being of billions of people, face a range of threats. These include climate change, over-exploitation, droughts, habitat loss and the impact of invasive species.

At RBG Kew, we document and conduct research into global plant and fungal diversity and its uses for humanity. We do this in partnership with many organisations around the world and we draw upon our own extensive scientific collections and those of partner institutions.

This publication is a wonderful example of our work and is one of the many invaluable outcomes of 'The MGU — Useful Plants Project' (UPP). Since 2007 this project has been working with partners and rural communities in Botswana, Kenya, Mali, Mexico and South Africa to conserve and sustainably use indigenous plants that are important to human well-being. This has been achieved through their conservation in seed banks, propagation in community nurseries and planting in community gardens, woodlands and forests, each supported by research.

As a result of the UPP, we have been able to compile multi-disciplinary information on 110 useful species (relating, for example, to seed conservation, propagation and traditional uses), all of which is presented with great clarity to optimise its value for practitioners working in agriculture, forestry and conservation.

None of this would have happened without the generosity of Maite García-Urtiaga, to whom we are all tremendously grateful. Equally vital was the invaluable support of our many partners around the world, without whom this project would have proved impossible. This publication is testament to that collaborative effort; it involved 13 editors and 40 authors drawn from Kew and five countries. I thank Maite García-Urtiaga and our many partners; we are greatly indebted to you.

This book is a perfect example of Kew's mission to be the global resource for plant and fungal knowledge. It contributes to our knowledge of plants and fungi, upon which all of our lives depend, and it directly supports the UN Sustainable Development Goals "to end poverty, protect the planet and ensure prosperity for all".

Richard Deverell

Director of the Royal Botanic Gardens, Kew

Symbols, abbreviations and acronyms

SYMBOLS

°	degrees
>	greater than
<	less than
±	more or less

ABBREVIATIONS

a.s.l.	above sea level (i.e. altitude)
c.	*circa*, about
IBA	indole-3-butyric acid (a rooting hormone)
MC	moisture content (of seeds)
PD	physiological dormancy (of seeds)
pH	potential of hydrogen, a measure of acidity or alkalinity, where pH7 is neutral, lower values represent increasing acidity, and higher values represent increasing alkalinity
ppm	parts per million
PY	physical dormancy (of seeds)
PY+PD	combinational dormancy, consisting of both physical and physiological dormancy (of seeds)
RH	relative humidity, a measure of the amount of water vapour in the air
sp.	species (singular)
spp.	species (plural)
subsp.	subspecies
var.	variety

ACRONYMS

BUAN	Botswana University of Agriculture and Natural Resources
CITES	Convention on International Trade in Endangered Species of Wild Fauna and Flora
CRRA	Centre Régional de Recherche Agronomique, Mali
DMT	Département de Médecine Traditionelle, Mali
DNEF	Direction Nationale des Eaux et Forêts, Mali
FESI-UNAM	Facultad de Estudios Superiores Iztacala, Universidad Nacional Autónoma de Mexico
ICDT	Ithuseng Community Development Trust, Botswana
IER	Institut d'Economie du Mali
IRNSP	Institut National de Recherche en Santé Publique (National Institute for Research in Public Health), Mali
IUCN	International Union for Conservation of Nature and Natural Resources
LNBG	Lowveld National Botanical Garden, SANBI, South Africa
MDRE	Ministère du Développement Rural et de l'Environnement, Mali
MEWT	Ministry of Environment, Wildlife and Tourism, Botswana
MSB	Millennium Seed Bank, RBG Kew, UK
MSBP	Millennium Seed Bank Partnership
OOAS	Organisation Ouest Africaine de la Santé, Burkina Faso
RBG Kew	Royal Botanic Gardens, Kew, UK
SANBI	South African National Biodiversity Institute
UNAM	Universidad Nacional Autónoma de Mexico
UNISA	University of South Africa
UPP	MGU — Useful Plants Project
WAHO	West African Health Organization, Burkina Faso

Introduction

by Efisio Mattana, Alex J. Hudson, Steve Davis and Tiziana Ulian

The MGU — Useful Plants Project

We live in the so-called Anthropocene age when human activities are driving global climate and environmental changes. Directly and indirectly, this is causing the rapid extinction of plant diversity, faster than during any period during the recent geological past. This extinction marks a global threat to many rural communities who, particularly in Africa and Latin America, still rely on wild plants for food, medicine and income generation. Local knowledge of traditional plant uses is disappearing as rapid cultural changes occur, strongly influenced by urbanisation and globalisation. The loss of these species and knowledge reduces the options for dealing with future environmental challenges. To counteract this effectively, the conservation of useful plants should not be limited to the preservation of genetic resources, but should include the gathering and cataloguing of traditional knowledge for the benefit of future generations, as well as the generation of information to enable the conservation and sustainable utilisation of these plants.

In order to address these critical challenges, the Royal Botanic Gardens, Kew (RBG Kew) documents and conducts research into global plant and fungal diversity and its uses for humanity [7] through a portfolio of international science projects, such as the MGU — Useful Plants Project (UPP). The UPP started in 2007 under the Millennium Seed Bank Partnership (MSBP) led by RBG Kew. The MSBP is a global network that focuses on conserving bankable (desiccation tolerant) seeds of wild plants that are most at risk and most useful for the future. The UPP was set up to conserve and use sustainably wild plants that are important for rural communities in Botswana, Kenya, Mali, South Africa and Mexico. A scientific approach was applied throughout the main components of the project: selecting useful plant species; *ex situ* conservation; plant propagation and planting; and supporting people's livelihoods. In each country, the project was led by institutional partners and brought together Kew scientists and a wide range of experts from different disciplines, including botanists, horticulturists, agronomists and foresters, who worked with the rural communities, local authorities and schools using a participatory approach [11].

The book

This book is one of the main dissemination achievements of the UPP and the result of a large collaborative effort working with international partners and rural communities. It is aimed at practitioners from governmental institutions and NGOs, working in Africa and Latin America. It aims to promote the conservation and sustainable use of wild multipurpose species in conservation, agriculture and forestry projects addressing the UN Sustainable Development Goals (https://sustainabledevelopment.un.org/sdgs). The book includes five chapters, one for each project country, featuring a total of 110 technical species profiles (20 species per country, except for 30 species in Kenya). These species were identified and selected for the UPP as being among the most important useful plants for the rural communities. Most of them are trees and/or shrubs (85 in total), with the rest being cacti (six species for Mexico), subshrubs or perennial or annual herbs. For Botswana, a species of truffle, a fungus, is also included. The overall habit of all the species included in the book is summarised in the Annex. The profiles have been compiled by experts in each country in collaboration with those at RBG Kew. The result is a synthesis of information for each species (taxonomy and nomenclature, plant description, fruits and seeds, distribution, habitat, uses, known hazards and safety, conservation status, seed conservation, propagation and trade) gathered within the framework of the project, and combined with information available in-country and in the scientific literature. The species profiles are structured in a consistent format, so they can be read individually and information can be easily accessed. The structure of the profiles was inspired by that of the 'seed leaflets' generated as an outcome of the 'Darwin Initiative Research Exercise on Community Tree Seeds (DIRECTS)' project (Ref. Number 162/12/001; 2003-2006), which was led and managed by RBG Kew [6]. These leaflets were published on the internet through the Working Programme of the DANIDA Forest Seed Centre (available at http://sl.ku.dk/rapporter/seed-leaflets/). A glossary of botanical and technical terms is included at the end of the book. Definitions of botanical terms mainly follow the second edition of *The Kew Plant Glossary* [2].

The arrangement of the profiles within each chapter is alphabetical according to genus and species names. The currently accepted scientific name of each species (at January 2018) follows the botanical nomenclature of the *World Checklist of Selected Plant Families* [13]. For plant families not yet included in the *World Checklist*, scientific names were checked using *The Plant List* [10]. Relevant Floras and monographic treatments were also consulted. In most cases, only the main synonyms (alternative names that are not the currently accepted names) are included. Full synonymy in each case can be found by consulting the *World Checklist* or *The Plant List*. The accepted scientific names are also highlighted in bold in the Index of scientific names at the end of the book.

The sub-family classification of legumes (Leguminosae) follows the Legume Phylogeny Working Group [5]. Names of *Acacia* sensu lato follow Kyalangalilwa *et al.* [4].

The main vernacular or common names of plants are given in the languages mostly used by the partner communities involved and in the regions concerned. English names are given where possible. The online version of *Ethnologue* [9] was used as the main reference for language names.

A brief botanical description provides the main characteristics of each species, including their life form, habit and the main distinctive characteristics of their leaves and flowers. Fruits and seeds are described in detail, along with their time of availability, to help practitioners identify and sustainably collect them at the best time. Whenever possible, photographs of the external views and internal sections of the diaspores are also provided.

Distribution and habitat details are included to help guide practitioners on the climatic and ecological conditions that each species requires; these are important considerations when planting into new localities. Unless otherwise stated, distribution, habitat and altitude ranges pertain to the full (global) range of the species, rather than to the individual country in whose chapter the species profile is presented.

Data on uses of the plants and their products focus on those of most relevance to the country concerned. The data are a summary of the main uses — the aim being to increase awareness, to support perpetuation of traditional uses, to promote those species of importance for local communities, and to encourage propagation of the plants and their use in reforestation and agriculture projects. Information on known hazards (e.g. toxicity) and safety is provided. For each species, the main categories of uses presented in the species profiles of this book (standardised according to Ulian *et al.* [11]) are summarised in the Annex.

The conservation status of each species is included using categories assigned in *The IUCN Red List of Threatened Species* [3] or in national or regional Red Lists at the time of our consultation (i.e. up to January 2018).

The section on seed conservation provides information on seed harvesting, processing and handling, seed storage and viability for each species, derived from literature, acquired by first hand experience by the authors and extracted from the RBG Kew Seed Bank Database (SBD). An explanation is given on how to collect the fruits and/or seeds, how best to separate seeds from fruits, or whether the whole fruit can be stored as the conservation unit. When available, the seed storage behaviour of the species is reported [8], identifying those species (i.e. desiccation-tolerant species) whose seed can be stored at sub-zero temperatures potentially for long-term conservation and those (desiccation-sensitive species) whose seeds can only be stored moist for short-term conservation at temperatures above 0°C. Seed dormancy classification follows the system proposed by Baskin & Baskin [1].

Information on how to propagate the species, by seeds and/or by vegetative propagation is provided, with particular focus on pre-treatments needed to break seed dormancy and optimum germination conditions in the laboratory and nursery.

For Mali, the results of the phytochemical analyses carried out by the Département de Médecine Traditionelle/Institut National de Recherche en Santé Publique (DMT/INRSP) as part of the UPP are also reported for the investigated species.

Finally, information on trade in the species and/or of their products at local, national and international levels is reported. If a species is CITES-listed (i.e. included on one of the CITES Appendices) this is also noted (UNEP-WCMC [12]).

Photographs were provided by many collaborators who are acknowledged throughout the book.

References

[1] Baskin, J. M. & Baskin, C. C. (2004). A classification system for seed dormancy. *Seed Science Research* 14 (1): 1–16.

[2] Beentje, H. (2016). *The Kew Plant Glossary: An Illustrated Dictionary of Plant Terms*, 2nd edition. Royal Botanic Gardens, Kew.

[3] IUCN (2017). *The IUCN Red List of Threatened Species.* Version 2017-3. http://www.iucnredlist.org/

[4] Kyalangalilwa, B., Boatwright, J. S., Daru, B. H., Maurin, O. & Van der Bank, M. (2013). Phylogenetic position and revised classification of *Acacia* s.l. (Fabaceae: Mimosoideae) in Africa, including new combinations in *Vachellia* and *Senegalia*. *Botanical Journal of the Linnean Society* 172 (4): 500–523.

[5] Legume Phylogeny Working Group (2017). A new subfamily classification of the Leguminosae based on a taxonomically comprehensive phylogeny. *Taxon* 66 (1): 44–77.

[6] Pritchard, H. W., Vautier, H., Sarkis, G. & Sacande, M. (2006). *Darwin Initiative Research Exercise on Community Tree Seeds (DIRECTS). Ref. Number 162/12/001. Final Report.* http://webarchive.nationalarchives.gov.uk/20130907181145/http://darwin.defra.gov.uk/wp-content/uploads/documents/12001/5231/12-001%20FR%20-%20edited.pdf

[7] Royal Botanic Gardens, Kew (2015). *A Global Resource for Plant and Fungal Knowledge — Kew's Science Strategy (2015–2020).* The Board of Trustees of the Royal Botanic Gardens, Kew. https://www.kew.org/science/who-we-are-and-what-we-do/kews-science-strategy

[8] Royal Botanic Gardens, Kew (2017). *Seed Information Database (SID).* Version 7.1. http://data.kew.org/sid/

[9] Simons, G. F. & Fennig, C. D. (eds) (2017). *Ethnologue: Languages of the World, Twentieth edition.* SIL International, Dallas, Texas, USA. http://www.ethnologue.com

[10] *The Plant List* (2013). Version 1.1. http://www.theplantlist.org/

[11] Ulian, T., Sacandé, M., Hudson, A. & Mattana, E. (2017). Conservation of indigenous plants to support community livelihoods: the MGU – Useful Plants Project. *Journal of Environmental Planning and Management* 60 (4): 668–683.

[12] UNEP-WCMC (2017). *The Checklist of CITES Species Website.* CITES Secretariat, Geneva, Switzerland. Compiled by UNEP-WCMC, Cambridge, UK. http://checklist.cites.org

[13] WCSP (2017). *World Checklist of Selected Plant Families.* Facilitated by the Royal Botanic Gardens, Kew. http://wcsp.science.kew.org/

Participants at 'The MGU — Useful Plants Project' (UPP) workshops in Bamako, Mali (2010), above, and at Kew's Millennium Seed Bank in the UK (2014), below. (Photos: IER and W. Stuppy, RBG Kew.)

Student learning how to identify wild plants with a traditional healer in the Tsetseng community.
(Photo: T. Ulian, RBG Kew.)

BOTSWANA

Adansonia digitata L.

TAXONOMY AND NOMENCLATURE

FAMILY. Malvaceae

SYNONYMS. *Adansonia bahobab* L., *A. baobab* Gaertn., *A. integrifolia* Raf., *A. scutula* Steud., *A. situla* (Lour.) Spreng., *A. somalensis* Chiov., *A. sphaerocarpa* A.Chev., *A. sulcata* A.Chev.

VERNACULAR / COMMON NAMES. Baobab, cream of tartar tree, dead-rat tree, monkey-bread tree, upside-down tree (English); ibozu, mbuyu, moana, mobuyu, mowana, muwane, (Tswana); omuzu (Herero, Namibia); mGuya (Kalanga); omukwa (Ndonga, Namibia); omikwa (Oshikwanyama, Namibia); mauyu (Shona); shimuwu (Tsonga).

PLANT DESCRIPTION

LIFE FORM. Tree.

PLANT. Deciduous, usually not more than 20 m high (but can reach 28 m), with a hugely swollen trunk <10 m in diameter (28 m for very old individuals or those in favourable sites); primary branches stout but gradually tapering, young branches often tomentose, rarely quite glabrous (Figure 1); bark smooth, sometimes heavily folded, grey brown to reddish brown. **LEAVES.** At ends of branches, compound, with 3–9 leaflets, each 5–15 cm long, elliptic to widely lanceolate, somewhat acuminate, smooth, dark green when mature. **FLOWERS.** Solitary, pendulous, large (<20 cm in diameter), showy, white, petals 5, curled, 6–10 x 7–12 cm, calyx lobes velvety and curved; stamens in a column dividing into many, showy filaments [5,7,13,14,19].

Figure 1 – *A. digitata* tree. (Photo: S. Ngwako, BUAN.)

Figure 2 – Fruit and seeds of *A. digitata*. (Photo: P. Gómez-Barreiro, RBG Kew.)

FRUITS AND SEEDS

Fruits (Figure 2) are hard woody capsules 8–20 cm long, densely covered in short, velvety, yellowish hairs [5,13], remaining on the tree for up to a year [13]. Seeds (Figure 3) many (>100 per fruit), surrounded by a powdery white pulp (cream of tartar) [20].

FLOWERING AND FRUITING. Flowering primarily occurs before the start of the rainy season (in Southern Africa the main flowering period is from October to December) [5]. The flowers are pollinated by bats and bushbabies [2]. Fruits mature 5–8 months after pollination [17].

Figure 3 – External views of the seeds and their longitudinal section showing the embryo.
(Photo: P. Gómez-Barreiro, RBG Kew.)

DISTRIBUTION

Widely distributed throughout sub-Saharan Africa; introduced to Madagascar, India and Indian Ocean islands [3,5,7].

HABITAT

Dry bushland, woodland and wooded grassland; near settlements and often left standing in cultivated areas [2,10,21]. Usually at low altitudes (<1,000 m), on well-drained soils, with mean annual rainfall of 100–1,000 mm, in frost-free areas with a mean annual temperature of 20–30°C [3].

USES

An important multipurpose tree. The fruits and seeds are eaten by local people and are a rich source of vitamin C [3,14]. Fruits are eaten directly after picking, or are sliced and mixed with fresh milk. Seeds can be roasted and used as a substitute for coffee. Subiya women chew the bark and use the resulting decoction to wash babies for the first three months of their life [11]. The white pulp is mixed with water to make a pleasant drink. It has also been used in traditional medicine to treat fevers, diarrhoea and haemoptysis [20]. The tree has many other traditional medicinal uses, and is culturally important in many communities. Bark fibre is used for making paper, rope and clothing [3,14]. Hollow trunks have been used as houses, prisons, storage barns and shelters. The leaves, fruits and bark are eaten by primates, elephants and many other animals [5,14].

KNOWN HAZARDS AND SAFETY

No known hazards.

CONSERVATION STATUS

Baobab is often protected by local communities and retained in cultivated areas on account of its social and economic value. In parts of its range, including Botswana, Namibia and South Africa, baobab is afforded protection under forestry legislation [4,6,16]. Assessed as Least Concern (LC) in South Africa according to IUCN Red List criteria [8], but yet to be assessed in Botswana and to have its status confirmed in the IUCN Red List [12,18].

SEED CONSERVATION

HARVESTING. The fruits can be collected from the crown of standing trees by using a hook to cut the hanging fruits or by climbing into the crown [3].

PROCESSING AND HANDLING. Seeds can be extracted by hitting the fruit on a hard surface to break it open. The mixture of seeds and pulp is then soaked in water to remove the pulp by gently squashing and floating in water [3]. The seeds can then be dried in the sun for two days.
STORAGE AND VIABILITY. X-ray analysis carried out on seed lots stored at the MSB highlighted high percentages of filled seeds (c. 90%). Seeds of this species are reported to be desiccation tolerant [15]. They can be stored for up to four years at room temperature without significant loss of viability [17].

PROPAGATION

SEEDS

Dormancy and pre-treatments: Seeds of this species exhibit physical dormancy [1], determined by the presence of a chalazal locule [9]. They therefore need to be scarified (filed) before sowing [15].
Germination, sowing and planting: Seeds can be sown in beds or containers. Germination is epigeal and takes 15–40 days. Seedlings need to be 3–4 months old and a height of 30–40 cm (Figure 4) before transplanting [17]. In the laboratory, filed seeds can be sown at temperatures of c. 25°C, with a photoperiod of 12 hours of darkness and 12 hours of light, to achieve >75% germination [15].

VEGETATIVE PROPAGATION

Can be propagated by stem cuttings and grafting [3].

TRADE

A small amount of informal trade of the fruits is carried out by local communities in Botswana.

Figure 4 – A 7-year-old tree in the BUAN garden, and a sapling at Tsetseng.
(Photos: S. Ngwako, BUAN; C. Varet, The Ivory Foundation, Paris.)

Authors

Samodimo Ngwako, Kebadire K. Mogotsi, Moctar Sacande, Tiziana Ulian, Steve Davis and Efisio Mattana.

References

[1] Baskin, C. C. & Baskin, J. M. (2014). *Seeds: Ecology, Biogeography, and Evolution of Dormancy and Germination*, 2nd edition. Academic Press, San Diego, USA.

[2] Beentje, H. & Goyder, D. (eds) (undated). *Adansonia digitata* (baobab). Royal Botanic Gardens, Kew. http://www.kew.org/science-conservation/plants-fungi/adansonia-digitata-baobab

[3] Bosch, C. H., Sié, K. & Asafa, B. A. (2004). *Adansonia digitata* L. In: *PROTA 2: Vegetables/ Légumes.* [CD-Rom]. Edited by G. J. H. Grubben & O. A. Denton. PROTA, Wageningen, The Netherlands.

[4] Botswana Government (2009). *Botswana Fourth National Report to the Convention on Biological Diversity*. Government of Botswana, Gaborone, Botswana.

[5] Coates Palgrave, K. & Coates Palgrave, M. (2002). *Trees of Southern Africa*, 3rd edition. Struik Publishers, Cape Town, South Africa.

[6] Curtis, B. A. & Mannheimer, C. A. (2005). *Tree Atlas of Namibia*. National Botanical Research Institute of Namibia, Windhoek, Namibia.

[7] Exell, A. W. & Wild, H. (eds) (1961). Bombacaceae. In: *Flora Zambesiaca*, Vol. 1, part 2. Flora Zambesiaca Managing Committee, Royal Botanic Gardens, Kew.

[8] Foden, W. & Potter, L. (2005). *Adansonia digitata* L. In: *National Assessment: Red List of South African Plants*. Version 2015.1. http://redlist.sanbi.org/

[9] Gama-Arachchige, N. S., Baskin, J. M., Geneve, R. L. & Baskin C. C. (2013). Identification and characterization of ten new water gaps in seeds and fruits with physical dormancy and classification of water-gap complexes. *Annals of Botany* 112 (1): 69–84.

[10] Hankey, A. (2004). *Adansonia digitata* L. South African National Biodiversity Institute (SANBI), South Africa. http://pza.sanbi.org/

[11] Heath, A. & Heath, R. (2010). *Field Guide to the Plants of Northern Botswana Including the Okavango Delta*. Royal Botanic Gardens, Kew.

[12] IUCN (2017). *The IUCN Red List of Threatened Species*. Version 2017-3. http://www.iucnredlist.org/

[13] Moll, E. (2011). *What's That Tree? A Starter's Guide to Trees of Southern Africa*. Struik Nature, Cape Town, South Africa.

[14] Royal Botanic Gardens, Kew (1999–2015). *Survey of Economic Plants for Arid and Semi-Arid Lands (SEPASAL) Database*. http://apps.kew.org/sepasalweb/sepaweb

[15] Royal Botanic Gardens, Kew (2017). *Seed Information Database (SID).* Version 7.1. http://data.kew.org/sid/

[16] SA Forestry Online (2015). *Protected Trees in South Africa (2015). List of Protected Trees Under The National Forest Act, 1998 (Act No. 84 of 1998).* http://saforestryonline.co.za/indigenous/protected-trees-in-south-africa/

[17] Sacande, M., Rønne, C., Sanon, M. & Jøker, D. (2006). *Adansonia digitata* L. *Seed Leaflet* No. 109. Forest & Landscape Denmark, Hørsholm, Denmark.

[18] Setshogo, M. P. & Hargreaves, B. (2002). Botswana. In: *Southern African Plant Red Data Lists.* Edited by J. S. Golding. Southern African Botanical Diversity Network Report No. 14, SABONET, Pretoria, South Africa. pp. 16–20.

[19] Van Wyk, B., Van Wyk, P. & Van Wyk, B.-E. (2000). *Photographic Guide to Trees of Southern Africa*. Struik Nature, Cape Town, South Africa.

[20] Van Wyk, B.-E., Van Oudtshoorn, B. & Gericke, N. (2009). *Medicinal Plants of South Africa*, 2nd edition. Briza Publications, Pretoria, South Africa.

[21] Varmah, J. C. & Vaid, K. M. (1978). Baobab — the historic African tree at Allahabad. *Indian Forester* 104 (7): 461–464.

Bauhinia macrantha Oliv.

TAXONOMY AND NOMENCLATURE

FAMILY. Leguminosae – Cercidoideae

SYNONYMS. *Bauhinia petersiana* subsp. *macrantha* (Oliv.) Brummitt & J.H.Ross, *B. petersiana* subsp. *serpae* (Ficalho & Hiern) Brummitt & J.H.Ross, *B. serpae* Ficalho & Hiern, *Perlebia macrantha* (Oliv.) A.Schmitz, *P. macrantha* subsp. *serpae* (Ficalho & Hiern) A.Schmitz

VERNACULAR / COMMON NAMES. Camel's foot, coffee bauhinia, coffee neat's foot, white bauhinia, wild coffee bean (English); mogose, mogotswe (Tswana); ngwa, ntkgwa, tkguntkkowa (Khoisan); omutuanuta (Oshikwanyama, Namibia); mungandu, mupondo (Shona).

PLANT DESCRIPTION

LIFE FORM. Shrub or small tree.

PLANT. Usually erect, 1–4 m high, or sometimes scrambling to 10 m by means of branch tendrils; occasionally a subshrub <40 cm high; young branches rather finely brown or grey brown, spreading or appressed pubescent. **LEAVES.** 1–4.5 cm long, glabrous, lobed from one third to two thirds of their length. **FLOWERS.** Inflorescence 1–3(4)-flowered; flowers white (rarely with pink blotches); petals 6 cm long (Figure 1) [7].

Figure 1 – *B. macrantha* shrub with flowers and buds. (Photo: T. Ulian, RBG Kew.)

FRUITS AND SEEDS

The fruits are dark brown pods, 10–24 cm long, 2–4 cm wide, flattened, woody, pointed at both ends and splitting along seams, with sides curling up like a corkscrew [4]. Each pod contains 5–6 dark purplish brown seeds, which are strongly compressed, oblong to circular, 1.3–3 cm long, 0.7–1.8 cm wide (Figure 2) [7].

FLOWERING AND FRUITING. Flowers at the end of the dry season (late October in Botswana). Fruits ripen 6–7 months after pollination. In the Kalahari, seeds are available from April to July [1].

Figure 2 – External views of the seeds and the excised embryo. (Photo: P. Gómez-Barreiro, RBG Kew.)

DISTRIBUTION

Found throughout South Tropical Africa (from Angola to Zimbabwe), Southern Africa (Botswana, Namibia and the Northern and Cape Provinces of South Africa) and East Africa (Tanzania). In Botswana, the species is quite frequent in the north-west sandy areas of Kweneng, Ngami and Kgalagadi. Introduced to Western Australia and cultivated in Florida (USA) [4].

HABITAT

Medium and high altitudes (550–1,800 m a.s.l.), with mean annual rainfall of 350–1,000 mm [4] and temperatures ranging from 25°C to 0°C during winters, and 35°C to 10°C in summers. Occurs in a variety of vegetation types, from open grassland to woodland [1,4], on deep, well-drained sandy soils. Can withstand long periods of drought, although sensitive to frost [4].

USES

Powdered seeds (Figure 3) are eaten raw or roasted as a nourishing meal and are a delicacy in parts of Botswana. The seeds are also used as a coffee substitute [1,8] and as animal food [1]. Traditional healers powder the seeds and mix them with other plant products to create a protein-rich tonic for children and the elderly. The seed oil is also extracted for local use [1].

Figure 3 – Preparation of coffee substitute for the market (in plastic sachets), at ICDT, Tsetseng. (Photos: K. K. Mogotsi, BUAN.)

KNOWN HAZARDS AND SAFETY

No known hazards.

CONSERVATION STATUS

Assessed as Least Concern (LC) in South Africa according to IUCN Red List criteria [2], but has yet to be assessed in Botswana and to have its status confirmed in the IUCN Red List [3,6].

SEED CONSERVATION

HARVESTING. The pods are picked directly from the tree as they do not fall to the ground at maturity.

PROCESSING AND HANDLING. The pods are easily opened and crushed by hand.

STORAGE AND VIABILITY. X-ray analysis carried out on seed lots stored at the MSB highlighted c. 100% filled seeds, although in some cases high percentages of infested seeds (>50%) have been found. Seeds are reported to be orthodox [5].

PROPAGATION

SEEDS

Dormancy and pre-treatments: Seeds exhibit physical dormancy. Scarifying seeds, by chipping a portion of the seed coat with a scalpel before sowing, improves their germination [5].

Germination, sowing and planting: Scarified seeds easily germinate (c. 100% germination) when sown at a wide range of temperatures (i.e. 16–26°C) and with a photoperiod of 12 hours of

Figure 4 – *B. macrantha* seedling raised at the BUAN nursery. (Photo: E. Mattana, RBG Kew.)

light per day [5]. Seeds can also be soaked in hot water overnight and then planted in seed trays filled with river sand. If kept moist, seeds will germinate after 7–20 days. Seedlings can be transplanted into nursery bags (Figure 4) when they reach the 2-leaf stage, taking care not to damage the taproot [1].

VEGETATIVE PROPAGATION

Bauhinia macrantha can be propagated by cuttings or layering [1].

TRADE

The species is of commercial value and is traded internationally [4].

Authors

Gloria N. Mashungwa, Kebadire K. Mogotsi, Cecilia Amosso, Moctar Sacande, Tiziana Ulian and Efisio Mattana.

References

[1] Bosch, C. H. (2006). *Bauhinia petersiana* Bolle. In: *PROTA 1: Cereals and Pulses/Céréales et Légumes Secs. [CD-Rom].* Edited by M. Brink & G. Belay, PROTA, Wageningen, The Netherlands.

[2] Foden, W. & Potter, L. (2005). *Bauhinia petersiana* Bolle subsp. *macrantha* (Oliv.) Brummitt & J.H.Ross. In: *National Assessment: Red List of South African Plants.* Version 2015.1. http://redlist.sanbi.org/

[3] IUCN (2017). *The IUCN Red List of Threatened Species.* Version 2017-3. http://www.iucnredlist.org/

[4] Royal Botanic Gardens, Kew (1999–2015). *Survey of Economic Plants for Arid and Semi-Arid Lands (SEPASAL) Database.* http://apps.kew.org/sepasalweb/sepaweb

[5] Royal Botanic Gardens, Kew (2017). *Seed Information Database (SID).* Version 7.1. http://data.kew.org/sid/

[6] Setshogo, M. P. & Hargreaves, B. (2002). Botswana. In: *Southern African Plant Red Data Lists.* Edited by J. S. Golding. Southern African Botanical Diversity Network Report No. 14, SABONET, Pretoria, South Africa. pp. 16–20.

[7] Timberlake, J. R., Pope, G. V., Polhill, R. M. & Martins, E. S. (eds) (2007). Leguminosae (Caesalpinioideae). In: *Flora Zambesiaca*, Vol. 3, part 2. Flora Zambesiaca Managing Committee, Royal Botanic Gardens, Kew.

[8] Van Wyk, B.-E. & Gericke, N. (2000). *People's Plants: A Guide to Useful Plants of Southern Africa.* Briza Publications, Pretoria, South Africa.

Cassia abbreviata Oliv.

TAXONOMY AND NOMENCLATURE

FAMILY. Leguminosae – Caesalpinioideae

SYNONYMS. *Cassia abbreviata* Oliv. subsp. *abbreviata*

VERNACULAR / COMMON NAMES. Long-tail cassia, sambokpeul, sjambok pod (English); mokwankusha, monepenepe, ngaganyama, omutangaruru, sifonkola (Tswana); omutangaruru (Herero, Namibia); ofothi (Oshikwanyama, Namibia).

PLANT DESCRIPTION

LIFE FORM. Tree.

PLANT. Deciduous, <15 m tall (Figure 1), with longitudinally ridged, brownish grey bark [7,10,16]. **LEAVES.** Paripinnate, with narrowly triangular stipules and 5–12 pairs of leaflets; leaflets c. 2–8 x 1–4 cm [7,9,16]. **FLOWERS.** Fragrant, borne on terminal, racemose inflorescences, pale yellow, c. 4.5 cm in diameter [9,16].

Figure 1 – *C. abbreviata* tree with pods. (Photo: S. Ngwako, BUAN.)

FRUITS AND SEEDS

Fruits are cylindric, smooth and silky, golden brown pods, up to c. 1 m long and 1.5–2 cm in diameter (Figure 1), containing many seeds embedded in pulp. Seeds are dark brown or black, 9–12 mm long and ellipsoid (Figure 2) [16,17].

FLOWERING AND FRUITING. Flowers from September to November; fruits ripen from December to April [11]. Ripened pods remain on the tree for several months before the seeds are released [9].

Figure 2 – External views of the seeds and longitudinal section showing the embryo.
(Photos: P. Gómez-Barreiro, RBG Kew.)

DISTRIBUTION

Throughout Southern Africa (including Botswana, Namibia, Mozambique, Zambia and Zimbabwe) to South Africa (Northern Provinces), and from Gabon east to Somalia and Kenya [2,7,16].

HABITAT

Woodland, *Acacia-Commiphora* bushland, hillsides, plains, watercourses and often on termite mounds in arid lowlands [4], from low to medium altitudes (220–1,520 m a.s.l.) [10,11]. Mature trees are drought and fire resistant and can withstand moderate frost [12].

USES

A multipurpose tree with a wide range of uses in traditional medicine, for example in treating dysentery, diarrhoea, schistosomiasis (bilharzia), skin diseases, cough, pneumonia, fever, gonorrhoea, heart diseases and snake bites [4,8]. An extract of the bark (Figure 3) is used to treat malaria and blackwater fever. Dried, powdered root cortex is taken as a mild, non-cramping laxative. An infusion of the bark and roots is used for relief of abdominal pains and toothache. The smoke from burning twigs and leaves is inhaled to cure headache [7]. Seeds are sucked as a general tonic. The wood is used to make furniture, charcoal and firewood [3]. The tree provides shade and is also used as an ornamental [12].

Figure 3 – Bark harvested in the Pilikwe community. (Photo: A. Hudson, RBG Kew.)

KNOWN HAZARDS AND SAFETY

No known hazards.

CONSERVATION STATUS

The demand for bark and roots for medicines is reported to be depleting wild populations at a high rate [7]. *Cassia abbreviata* subsp. *beareana* (Holmes) Brenan is assessed as Least Concern (LC) in South Africa according to IUCN Red List criteria [5]. *Cassia abbreviata* has yet to be assessed in Botswana and to have its status confirmed in the IUCN Red List [6,14].

SEED CONSERVATION

HARVESTING. Pods are picked from the tree as soon as they are ripe [9] to avoid damage by insects.
PROCESSING AND HANDLING. Pods can be opened by hand to enable the seeds to be separated from the green pulp [9].
STORAGE AND VIABILITY. Seeds of this species are reported to be likely orthodox [13]. Locally, they can be stored at room temperature if kept dry and free from insects.

PROPAGATION

SEEDS

Dormancy and pre-treatments: Seeds of this species are reported to exhibit physical dormancy [1] due to the low germination percentage (10%) achieved on seeds incubated without any pre-treatments [15]. Soaking seeds in hot water improves germination [7]. However, more studies are needed to confirm the type of dormancy present.
Germination, sowing and planting: Seed germination in moist conditions takes place in 4–10 days. For optimum results, seeds should be sown in a mixture of equal parts by volume of sand and compost [7].

VEGETATIVE PROPAGATION

Can be propagated from wildings [12].

TRADE

The Pilikwe community processes various parts of the plant for income generation. Informal trade in bark powder is increasing in both urban and rural areas of Botswana.

Authors

Bamphitlhi Tiroesele, Cecilia Amosso, Kebadire K. Mogotsi, Moctar Sacande, Tiziana Ulian and Efisio Mattana.

References

[1] Baskin, C. C. & Baskin, J. M. (2014). *Seeds: Ecology, Biogeography, and Evolution of Dormancy and Germination*, 2nd edition. Academic Press, San Diego, USA.

[2] Brenan, J. P. M. (1967). Leguminosae Subfamily Caesalpinioideae. In: *Flora of Tropical East Africa*. Edited by E. Milne-Redhead & R. M. Polhill. Crown Agents for Oversea Governments and Administrations, London.

[3] Dharani, M., Rukunga, G., Yenesew, A., Mbora, A., Mwaura, L., Dawson, I. & Jamnadass, R. (2010). *Common Antimalarial Trees and Shrubs of East Africa: A Description of Species and a Guide to Cultivation and Conservation Through Use*. World Agroforestry Centre, Nairobi, Kenya.

[4] Erasto, P. & Majinda, R. R. T. (2011). Bioactive proanthocyanidins from the root bark of *Cassia abbreviata. International Journal of Biological and Chemical Sciences* 5 (5): 2170–2179.

[5] Foden, W. & Potter, L. (2005). *Cassia abbreviata* Oliv. subsp. *beareana* (Holmes) Brenan. In: *National Assessment: Red List of South African Plants*. Version 2015.1. http://redlist.sanbi.org/

[6] IUCN (2017). *The IUCN Red List of Threatened Species.* Version 2017-3. http://www.iucnredlist.org/

[7] Kawanga, V. (2008). *Cassia abbreviata* Oliv. In: *Plant Resources of Tropical Africa 11 (1): Medicinal Plants 1*. Edited by G. H. Schmelzer & A. Gurib-Fakim, PROTA Foundation/Backhuys Publishers/CTA, Wageningen, The Netherlands. pp. 144–146.

[8] Leteane, M. M., Ngwenya, B. N., Muzila, M., Namushe, A., Mwinga, J., Musonda, R., Moyo, S., Mengestu, Y. B., Abegaz, B. M. & Andrae-Marobela, K. (2012). Old plants newly discovered: *Cassia sieberiana* D.C. and *Cassia abbreviata* Oliv. Oliv. [*sic*] root extracts inhibit *in vitro* HIV-1c replication in peripheral blood mononuclear cells (PBMCs) by different modes of action. *Journal of Ethnopharmacology* 141 (1): 48–56.

[9] Mojeremane, W., Legwaila, G. M., Mogotsi, K. K. & Tshwenyane, S. O. (2005). Monepenepe (*Cassia abbreviata*): a medicinal plant in Botswana. *International Journal of Botany* 1 (2): 108–110.

[10] Mongalo, N. I. & Mafoko, B. J. (2013). *Cassia abbreviata* Oliv. A review of its ethnomedicinal uses, toxicology, phytochemistry, possible propagation techniques and pharmacology. *African Journal of Pharmacy and Pharmacology* 7 (45): 2901–2906.

[11] Mulofwa, J., Simute, S. & Tengnäs, B. (1994). *Agroforestry: Manual for Extension Workers in Southern Province, Zambia*. Regional Soil Conservation Unit/Swedish International Development Authority, Nairobi, Kenya.

[12] Orwa, C., Mutua, A., Kindt, R., Jamnadass, R. & Simons, A. (2009). *Agroforestree Database: A Tree Reference and Selection Guide*. Version 4.0. World Agroforestry Centre, Kenya. http://www.worldagroforestry.org/output/agroforestree-database

[13] Royal Botanic Gardens, Kew (2017). *Seed Information Database (SID).* Version 7.1. http://data.kew.org/sid/

[14] Setshogo, M. P. & Hargreaves, B. (2002). Botswana. In: *Southern African Plant Red Data Lists*. Edited by J. S. Golding. Southern African Botanical Diversity Network Report No. 14, SABONET, Pretoria, South Africa. pp. 16–20.

[15] Tietema, T., Merkesdal, E. & Schroten J. (1992). *Seed Germination of Indigenous Trees in Botswana*. ACTS Press, African Centre for Technology Studies, Nairobi, Kenya.

[16] Timberlake, J. R., Pope, G. V., Polhill, R. M. & Martins, E. S. (eds) (2007). Leguminosae (Caesalpinioideae). In: *Flora Zambesiaca*, Vol. 3, part 2. Flora Zambesiaca Managing Committee, Royal Botanic Gardens, Kew.

[17] Venter, F. & Venter, J.-A. (2016). *Making The Most of Indigenous Trees*, 3rd edition. Briza Publications, Pretoria, South Africa.

Citrullus lanatus (Thunb.) Matsum. & Nakai

TAXONOMY AND NOMENCLATURE

FAMILY. Cucurbitaceae

SYNONYMS. *Anguria citrullus* Mill., *Citrullus lanatus* (Thunb.) Mansf., *C. vulgaris* Schrad. ex Eckl. & Zeyh., *Colocynthis citrullus* (L.) Kuntze, *Cucurbita citrullus* L., *Momordica lanata* Thunb.

VERNACULAR / COMMON NAMES. Bitter melon, colocynth, common wild melon, desert melon, khama melon, sweet melon, watermelon (English); kgengwe, tsama (Tswana); etanga (Herero, Namibia); kgeme (Kgalagadi); t'sama (Khoisan); ontanga (Ndonga); bawora (Shona).

PLANT DESCRIPTION

LIFE FORM. Annual herb.

PLANT. Climbing or prostrate; stems <10 m, with forked tendrils (Figure 1); monoecious.

LEAVES. Usually deeply palmately 3–5-lobed; lobes elliptic, usually deeply pinnately lobulate, sub-entire; lamina 5–20 cm long, 3.5–19 cm wide, ovate or narrowly ovate; leaf surface ± hairy, especially on the veins beneath; petioles 2–19 cm long, hairy (Figure 1). **FLOWERS.** Axillary, light yellow, 5-lobed corolla, greenish beneath [6,13].

Figure 1 – Plant with fruit, growing in the ICDT garden (Tsetseng), and a fruit eaten by livestock.
(Photos: C. Varet, The Ivory Foundation, Paris.)

FRUITS AND SEEDS

Fruits usually 2–20 cm in diameter (up to 60 x 30 cm in cultivated plants), subglobose, greenish mottled with darker green lines, succulent, with a shiny or powdery look. Inner contents fleshy or pulpy, containing multiple seeds in rows (Figure 1). Seeds ovate-elliptic, usually 9–11 x 5–6 x 2.5–2.7 mm, smooth or slightly verrucose, dark or pale-coloured, often mottled, and sometimes bordered (Figure 2) [6].

FLOWERING AND FRUITING. In Southern Africa fruits ripen during winter months [9].

Figure 2 – External views of the seeds, and longitudinal and cross-sections showing the embryo.
(Photos: P. Gómez-Barreiro, RBG Kew.)

DISTRIBUTION

Native of the western Kalahari region and elsewhere in Southern and South Tropical Africa, but now found throughout the African continent due to cultivation – a process that started more than 5,000 years ago. Subsequently introduced to all tropical and subtropical countries, where it is cultivated and sometimes naturalised [6,9,12].

HABITAT

Citrullus lanatus is found throughout Botswana [8], especially in the southern and western parts of the country. It prefers deep sands, such as those of the Kalahari, but is also found on shallow sands of the Kweneng, North-West and Central districts. It also grows alongside dry watercourses and in disturbed areas, where it can be a pioneer. It is sometimes a troublesome weed of cultivated land. It is extremely resistant to drought [9], but it is not frost tolerant [7].

USES

The succulent fruits (Figure 1) are used as a source of water during hunting or when found far from conventional water sources. Most livestock will also eat the fruits as a water source, especially during drought or extremely high temperatures which are frequent in the Kalahari (Figure 1). Water content remains in the fruits for several months after abscission. The fruit pulp can be sliced and sun-dried or cooked and mixed with maize meal to create a porridge served with creamy milk. A sweet jam is also prepared. Seeds are roasted and eaten, or are dried and pounded into flour to make a type of bread. Seeds are a useful source of protein and fat for people living in sub-Saharan Africa and savanna regions. In Southern Africa, watermelon is often cultivated in home gardens as a seasonal crop with maize (*Zea mays* L.) and kaffir corn (*Sorghum*). Cultivated forms have a high yield even in drought years [9,13].

KNOWN HAZARDS AND SAFETY

Some fruits (known as 'karkoer' or 'bitterwaatlemoen') contain high levels of cucurbitacin, making them bitter tasting and unfit for consumption [13]. Some people experience an allergic reaction on ingestion of watermelon, including swelling of the mouth and throat [1].

CONSERVATION STATUS

Assessed as Least Concern (LC) in South Africa according to IUCN Red List criteria [4], but yet to be assessed in Botswana and to have its status confirmed in the IUCN Red List [5,11].

SEED CONSERVATION

HARVESTING. Seed maturity is indicated by withering of the fruit stalk and ancillary tendril, and the brown colour of the fruit. Ripe fruits make a different sound to unripe fruits when knocked. Collected fruits can be stored in large containers until their seeds are extracted.

PROCESSING AND HANDLING. The seeds can be separated from the pulp during consumption, or removed earlier by splitting open the fruits (either by hitting them on the ground or using a sharp blade to cut them open) and then spreading the open fruits on sacks to allow them to dry (Figure 3). They can be left to dry for several days (usually seven days), depending on air temperature and season. Exposure to direct sunlight should be avoided.

STORAGE AND VIABILITY. Seeds are reported to be orthodox and therefore after appropriate drying (Figure 4) can be stored long-term. Seeds are not damaged by exposure to liquid nitrogen so cryoconservation may also be carried out [10]. X-ray analysis on seed lots stored at the MSB highlighted a high percentage of filled seeds (c. 100%). Seed viability is halved after five years of storage at room temperature [10]; the most common method of storage used by local people is to store seeds in airtight containers (mostly recycled transparent glass jars).

PROPAGATION

SEEDS

Dormancy and pre-treatments: Cucurbit seeds are reported to be physiologically dormant [3]. Although high germination percentages (>80%) can be achieved without any pre-treatments, germination of *C. lanatus* can be enhanced by removing the testa [10]. This suggests the occurrence of non-deep physiological dormancy.

Germination, sowing and planting: Germination is promoted by high temperatures (≥25°C) [10].

VEGETATIVE PROPAGATION

Not normally propagated vegetatively, but cuttings of commercial watermelon cultivars have been rooted successfully under experimental conditions [2].

Figure 3 – Tsetseng village chief's wife with fruits of *C. lanatus* in the garden. (Photo: T. Ulian, RBG Kew.)

Figure 4 – *C. lanatus* seeds for planting in the Tsetseng community garden. (Photo: K. K. Mogotsi, BUAN.)

TRADE

A major commercial crop which is internationally traded. The fruits and seeds are also sold locally along roadsides and in markets [9,12].

Authors

Kebadire K. Mogotsi, Moctar Sacande, Tiziana Ulian, Efisio Mattana, Cecilia Amosso and Albertinah T. Matsika.

References

[1] Asero, R., Mistrello, G. & Amato, S. (2011). The nature of melon allergy in ragweed-allergic subjects: a study of 1000 patients. *Allergy & Asthma Proceedings* 32 (1): 64–67.

[2] El-Eslamboly, A. A. S. A. (2014). Effect of watermelon propagation by cuttings on vegetative growth, yield and fruit quality. *Egyptian Journal of Agricultural Research* 92 (2): 553–579.

[3] Finch-Savage, W. E. & Leubner-Metzger, G. L. (2006). Seed dormancy and the control of germination. *New Phytologist* 171 (3): 501–523.

[4] Foden, W. & Potter, L. (2005). *Citrullus lanatus* (Thunb.) Matsum. & Nakai. In: *National Assessment: Red List of South African Plants.* Version 2015.1. http://redlist.sanbi.org/

[5] IUCN (2017). *The IUCN Red List of Threatened Species.* Version 2017-3. http://www.iucnredlist.org/

[6] Launert, E. (ed.) (1978). Cucurbitaceae. In: *Flora Zambesiaca*, Vol. 4. Flora Zambesiaca Managing Committee, Royal Botanic Gardens, Kew.

[7] Purseglove, J. W. (1968). *Tropical Crops: Dicotyledons.* Longman, London.

[8] Renew, A. (1968). Some edible wild cucumbers (Cucurbitaceae) of Botswana. *Botswana Notes and Records* 1: 5–8.

[9] Royal Botanic Gardens, Kew (1999–2015). *Survey of Economic Plants for Arid and Semi-Arid Lands (SEPASAL) Database.* http://apps.kew.org/sepasalweb/sepaweb

[10] Royal Botanic Gardens, Kew (2017). *Seed Information Database (SID).* Version 7.1. http://data.kew.org/sid/

[11] Setshogo, M. P. & Hargreaves, B. (2002). Botswana. In: *Southern African Plant Red Data Lists.* Edited by J. S. Golding. Southern African Botanical Diversity Network Report No. 14, SABONET, Pretoria, South Africa. pp. 16–20.

[12] van der Vossen, H. A. M., Denton, O. A. & El Tahir, I. M. (2004). *Citrullus lanatus* (Thunb.) Matsum. & Nakai. [Internet] Record from PROTA4U. In: *PROTA (Plant Resources of Tropical Africa/Ressources Végétales de l'Afrique Tropicale).* Edited by G. J. H. Grubben & A. A. Denton. Wageningen, The Netherlands. http://www.prota4u.org/search.asp

[13] Van Wyk, B.-E. & Gericke, N. (2000). *People's Plants: A Guide to Useful Plants of Southern Africa.* Briza Publications, Pretoria, South Africa.

Colophospermum mopane (J.Kirk ex Benth.) J.Léonard

TAXONOMY AND NOMENCLATURE

FAMILY. Leguminosae – Detarioideae

SYNONYMS. *Copaiba mopane* (J.Kirk ex Benth.) Kuntze, *Hardwickia mopane* (J.Kirk ex Benth.) Breteler

VERNACULAR / COMMON NAMES. Balsam tree, butterfly tree, turpentine tree (English); mopane, mopani, mophane (Tswana); omuntati (Herero, Namibia); tsaurahais (Khoekhoe, Namibia); omusati (Ndonga, Namibia); musharo (Shona); nxanatsi (Tsonga, South Africa).

PLANT DESCRIPTION

LIFE FORM. Tree.

PLANT. Deciduous small to medium-sized tree, <20 m tall (Figure 1); bark deeply vertically fissured, often in an elongated reticulate pattern. **LEAVES.** Leaflets 2–13 x 2–6.5 cm, stalkless, resembling two butterfly wings, hairless, smelling strongly of turpentine when crushed; petioles 2–5 cm long (Figure 1). **FLOWERS.** Small, greenish and inconspicuous; inflorescence racemose, slender and spike-like [8,9,14–17].

Figure 1 – *C. mopane* woodland and details of leaves and fruits. (Photos: S. Ngwako, BUAN and B. Wursten, Flora of Zimbabwe: https://www.zimbabweflora.co.zw/index.php.)

FRUITS AND SEEDS

The pods (Figures 1 and 2) are c. 6 x 3 cm, pale brown, flattened, oval and indehiscent [17]. The seeds have small, reddish glands, exuding a sticky fluid on both surfaces (Figure 2) [16].

FLOWERING AND FRUITING. Flowering is erratic, occurring between October and March. Fruits mature from March to June [3].

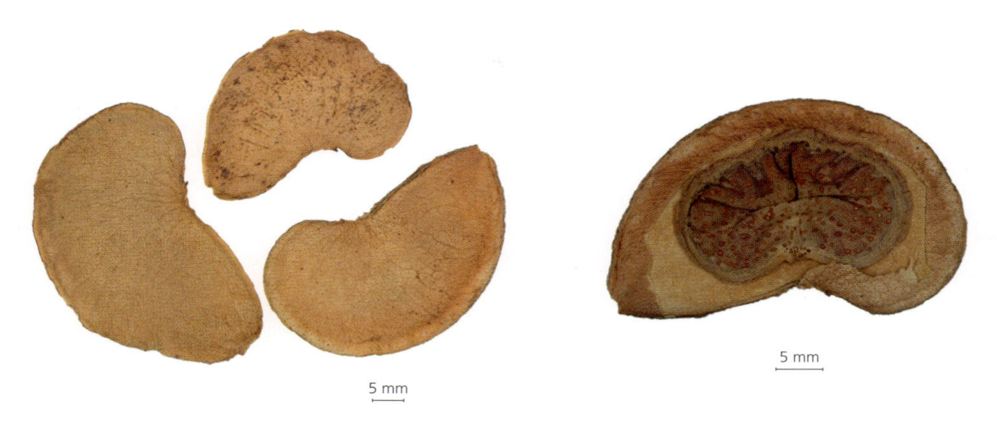

Figure 2 – External views of the indehiscent one-seeded pods and longitudinal section showing the seed with reddish glands. (Photos: P. Gómez-Barreiro, RBG Kew.)

DISTRIBUTION

Angola, Botswana, Namibia, Zimbabwe, Zambia, Malawi, South Africa and Mozambique [8,9,14]. In Botswana, it is found in the central, northern and eastern parts of the country [16].

HABITAT

Dominant in mopane woodland in south-central Africa, occurring in hot, low-lying, mainly alluvial or lime-rich soils at 60–1,000 m a.s.l. (occasionally up to 1,300 m) [14], in areas with 200–800 mm annual rainfall [9,15,17] and mean maximum temperature of 30°C [8]. It does not tolerate cold or frost [1].

USES

The wood is used for making furniture and ornaments, buildings, fences, mine props and railway sleepers [8,10,15], and is considered to be a good fuel because it burns slowly with a constant high heat [8]. The caterpillars of *Imbrasia belina* (mopane moth; Tswana: phane) feed on the leaves and are hand-picked and eaten raw or fried [6,10,15]. The leaves provide fodder and have high protein content; many animals, including cattle, browse the leaves and eat dry pods that have fallen to the ground [9]. *Colophospermum mopane* is used in traditional medicine to treat a wide range of ailments [7,10].

KNOWN HAZARDS AND SAFETY

No known hazards.

CONSERVATION STATUS

A widespread tree, assessed as Least Concern (LC) in South Africa according to IUCN Red List criteria [4], but yet to be assessed in Botswana and to have its status confirmed in the IUCN Red List [5,12].

SEED CONSERVATION

HARVESTING. Pods are picked from the tree from March to June [3].

PROCESSING AND HANDLING. The seeds are very soft and easily damaged, so are best kept in the pods during storage. The fruit is therefore the conservation unit.

STORAGE AND VIABILITY. X-ray analysis carried out on seed lots stored at the MSB highlighted high percentages of filled seeds (c. 85–100%) with no sound seeds being infested. Seeds are reported to be desiccation tolerant [7,11].

PROPAGATION

SEEDS

Dormancy and pre-treatments: A high proportion of seeds will germinate without any pre-treatments. However, soaking in cold water for 24 hours speeds up germination [3,7]. The species is reported as either being non-dormant or having physical dormancy [2]. Removing the seeds from the fruit, or chipping, is recommended when testing the seeds in the laboratory [11].

Germination, sowing and planting: Chipped seeds (as above) germinate to high percentages (>75%) under laboratory conditions when incubated in the light at constant temperatures of >20°C [11]. A recent study showed the following range of temperatures for germination: minimum 11.6°C, optimum 34.1°C, maximum 43.6°C [13]. Under nursery conditions, seeds should be planted on moist river sand [1]. Seedlings (Figure 3) grow slowly at first but their growth rate increases when they are c. 20 cm tall [1].

VEGETATIVE PROPAGATION

Can be propagated by wildings and cuttings [7].

Figure 3 – Seedlings of *C. mopane* at BUAN. (Photo: S. Ngwako, BUAN.)

TRADE

Production of timber and firewood for local markets is extensive. 'Phane' is an important food source in Botswana; the caterpillars are a highly sought-after and protein-rich delicacy, and are the basis of a very lucrative market in Southern Africa [7].

Authors

Samodimo Ngwako, Kebadire K. Mogotsi, Moctar Sacande, Tiziana Ulian and Efisio Mattana.

References

[1] Aubrey, A. (2004). *Colophospermum mopane*. South African National Biodiversity Institute (SANBI), South Africa. http://pza.sanbi.org/

[2] Baskin, C. C. & Baskin, J. M. (2014). *Seeds: Ecology, Biogeography, and Evolution of Dormancy and Germination*, 2nd edition. Academic Press, San Diego, USA.

[3] Coates Palgrave, K. & Coates Palgrave, M. (2002). *Trees of Southern Africa*, 3rd edition. Struik Publishers, Cape Town, South Africa.

[4] Foden, W. & Potter, L. (2005). *Colophospermum mopane* (J.Kirk ex Benth.) J.Kirk ex J.Léonard. In: *National Assessment: Red List of South African Plants*. Version 2015.1. http://redlist.sanbi.org/

[5] IUCN (2017). *The IUCN Red List of Threatened Species*. Version 2017-3. http://www.iucnredlist.org/

[6] Mackinder, B. (ed.) (undated). *Colophospermum mopane* (mopane). Royal Botanic Gardens, Kew. http://www.kew.org/science-conservation/plants-fungi/colophospermum-mopane-mopane

[7] Manyuela, F., Tsopito, C., Kamau, J., Mogotsi, K. K., Nsoso, S. J. & Moreki, J. C. (2013). Effect of *Imbrasia belina* (westwood), *Tylosema esculentum* (Burchell) Schreiber and *Vigna subterranea* (L) Verde [*sic*] as protein sources on growth and laying performance of Tswana hens raised under intensive production system. *Scientific Journal of Animal Science* 2 (1): 1–8.

[8] Melusi, R. & Mojeremane, W. (2012). *Colophospermum mopane* (Benth.) J.Léonard. [Internet] Record from PROTA4U. In: *PROTA (Plant Resources of Tropical Africa/Ressources Végétales de l'Afrique Tropicale)*. Edited by R. H. M. J. Lemmens, D. Louppe & A. A. Oteng-Amoako. Wageningen, The Netherlands. https://www.prota4u.org

[9] Orwa, C., Mutua, A., Kindt, R., Jamnadass, R. & Simons, A. (2009). *Agroforestree Database: A Tree Reference and Selection Guide*. Version 4.0. World Agroforestry Centre, Kenya. http://www.worldagroforestry.org/output/agroforestree-database

[10] Royal Botanic Gardens, Kew (1999–2015). *Survey of Economic Plants for Arid and Semi-Arid Lands (SEPASAL) Database*. http://apps.kew.org/sepasalweb/sepaweb

[11] Royal Botanic Gardens, Kew (2017). *Seed Information Database (SID).* Version 7.1. http://data.kew.org/sid/

[12] Setshogo, M. P. & Hargreaves, B. (2002). Botswana. In: *Southern African Plant Red Data Lists*. Edited by J. S. Golding. Southern African Botanical Diversity Network Report No. 14, SABONET, Pretoria, South Africa. pp. 16–20.

[13] Stevens, N., Seal, C. E., Archibald, S. & Bond, W. (2014). Increasing temperatures can improve seedling establishment in arid-adapted savanna trees. *Oecologia* 175 (3): 1029–1040.

[14] Timberlake, J. R., Pope, G. V., Polhill, R. M. & Martins, E. S. (eds) (2007). Leguminosae (Caesalpinioideae). In: *Flora Zambesiaca*, Vol. 3, part 2. Flora Zambesiaca Managing Committee, Royal Botanic Gardens, Kew.

[15] Van Wyk, B. & Van Wyk, P. (1997). *Field Guide to Trees of Southern Africa*. Struik Nature, Cape Town, South Africa.

[16] Van Wyk, B., Van Wyk, P. & Van Wyk, B.-E. (2000). *Photographic Guide to Trees of Southern Africa*. Struik Nature, Cape Town, South Africa.

[17] Van Wyk, B.-E. & Gericke, N. (2000). *People's Plants: A Guide to Useful Plants of Southern Africa*. Briza Publications, Pretoria, South Africa.

Cucumis africanus L.f.

TAXONOMY AND NOMENCLATURE

FAMILY. Cucurbitaceae

SYNONYMS. *Cucumis arenarius* Schrad., *C. hookeri* Naudin

VERNACULAR / COMMON NAMES. Bitter apple, bitter wild cucumber, horned cucumber, jelly melon, thorn cucumber, wild cucumber (English); konkomba, lekatane, magabala, mosumo (Tswana); etanga (Oshikwanyama, Namibia).

PLANT DESCRIPTION

LIFE FORM. Annual (sometimes perennial) herb.

PLANT. Up to 20 m long (Figure 1), sometimes developing a woody rootstock; stems finely ridged, roughly whitish hairy, usually prostrate, but also climbing on bushes and other supports; monoecious. **LEAVES.** Broadly ovate, deeply palmately 3(5)-lobed, 1.5–11.5 cm long, dull green, roughly hairy on both sides, hairs bulbous-based. **FLOWERS.** Pale to dark yellow with green veins; male flowers in groups of 5–10, petals <9 mm long; female flowers solitary, petals <11 mm long [3,7,11].

Figure 1 – *C. africanus* growing naturally in the ICDT garden, Tsetseng, showing an open fruit (left).
(Photos: A. Hudson, RBG Kew and C. Varet, The Ivory Foundation, Paris.)

FRUITS AND SEEDS

Ripe fruits (Figure 1) c. 30 mm long, prickly, with prickles <10 mm long, fleshy to hard, base flat; unripe fruits longitudinally striped dark and light green, maturing to greenish yellow or white, striped purplish brown or yellow, prickles mainly on dark bands; occurring in two distinct forms, either small, ellipsoid, bitter and poisonous when ripe, or large, cylindric, non-bitter and edible when ripe; flesh green and translucent, many-seeded. Seeds elliptic, c. 4.5 x 2.5 mm, and 1 mm thick, cream or light brown in colour (Figure 2) [7,11].

FLOWERING AND FRUITING. Flowers and fruits from about September to May, but mostly in March [11].

Figure 2 – External views of the extracted seeds.
(Photo: P. Gómez-Barreiro, RBG Kew.)

DISTRIBUTION

Native to Angola, Botswana, Namibia, South Africa and Zimbabwe. It is also found in Madagascar where it is probably introduced [3,11].

HABITAT

Bushland, near non-permanent watercourses and in floodplains, on a variety of soils, including deep and well-drained sandy soils, loams, clays and rock outcrops, at 150–2,115 m a.s.l., in areas of low to moderately high rainfall (100–800 mm per year) [11].

USES

The fresh young leaves are rich in calcium, iron, nicotinic acid and vitamin C and are eaten as a pot herb or cooked as a side dish [1]. The fruits of non-bitter forms are a source of water and are eaten as vegetables by Basarwa people in the Kalahari and in other dry areas [1]. The fruit, leaf and root are used as an emetic, purgative or enema for various ailments; the boiled leaf is used as a poultice in traditional medicine for humans and livestock [11]. *Cucumis africanus* has potential for domestication to improve related cultivated species [3].

KNOWN HAZARDS AND SAFETY

The poisonous form of the fruit contains the bitter principle curcumin and can cause fatal human and cattle poisoning if eaten [10].

CONSERVATION STATUS

Assessed as Least Concern (LC) in South Africa according to IUCN Red List criteria [4], but yet to be assessed in Botswana and to have its status confirmed in the IUCN Red List [6,9].

SEED CONSERVATION

HARVESTING. Fresh fruits are collected from wild or semi-wild plants [3].

PROCESSING AND HANDLING. The fruits (Figure 3) can be opened using a knife to slice the pulp and extract the seeds.

STORAGE AND VIABILITY. X-ray analysis on seed lots stored at the MSB highlighted high percentages (almost 100%) of filled seeds, although up to 40% of empty seeds have also been detected. The seeds are reported to be orthodox [8].

Figure 3 – Fruits gathered during seed collection. (Photo: K. K. Mogotsi, BUAN.)

PROPAGATION

SEEDS

Dormancy and pre-treatments: Seeds of other *Cucumis* spp. are reported to be physiologically dormant (PD) [2]. Scarification (by removing the testa) is reported to improve seed germination suggesting the occurrence of non-deep PD [8]. However, further studies are needed to characterise seed dormancy in *C. africanus*, particularly as the family Cucurbitaceae is listed among those with seeds having physical dormancy [5].

Germination, sowing and planting: Scarified seeds germinate at high percentages (100%) when incubated at c. 25°C in the light. Sterilisation of the seeds, by soaking them in a solution (10%) of commercial bleach for five minutes, can help to avoid contamination [8].

VEGETATIVE PROPAGATION

No protocols available.

TRADE

No data available.

Authors

Gloria N. Mashungwa, Samodimo Ngwako, Kebadire K. Mogotsi, Moctar Sacande, Tiziana Ulian and Efisio Mattana.

References

[1] Arnold, T. H., Wells, M. J. & Wehmeyer, A. S. (1985). Khoisan food plants: taxa with potential for future economic exploitation. In: *Plants for Arid Lands: Proceedings of the Kew International Conference on Economic Plants for Arid Lands held in the Jodrell Laboratory, Royal Botanic Gardens, Kew, England, 23–27 July 1984*. Edited by G. E. Wickens, J. R. Goodin & D. V. Field. Allen & Unwin, London. pp. 69–86.

[2] Baskin, C. C. & Baskin, J. M. (2014). *Seeds: Ecology, Biogeography, and Evolution of Dormancy and Germination*, 2nd edition. Academic Press, San Diego, USA.

[3] Bosch, C. H. (2004). *Cucumis africanus* L.f. In: *Plant Resources of Tropical Africa 2. Vegetables*. Edited by G. J. H. Grubben & O. A. Denton. PROTA Foundation, Wageningen, The Netherlands; Backhuys Publishers, Leiden, The Netherlands; CTA, Wageningen, The Netherlands. pp. 237–238.

[4] Foden, W. & Potter, L. (2005). *Cucumis africanus* L.f. In: *National Assessment: Red List of South African Plants*. Version 2015.1. http://redlist.sanbi.org/

[5] Gama-Arachchige, N. S., Baskin, J. M., Geneve, R. L. & Baskin C. C. (2013). Identification and characterization of ten new water gaps in seeds and fruits with physical dormancy and classification of water-gap complexes. *Annals of Botany* 112 (1): 69–84.

[6] IUCN (2017). *The IUCN Red List of Threatened Species*. Version 2017-3. http://www.iucnredlist.org/

[7] Launert, E. (ed.) (1978). Cucurbitaceae. In: *Flora Zambesiaca*, Vol. 4. Flora Zambesiaca Managing Committee, Royal Botanic Gardens, Kew.

[8] Royal Botanic Gardens, Kew (2017). *Seed Information Database (SID)*. Version 7.1. http://data.kew.org/sid/

[9] Setshogo, M. P. & Hargreaves, B. (2002). Botswana. In: *Southern African Plant Red Data Lists*. Edited by J. S. Golding. Southern African Botanical Diversity Network Report No. 14, SABONET, Pretoria, South Africa. pp. 16–20.

[10] Watt, J. M. & Breyer-Brandwijk, M. G. (1962). *The Medicinal and Poisonous Plants of Southern and Eastern Africa*, 2nd edition. E. & S. Livingstone, Edinburgh & London.

[11] Welman, M. (2005). *Cucumis africanus* L.f. South African National Biodiversity Institute (SANBI), South Africa. http://pza.sanbi.org/

Grewia flava DC.

TAXONOMY AND NOMENCLATURE

FAMILY. Malvaceae

SYNONYMS. *Grewia cana* Sond., *G. hermannioides* Harv.

VERNACULAR / COMMON NAMES. Brandybush, raisin tree, wild currant, wild plum, velvet raisin (English); maphokwe, maretlwa, monabo, moreswe, moretlwa, moseme, ntewa, phomphokwe (Tswana); omundjembere (Herero, Namibia); ehonga (Ndonga, Namibia).

PLANT DESCRIPTION

LIFE FORM. Shrub.

PLANT. Up to 2 m tall (occasionally 3 m), with greyish or greyish brown young branches and dark purplish black old branches. **LEAVES.** Elliptic or oblanceolate, c. 60 mm long, upper surface silvery grey green, underside quite markedly paler, finely hairy above, more densely hairy below. **FLOWERS.** Usually yellow, single and axillary, c. 1–1.5 cm diameter (Figure 1) [2,3,6].

Figure 1 – *G. flava* with flowers in the Tsetseng community garden (Photo: C. Amosso, BUAN) and details of flower and fruit. (Photos: R. and A. Heath, Plants and People Africa.)

FRUITS AND SEEDS

Fruits globose or 2-lobed, becoming reddish when ripe (Figure 2), <8 mm in diameter, with sparse minute bristles [2,3]. Seeds c. 3 mm long, brown, laterally compressed, and elliptic in outline.

FLOWERING AND FRUITING. Flowers from October to March; fruits available from December to April [6].

Figure 2 – External views of the fruits on the left; longitudinal section showing the seeds and the embryo on the right. (Photos: P. Gómez-Barreiro, RBG Kew.)

DISTRIBUTION

Widely distributed throughout Southern Africa (Botswana, Namibia, Swaziland and Cape Provinces, Free State, KwaZulu-Natal and Northern Provinces of South Africa) and South Tropical Africa (Zimbabwe). In Botswana, it is mostly found in eastern, central and northern areas; in particular, it is widespread in the dry habitats of Okavango Delta [3,6].

HABITAT

Grewia flava occurs from low to high altitudes (155–1,830 m a.s.l.), in areas with a mean annual rainfall of 300–350 mm (mainly from November to April). It occurs on Kalahari sand and on limestone ridges, and is common in open deciduous woodlands and Kalahari veld (grasslands with low scrub), often associated with mopane woodland dominated by *Colophospermum mopane* (Benth.) J.Léonard [6].

USES

The fruits have a sweet edible pulp which is particularly high in sugar content. They are an important source of food in desert regions and may be dried and stored for future use. An alcoholic beverage is made by soaking fruits in water for about 24 hours. The San distil liquor from the fruits. In traditional medicine, powdered bark and roots mixed with milk are given to babies to treat stomach problems. Tea made from twigs and leaves is drunk frequently in small quantities, and is said by local people to be beneficial for the kidneys. Larger and longer stems are used to make lightweight, sturdy walking sticks. The wood is also highly valued for making bows and knobkerries (a form of club traditionally used as a weapon), as well as other tools. The bark can be used for weaving after softening it in water. The roots are used as a remedy for pulmonary diseases in cattle, and the plant is eaten by cattle and game, especially in early spring and later summer when green grass is scarce [6].

KNOWN HAZARDS AND SAFETY

Excessive consumption of the fruit can cause constipation [2].

CONSERVATION STATUS

Assessed as Least Concern (LC) in South Africa according to IUCN Red List criteria [4], but yet to be assessed in Botswana and to have its status confirmed in the IUCN Red List [5,8].

SEED CONSERVATION

HARVESTING. Ripe fruits (which are indehiscent) are picked directly from the plant to fill baskets or sacks, or an open sack is laid on the ground and the branches are shaken to dislodge the fruit.
PROCESSING AND HANDLING. Locally, fruits are eaten to obtain the seeds, or the fruit flesh is separated using a knife when the fruit is still fresh. Seeds are then cleaned and dried in the shade. However, whole fruits are also stored for long-term conservation [7].
STORAGE AND VIABILITY. Seeds exhibit orthodox storage behaviour [7]. X-ray analysis carried out on seed lots stored at the MSB highlighted a high proportion (c. 50%) of empty seeds.

PROPAGATION

SEEDS

Dormancy and pre-treatments: Seeds of *Grewia* spp. are reported to be physically dormant [1], but no tests have been carried out to determine the presence of this type of dormancy in *G. flava*.
Germination, sowing and planting: High germination rates have been achieved under controlled conditions when sowing the seeds at 30–35°C with a photoperiod of 12 hours of light per day [7]. In nurseries, seeds can be planted in trays filled with river sand after soaking them in hot water overnight. If kept moist, c. 50–70% germination can be achieved. Seedlings can be transplanted into nursery bags when they reach the 2-leaf stage.

VEGETATIVE PROPAGATION

No protocols available.

TRADE

Communities in Botswana sell the fruits in the local informal markets in cities, towns and rural areas. In Tsetseng, ICDT acts as middle trader between the community and the final consumers.

Authors

Gloria N. Mashungwa, Kebadire K. Mogotsi, Cecilia Amosso, Moctar Sacande, Tiziana Ulian and Efisio Mattana.

References

[1] Baskin, C. C. & Baskin, J. M. (2014). *Seeds: Ecology, Biogeography, and Evolution of Dormancy and Germination*, 2nd edition. Academic Press, San Diego, USA.

[2] Curtis, B. A. & Mannheimer, C. A. (2005). *Tree Atlas of Namibia*. National Botanical Research Institute of Namibia, Windhoek, Namibia.

[3] Exell, A. W., Fernandes, A. & Wild, H. (eds) (1963). Tiliaceae. In: *Flora Zambesiaca*, Vol. 2, part 1. Flora Zambesiaca Managing Committee, Royal Botanic Gardens, Kew.

[4] Foden, W. & Potter, L. (2005). *Grewia flava* DC. In: *National Assessment: Red List of South African Plants*. Version 2015.1. http://redlist.sanbi.org/

[5] IUCN (2017). *The IUCN Red List of Threatened Species.* Version 2017-3. http://www.iucnredlist.org/

[6] Royal Botanic Gardens, Kew (1999–2015). *Survey of Economic Plants for Arid and Semi-Arid Lands (SEPASAL) Database*. http://apps.kew.org/sepasalweb/sepaweb

[7] Royal Botanic Gardens, Kew (2017). *Seed Information Database (SID).* Version 7.1. http://data.kew.org/sid/

[8] Setshogo, M. P. & Hargreaves, B. (2002). Botswana. In: *Southern African Plant Red Data Lists*. Edited by J. S. Golding. Southern African Botanical Diversity Network Report No. 14, SABONET, Pretoria, South Africa. pp. 16–20.

Guibourtia coleosperma (Benth.) J.Léonard

TAXONOMY AND NOMENCLATURE

FAMILY. Leguminosae – Detarioideae

SYNONYMS. *Copaifera coleosperma* Benth.

VERNACULAR / COMMON NAMES. African rosewood, bastard mopane, bastard teak, large false mopane, large mock mopane, Rhodesian mahogany, Rhodesian teak, (English); tsaudi (Tswana); omsii (Ndonga); mungenge (Shona).

PLANT DESCRIPTION

LIFE FORM. Tree.

PLANT. Semi-evergreen, <30 m tall (Figure 1); bark smooth (sometimes flaking) and grey (sometimes pale reddish brown to black). **LEAVES.** Paripinnate, arranged spirally, with 1 pair of leaflets shaped like a butterfly; leaflets ovate, curved, 3–10 x 2–4 cm, markedly asymmetric, dark glossy green above, paler below. **FLOWERS.** Small, fragrant, creamy white, star-shaped, c. 1 cm in diameter, produced in axillary and terminal paniculate inflorescences; petals absent [1,5,9].

Figure 1 – *G. coleosperma* tree with ripe fruits and close-up of flowers. (Photos: M. Legwaila, MEWT, Botswana and B. Wursten, Flora of Zimbabwe: https://www.zimbabweflora.co.zw/index.php.)

FRUITS AND SEEDS

Fruits are obliquely elliptical flattened pods; c. 2–3.5 x 1.5–2 cm, glabrous, wrinkled and brown, tardily dehiscent, single-seeded [1]. Seeds ellipsoid, somewhat flattened, c. 1 x 2 cm, shiny red to reddish brown (Figure 2), completely enclosed by a red aril [1,5,9].

FLOWERING AND FRUITING. Flowering takes place from December to March (rarely to May); fruits ripen mainly from May to July [2,5]. The seeds are harvested from June to October [1,5].

Figure 2 – External views of the dark brown seeds, almost completely enclosed in the pale brown dry aril. (Photo: P. Gómez-Barreiro, RBG Kew.)

DISTRIBUTION

Angola, Botswana, Democratic Republic of Congo, Namibia, Zambia and Zimbabwe [9].

HABITAT

Woodlands and dry forest, often along rivers, at altitudes of 750–1,400 m a.s.l. Predominantly found on Kalahari sands, with mean annual rainfall of 45–1,100 mm and mean annual temperature of 20–28°C [5,9].

USES

The wood is used for flooring, furniture, railway sleepers, mine props, tool handles, toys and ornaments [5]. Traditionally, it is used in the construction of canoes. The bark is used for tanning and dyeing [5]. The seeds and arils contain oil which is used in cooking [5]. The seeds are eaten after roasting and pounding, or can be eaten boiled (as beans or mixed with meat) (Figure 3). The red dye from the aril is used to stain furniture. The arils can be removed with warm water and made into a red paste, which is then cooked with meat or made into a nourishing drink [5]. The roots are used in traditional medicine for promoting wound healing [4]. A root decoction is used to treat venereal diseases. Roots and bark are used as a vapour bath to treat headache; a mixture of the roots and leaves is used to treat fever and mental problems. A leaf decoction is used to treat stomach complaints; young leaves are used to treat coughs. The tree makes an attractive ornamental subject on account of its striking flowers and fruits, and it is also a good shade tree [5].

Figure 3 – Traditional dish prepared with *G. coleosperma* seeds used as relish. (Photo: O. Morebodi, BUAN.)

KNOWN HAZARDS AND SAFETY

No known hazards.

CONSERVATION STATUS

A widespread species, protected by forestry legislation in Botswana and Namibia [2,5], but yet to be assessed according to IUCN Red List criteria in Botswana and to have its status confirmed in the IUCN Red List [3,7].

SEED CONSERVATION

HARVESTING. Dehiscent fruits can be harvested directly from the tree.

PROCESSING AND HANDLING. Seeds are easily extracted from the pods and the red aril can be removed before storing.

STORAGE AND VIABILITY. X-ray analysis carried out on a seed lot stored at the MSB highlighted that 100% of seeds were filled. Seeds of this species are reported to be likely orthodox [6].

PROPAGATION

SEEDS

Dormancy and pre-treatments: Untreated seeds germinated just as well (95%) as those treated with boiling water for one minute or with cold sulphuric acid for 10 minutes, suggesting that seeds of this species do not have dormancy [8].

Germination, sowing and planting: No information is available on the optimum germination requirements in the laboratory. In the nursery, germination usually starts within 10 days of sowing [5].

VEGETATIVE PROPAGATION

Can be propagated by cuttings and suckers.

TRADE

The timber is traded internationally [5].

Authors

Seoleseng O. Tshwenyane, Kebadire K. Mogotsi, Moctar Sacande, Tiziana Ulian and Efisio Mattana.

References

[1] Coates Palgrave, K. (1988). *Trees of Southern Africa*, 2nd edition. Struik Publishers, Cape Town, South Africa.

[2] Curtis, B. A. & Mannheimer, C. A. (2005). *Tree Atlas of Namibia*. National Botanical Research Institute of Namibia, Windhoek, Namibia.

[3] IUCN (2017). *The IUCN Red List of Threatened Species.* Version 2017-3. http://www.iucnredlist.org/

[4] Letloa LLHRC (Lands, Livelihoods and Heritage Resource Centre) (2007). *The Khwe of the Okavango Panhandle: the Use of Veld Plants for Food and Medicine.* CTP Book Printers, Cape Town, South Africa.

[5] Mojeremane, W. & Kopong, I. (2012). *Guibourtia coleosperma* (Benth.) J.Léonard. In: *Plant Resources of Tropical Africa 7 (2): Timbers 2*. Edited by R. H. M. J. Lemmens, D. Louppe & A. A Oteng-Amoako. PROTA Foundation, Wageningen, The Netherlands/CTA, Wageningen, The Netherlands. pp. 370–372.

[6] Royal Botanic Gardens, Kew (2017). *Seed Information Database (SID).* Version 7.1. http://data.kew.org/sid/

[7] Setshogo, M. P. & Hargreaves, B. (2002). Botswana. In: *Southern African Plant Red Data Lists*. Edited by J. S. Golding. Southern African Botanical Diversity Network Report No. 14, SABONET, Pretoria, South Africa. pp. 16–20.

[8] Tietema, T., Merkesdal, E. & Schroten J. (1992). *Seed Germination of Indigenous Trees in Botswana*. ACTS Press, African Centre for Technology Studies, Nairobi, Kenya.

[9] Timberlake, J. R., Pope, G. V., Polhill, R. M. & Martins, E. S. (eds) (2007). Leguminosae (Caesalpinioideae). In: *Flora Zambesiaca*, Vol. 3, part 2. Flora Zambesiaca Managing Committee, Royal Botanic Gardens, Kew.

Harpagophytum procumbens (Burch.) DC. ex Meisn.

TAXONOMY AND NOMENCLATURE

FAMILY. Pedaliaceae

SYNONYMS. *Harpagophytum burchellii* Decne., *Uncaria procumbens* Burch.

VERNACULAR / COMMON NAMES. Devil's claw, grapple plant, grapple thorn, wood spider (English); sengaparile (Tswana); otijhengatene (Herero, Namibia); elyata (Oshikwanyama, Namibia).

PLANT DESCRIPTION

LIFE FORM. Perennial herb.

PLANT. Prostrate (Figure 1), stems <2 m long, grow annually from a primary (or 'mother') tuber whose succulent taproot can grow to a depth of 2 m; secondary lateral tubers ('babies') develop on fleshy lateral roots, <25 cm long and 6 cm thick. **LEAVES.** Variable in shape, 6.5 x 4 cm, entire or deeply lobed, with 3 or 5 main lobes. **FLOWERS.** Solitary, showy, tubular, tube constricted at base; outside of tube usually pink, but can be white, yellow or purple, occasionally dark red; inside of tube yellow [7,8,18].

Figure 1 – Flowering plant.
(Photo: T. Ulian, RBG Kew.)

Figure 2 – X-ray image of the indehiscent fruit, showing the rows of seeds in two of the four loculi. (Photo: P. Gómez-Barreiro, RBG Kew.) .

Figure 3 – External views of the seeds and longitudinal and cross-sections showing outer and inner teguments and the spatulate embryo. (Photos: P. Gómez-Barreiro, RBG Kew.) .

FRUITS AND SEEDS

Fruits are large (7–20 x 6 cm), laterally compressed, woody indehiscent capsules with four rows of curved arms bearing recurved spines, the length of the longest arm exceeding the width of the capsule (Figure 2). The genus name and vernacular names refer to the claw-like shape of the fruit. Seeds black or brown, obovate with a rough and fibrillar outer region on the seed coat (Figure 3), c. 50 seeds per fruit [6,7,8].

FLOWERING AND FRUITING. Flowers from November to April; fruits ripen in January [16].

DISTRIBUTION

There are two subspecies: *H. procumbens* subsp. *procumbens* occurs in Botswana, Namibia and South Africa; and *H. procumbens* subsp. *transvaalense* Ihlenf. & H.E.K.Hartm. in South Africa and Zimbabwe. The species also occurs in Mozambique and Zambia [4,18].

HABITAT

Mainly associated with deep sandy soils, open plains and dunes, but also occurs on rocky soils and clay pans. *Harpagophytum procumbens* does not like competition from other plants, so it is often found in grazed (and overgrazed) areas or along tracks. However, it can also be found in dry savanna and open woodland. It grows well at 17–30°C and can withstand mild frost [11,16].

USES

The tubers are used in traditional medicine to treat a wide range of disorders [12,14]. The Khwe people use *H. procumbens* for treating sexually transmitted diseases, particularly gonorrhoea and syphilis, and for treating intestinal worms [9]. It is also used to treat osteoarthritis, fibrosis, rheumatism, atherosclerosis, lumbago, digestive tract disorders, and diseases of the liver, kidney and bladder, as well as blood and heart-related disorders [18]. Small doses are used for menstrual cramps, whereas higher doses are used to expel a retained placenta. It is also used as a postpartum analgesic, and for uterine contraction. It is used to relieve joint and muscle pain, including gout and back pain. Taken on a regular daily basis, it has a subtle laxative effect [18]. The Tsetseng community sustainably harvest the tubers for income generation (Figure 4).

KNOWN HAZARDS AND SAFETY

In high doses, extracts of devil's claw used in medicines could be hazardous. Devil's claw can be allergenic, and in Western medicine it is contraindicated for diabetics and people with duodenal and gastric ulcers. Pregnant and breastfeeding women, and people with heart disease, or high or low blood pressure, should seek medical advice before taking devil's claw [10]. The fruits should be harvested and processed with caution to avoid cuts from the sharp 'claws'. In Southern Africa, grazing animals can be injured if they tread on the fruit, or they may starve if the fruit gets caught in their mouths [16,20].

Figure 4 – Semi-processed tubers displayed by ICDT, as part of the dissemination activities of the UPP carried out at the Vision 2016 Commemoration Day.
(Photo: K. K. Mogotsi, BUAN.)

CONSERVATION STATUS

Large quantities of the tubers are harvested directly from the wild. For conservation reasons, the primary root should be replanted and only the secondary roots used [19]. Assessed as Least Concern (LC) according to IUCN Red List criteria in South Africa, and Lower Risk-Near Threatened (LR-nt) in Botswana [15,20], but its status has yet to be confirmed in the IUCN Red List [5].

SEED CONSERVATION

HARVESTING. The fruits are harvested directly from the plant or are gathered from the ground.

PROCESSING AND HANDLING. Seed extraction from the thorny and woody fruit is difficult and laborious, requiring the use of tools, such as secateurs, to remove the hooks and to crack open the hard fruit. Care is required to avoid crushing the seed inside the fruit.

STORAGE AND VIABILITY. X-ray analysis carried out on seed lots stored at the MSB highlighted a high percentage of filled seeds (70–100%) with the majority of unsound seeds being empty. No information on seed storage behaviour is available. However, seeds of the closely related *H. zeyheri* Decne. are reported to be desiccation tolerant [13], suggesting that seeds of *H. procumbens* are likely to be orthodox.

PROPAGATION

SEEDS

Dormancy and pre-treatments: Pedaliaceae is not listed among families showing physical dormancy. Although no information is available in the literature for *H. procumbens*, physiological dormancy is reported in other Pedaliaceae spp. [1]. There is evidence that the inner seed coat may be involved in morphological and possibly chemical dormancy [6]. In particular, there may be a barrier to full imbibition of the embryo and the endosperm [3], and the presence of phenolic compounds and other dark-staining substances may contribute towards dormancy [6]. Addition of gibberellic acid promoted germination, but concentrated sulphuric acid gave no germination response [3]. Further studies are needed to characterise seed dormancy in this species fully.

Germination, sowing and planting: Very low germination (c. 4%) is reported for seeds incubated at 45°/10°C, without any pre-treatments [3]. Cultivation is difficult, so most tubers come from plants growing in the wild [17].

VEGETATIVE PROPAGATION

Can be propagated by primary or secondary tubers. The tubers should be planted 10 cm deep and 50 cm apart [18].

TRADE

Devil's claw is widely used as a herbal medicine and is internationally traded for its anti-inflammatory and analgesic properties. The local market in Botswana is managed by Thusano Lefatsheng, an NGO that also sells devil's claw products on the international market. Semi-processed products generate income for local communities in the Kgalagadi, Kweneng, Southern and Ghanzi districts [2]. In 2000, recommendations were made to the Convention on International Trade in Endangered Species of Wild Fauna and Flora (CITES) to add devil's claw to Appendix II. In 2004, the proposal was withdrawn as a result of efforts made by range states to address sustainability issues [17]. Harvesting is regulated in Botswana under the Agricultural Resources Regulations (2006). Permits to harvest, transport, process and export devil's claw are also required in Namibia and South Africa [16].

Authors

Seoleseng O. Tshwenyane, Kebadire K. Mogotsi, Moctar Sacande, Tiziana Ulian, Steve Davis and Efisio Mattana.

References

[1] Baskin, C. C. & Baskin, J. M. (2014). *Seeds: Ecology, Biogeography, and Evolution of Dormancy and Germination*, 2nd edition. Academic Press, San Diego, USA.

[2] Department of Forestry and Range Resources (DFRR) (2009). *Grapple Plant.* DFRR Veldt Product Series, number 2, Gaborone, Botswana.

[3] Ernst, W. H. O., Tietema, T., Veenendaal, E. M. & Masene, R. (1988). Dormancy, germination and seedling growth of two Kalaharian perennials of the genus *Harpagophytum* (Pedaliaceae). *Journal of Tropical Ecology* 4 (2): 185–198.

[4] Ihlenfeldt, H.-D. & Hartmann, H. (1970). Die Gattung *Harpagophytum* (Burch.) DC. ex Meissn. (Monographien der afrikanischen Pedaliaceae). *Mitt. Staatsinst. Allg. Bot. Hamburg* 13: 15–69.

[5] IUCN (2017). *The IUCN Red List of Threatened Species.* Version 2017-3. http://www.iucnredlist.org/

[6] Jordaan, A. (2011). Seed coat development, anatomy and scanning electron microscopy of *Harpagophytum procumbens* (devil's claw), Pedaliaceae. *South African Journal of Botany* 77 (2): 404–414.

[7] Kadereit, J. W. (ed.) (2004). *The Families and Genera of Vascular Plants*, Vol. VII: *Flowering Plants. Dicotyledons: Lamiales (except Acanthaceae including Avicenniaceae)*. Edited by K. Kubitzki. Springer, Berlin.

[8] Launert, E. (ed.) (1988). Pedaliaceae. In: *Flora Zambesiaca*, Vol. 8, part 3. Flora Zambesiaca Managing Committee, Royal Botanic Gardens, Kew.

[9] Letloa LLHRC (Lands, Livelihoods and Heritage Resource Centre) (2007). *The Khwe of the Okavango Panhandle: the Use of Veld Plants for Food and Medicine.* CTP Book Printers, Cape Town, South Africa.

[10] MedlinePlus (2015). *Devil's Claw.* National Library of Medicine (US), Bethesda, Maryland, USA. https://medlineplus.gov/

[11] Nott, K. (1986). A survey of the harvesting and export of *Harpagophytum procumbens* and *Harpagophytum zeyheri* in SWA/Namibia. Unpublished report, Etosha Ecological Institute, Okaukuejo, Namibia.

[12] Royal Botanic Gardens, Kew (1999–2015). *Survey of Economic Plants for Arid and Semi-Arid Lands (SEPASAL) Database.* http://apps.kew.org/sepasalweb/sepaweb

[13] Royal Botanic Gardens, Kew (2017). *Seed Information Database (SID).* Version 7.1. http://data.kew.org/sid/

[14] Schneider, E., Sanders, J. & Von Willert, D. (2006). Devil's claw (*Harpagophytum procumbens*) from Southern Africa. In: *Medicinal and Aromatic Plants: Agricultural, Commercial, Ecological, Legal, Pharmacological and Social Aspects.* Edited by R. J. Bogers, L. E. Cracker & D. Lange. Springer, Dordrecht, The Netherlands. pp. 181–202.

[15] Setshogo, M. P. & Hargreaves, B. (2002). Botswana. In: *Southern African Plant Red Data Lists.* Edited by J. S. Golding. Southern African Botanical Diversity Network Report No. 14, SABONET, Pretoria, South Africa. pp. 16–20.

[16] Smithies, S. (2006). *Harpagophytum procumbens* (Burch.) DC. ex Meisn. subsp. *procumbens* and subsp. *transvaalense* Ihlenf. & H.E.K.Hartmann. South African National Biodiversity Institute (SANBI), South Africa. http://pza.sanbi.org/

[17] Stewart, K. M. & Cole, D. (2005). The commercial harvest of devil's claw (*Harpagophytum* spp.) in Southern Africa: the devil's in the details. *Journal of Ethnopharmacology* 100 (3): 225–236.

[18] Van Wyk, B.-E. & Gericke, N. (2000). *People's Plants: A Guide to Useful Plants of Southern Africa.* Briza Publications, Pretoria, South Africa.

[19] Van Wyk, B.-E., Van Oudtshoorn, B. & Gericke, N. (2009). *Medicinal Plants of South Africa*, 2nd edition. Briza Publications, Pretoria, South Africa.

[20] York, E. (ed.) (undated). *Harpagophytum procumbens* (devil's claw). Royal Botanic Gardens, Kew. http://www.kew.org/science-conservation/plants-fungi/harpagophytum-procumbens-devils-claw

Hyphaene petersiana Klotzsch ex Mart.

TAXONOMY AND NOMENCLATURE

FAMILY. Arecaceae

SYNONYMS. *Hyphaene aurantiaca* Dammer, *H. benguelensis* Welw. ex H.Wendl., *H. bussei* Dammer, *H. goetzei* Dammer, *H. obovata* Furtado, *H. ovata* Furtado, *H. plagiocarpa* Dammer, *H. ventricosa* Kirk

VERNACULAR / COMMON NAMES. Doum, dum palm, ilala palm, makola palm, mulala palm, northern ilala, real fan palm, vegetable ivory palm (English); mokolane, mokolwane (Tswana); evare (Herero); omulunga (Ndonga, Namibia); epokola (Oshikwanyama, Namibia); inala, muchindwi, mulala, munganda, mungwenji, murara (Shona).

PLANT DESCRIPTION

LIFE FORM. Palm tree.

PLANT. Solitary or rarely clustered (by branching below ground), dioecious palm, <20 m tall, with an almost always unbranched trunk, marked with horizontal leaf scars. **LEAVES.** Bluish grey, fan-shaped, <2 m long, with black thorns along the stalks. **FLOWERS.** Male inflorescences pendulous, somewhat elongate, <2 m long; female inflorescences smaller (<1.25 m long) (Figure 1) [4,6,7].

Figure 1 – *H. petersiana* palm tree. (Photo: T. Ulian, RBG Kew.)

FRUITS AND SEEDS

Fruits variable in shape but always rounded, ovoid, or near globose, typically 5–7 cm in diameter. The epicarp is very smooth and polished with only slight dimpling; reddish brown in colour (Figure 2). When ripe the epicarp parts easily from the aromatic, yellow orange flesh of the fibrous mesocarp [4]. Seeds flat at the base, nearly round in horizontal section, c. 2.9 cm long, 3.2 cm wide, with the apex slightly peaked at the embryo [4]. The hard white kernel of the seed (known as 'vegetable ivory') is surrounded by a fibrous outer husk. Inside the seed there is a small quantity of sweet liquid similar to coconut milk (Figure 2) [7].

FLOWERING AND FRUITING. Flowers from September to October. Fruiting can take place at any time of the year. Fully grown fruits can remain on the tree for more than two years. Fruits fall directly below the parent tree and are dispersed by seasonal flood waters or by elephants and baboons [4].

Figure 2 – External views of the fruits and section showing the hard white kernel and the empty space where sweet liquid was contained. (Photos: P. Gómez-Barreiro, RBG Kew.)

DISTRIBUTION

From Tanzania (Lake Manyara) and Mozambique in the east, to Angola and Namibia in the west, and from the Democratic Republic of Congo to South Africa. In Botswana, it occurs in the north of the country, where it is very common in the Okavango Delta [4,8].

HABITAT

Alluvial dry soils, from sandy to clays, with high water table, usually at 275–1,300 m a.s.l. [4].

USES

Hyphaene petersiana is used extensively by local people. Together with *H. coriacea* Gaertn., it is an important source of palm wine and contributes significantly to the rural economy. The sap of the palm is collected by burning stem clumps to make the stems accessible and to remove the leaf spines. The leaves are then trimmed with a sharp knife to start sap flow. The wine is usually consumed 36 hours after collection when the sap has been sufficiently fermented by natural yeasts. The wine is normally used in undiluted form, but may also be diluted (and sugar added) to increase profits. Palm wine is an important dietary supplement, especially for men in rural areas, adding substantial quantities of nicotinic acid, vitamin C and potassium to the normal starchy diet. A tree can produce 60–70 litres of sap (although repeated cutting may kill the plant). The fruits are also edible, as is the pith of young stems and leaves. The fibrous outer layer of the fruit is sweet and the fluid within the seed is similar to coconut milk in taste and appearance. The leaves are used for roofing and basket weaving. The hard, white nut is carved to make hand-crafted figures and ornaments. Juvenile leaves are used to make baskets, especially in Botswana (Figure 3). Palm fruits are a favourite food of elephants. These animals bump trees to dislodge the fruits which are swallowed whole, the flesh digested, and the seed excreted in a pile of manure, ready to germinate. In the absence of alternative fodder, young unopened leaves are browsed by domestic livestock, hippopotamuses and elephants [4].

KNOWN HAZARDS AND SAFETY

No known hazards.

Figure 3 – Baskets made from juvenile leaves of *H. petersiana* and a basket shop in Etsha, Ngamiland, Botswana selling products made from *H. petersiana* leaves. (Photos: L. Mogotsi-Yimbo and T. T. Mogotsi, BUAN.)

CONSERVATION STATUS

Assessed as Least Concern (LC), but locally threatened by overexploitation of the leaves [2,3]. Sap collection (for palm wine production) is often fatal to the tree, so the palm is disappearing from areas where exploitation is more intense. Surveys are needed in Botswana [5], South Africa and Zimbabwe to better understand current harvest pressures and to identify whether conservation measures are needed in some areas [3].

SEED CONSERVATION

HARVESTING. Ripe fruits that fall to the ground are easily collected. Alternatively, ripe fruits can be detached from the tree with the help of a long stick or by throwing stones.

PROCESSING AND HANDLING. Fruits are chewed to obtain the seed, or the fruit flesh is separated from the seed using a knife when the fruit is still fresh. In the laboratory, seeds may be extracted from the fruits by sawing through the endocarp [1].

STORAGE AND VIABILITY. Seeds are long-lived in dry, warm storage (21°C and 55% RH), with viability being retained for 2–5 years. Some seeds showed sensitivity to desiccation at low humidity (c. 20–30% RH) and susceptibility to –20°C freezing. Thus, conservation under conventional seed bank conditions is not yet guaranteed for all seeds [1].

PROPAGATION

SEEDS

Dormancy and pre-treatments: Soaking seeds in water before sowing reduced emergence time from 24–50 to 17–22 days but did not improve the germination percentage [1].

Germination, sowing and planting: Seeds incubated at 26°C, with a photoperiod of 12 hours of darkness and 12 hours of light, reached germination percentages of up to 70%, with mean emergence time of c. 30 days [1]. Seeds do not germinate easily, and plants are slow growing. The massive taproot makes it almost impossible to transplant the trees once they are established.

VEGETATIVE PROPAGATION

No protocols available.

TRADE

Baskets made in Botswana from juvenile leaves are sold locally and are in great demand in Europe and the USA [4].

Authors

Kebadire K. Mogotsi, Cecilia Amosso, Moctar Sacande, Gloria N. Mashungwa, Tiziana Ulian and Efisio Mattana.

References

[1] Davies, R. I. & Pritchard, H. W. (1998). Seed storage and germination of the palms *Hyphaene thebaica*, *H. petersiana* and *Medemia argun*. *Seed Science and Technology* 26: 823–828.

[2] Foden, W. & Potter, L. (2005). *Hyphaene petersiana* Klotzsch ex Mart. In: *National Assessment: Red List of South African Plants*. Version 2015.1. http://redlist.sanbi.org/

[3] IUCN (2017). *The IUCN Red List of Threatened Species.* Version 2017-3. http://www.iucnredlist.org/

[4] Royal Botanic Gardens, Kew (1999–2015). *Survey of Economic Plants for Arid and Semi-Arid Lands (SEPASAL) Database*. http://apps.kew.org/sepasalweb/sepaweb

[5] Setshogo, M. P. & Hargreaves, B. (2002). Botswana. In: *Southern African Plant Red Data Lists*. Edited by J. S. Golding. Southern African Botanical Diversity Network Report No. 14, SABONET, Pretoria, South Africa. pp. 16–20.

[6] Timberlake, J. R. & Martins, E. S. (eds) (2010). Arecaceae (Palmae). In: *Flora Zambesiaca*, Vol. 13, part 2. Flora Zambesiaca Managing Committee, Royal Botanic Gardens, Kew.

[7] Van Wyk, B.-E. & Gericke, N. (2000). *People's Plants: A Guide to Useful Plants of Southern Africa*. Briza Publications, Pretoria, South Africa.

[8] WCSP (2017). *World Checklist of Selected Plant Families.* Royal Botanic Gardens, Kew. http://wcsp.science.kew.org/

Kalaharituber pfeilii (Henn.) Trappe & Kagan-Zur

TAXONOMY AND NOMENCLATURE

FAMILY. Pezizaceae

SYNONYMS. *Terfezia pfeilii* Henn., *Tuber pfeilii* Henn.

VERNACULAR / COMMON NAMES. Kalahari truffle (English); mahupu (Tswana); omatumbula (Bantu, Ndonga); kgaboyamahupu, magupu, mosasawe (Kgalagadi); haban, hawan, n/abba (Khoekhoe); kuutse (Khoisan); dcoodcoò (Khoisan, Jul'hoansi).

DESCRIPTION

LIFE FORM. Hypogeous fungus (truffle).

FUNGUS. Ascomata (fruiting bodies) hypogeous, smooth, minutely pubescent, turbinate to obpyriform or subglobose, 7(–11) x 7(–12) cm in diameter; peridium minutely pubescent and covered with soil or sand up to 1 cm thick, usually pale to dark brown with yellowish wrinkles or cracks (Figure 1); gleba solid, fleshy, marbled with white veins and fertile pockets of yellowish white to brown; odour strongly fungoid; taste mild. Ascospores globose 16–22(26) µm in diameter, hyaline (translucent), turning light brown with age, ornamentation organised in dense acute spines. Asci with 5–8 ascospores, globose to ellipsoid or obovoid, 70–100 x 50–80 µm, arranged randomly in fertile pockets, not reactive to iodine solution. Peridial and glebal tissues with many thin-walled and inflated cells [2,9,11].

Figure 1 – *K. pfeilii* fruiting body.
(Photo: T. Mukwevho, UNISA.)

Desert truffles may vary in colour due to sand sticking to their surfaces. Mature truffles weigh up to c. 200 g depending on rainfall, but are usually 25–45 g [6,14,16]. Truffles weighing 400 g and 500 g have been found in South Africa and Botswana, respectively [6]. The fruiting bodies are carried on a 'stalk', 2–10 cm in length, composed of entangled hyphae, plant roots and soil or sand particles (Figure 2). The base of the stalk is connected to rhizomorphs and hyphae emanating from adjacent plant roots. The rhizomorphs extend from colonised roots to the fruiting body, which may be up to 40 cm away [6,14]. All truffles, including desert truffles, are mycorrhizal. *Kalaharituber pfeilii* has been reported to form endomycorrhizae lacking Hartig net and mantle, but displaying undifferentiated intracellular hyphae [5].

Figure 2 – A 'family' of four truffles in the sand.
(Photo: T. Mukwevho, UNISA.)

SEASON. Truffles mature from March to July, or sometimes up to August, at the end of the rainy season in Botswana (Kalahari), depending heavily on rainfall and region, during which the thin-walled cells of immature fruiting bodies take up large amounts of water to sustain spore formation. The process of swelling and expansion of the fruiting body results in cracked mounds appearing on the surface of the soil [16].

DISTRIBUTION

Kalahari Desert: Botswana, Namibia, South Africa and adjacent arid areas [9].

HABITAT

Kalahari truffles are found in arid and semi-arid areas of the Kalahari Desert in Southern Africa that have at least 200–250 mm rainfall per season [6], in fairly compact, pink or infrequently white, slightly calcareous sands (pH 5.5–7.2) [13], or in red sands [3,6]. They can also occur in the dips between sand dunes [13] and in arenosols [16]. Associated plants establishing symbiosis with *K. pfeilii* include mostly woody species, such as *Vachellia hebeclada* (DC.) Kyal. & Boatwr. [syn. *Acacia hebeclada* DC.], *Senegalia mellifera* (Vahl) Seigler & Ebinger [syn. *A. mellifera* (Vahl) Benth.], *Acacia uncinata* Lindl., *Boscia albitrunca* (Burch.) Gilg & Benedict, *Dichrostachys cinerea* (L.) Wight & Arn., *Grewia flava* DC. and *Terminalia sericea* Burch. ex DC., as well as herbaceous species such as the grasses *Aristida* spp. and *Eragrostis* spp., *Enneapogon cenchroides* (Licht. ex Roem. & Schult.) C.E.Hubb. and *Stipagrostis uniplumis* (Licht.) De Winter [13,14], and species of *Citrullus* (watermelon) [5] and *Tylosema* (morama) [11,16], *Rhigozum brevispinosum* Kuntze and *R. trichotomum* Burch. Kalahari truffles can also been found in cultivated fields of various food crops: e.g. pearl millet (*Pennisetum typhoides* (Burm.f.) Stapf & C.E.Hubb.) [16] and sorghum (*Sorghum bicolor* (L.) Moench) [8,11]. Records of Kalahari truffle spores being dispersed by animals are scarce [15,16].

USES

The truffles (Figure 3) are eaten raw or cooked (either boiled, roasted over a fire or buried in hot ashes [14], or fried). They are an important part of the diet of people living in arid and semi-arid areas [1,12]. They are highly nutritious with higher protein, fat, fibre, magnesium, potassium and phosphate content than many vegetables [8]. On a weight-for-weight basis, Kalahari truffles have been shown to be second only to cooked maize among 24 vegetables analysed for energy content (Nutrition Information Center, University of Stellenbosch, 2008, cited in [14]). They also have good antioxidant properties and are considered to be an aphrodisiac. The fungus is used in traditional Khoisan medicine as an antidote to some poisons [14].

Figure 3 – Fruiting bodies collected, peeled, sun-dried and ready for market in Tsetseng.
(Photos: A. Hudson, RBG Kew; K. K. Mogotsi, BCA; T. Mukwevho, UNISA.)

KNOWN HAZARDS AND SAFETY

No known hazards (see [6]).

CONSERVATION STATUS

Assessed as Vulnerable (VU) according to IUCN criteria [10], but its status has yet to be confirmed in the IUCN Red List [4]. It is currently under assessment in the Global Fungal Red List Initiative [15]. The population is decreasing due to commercial harvesting, land use changes and overgrazing by livestock. Climate change is considered to be a potential future threat.

CONSERVATION

HARVESTING. The presence of *K. pfeilii* is commonly determined by the appearance of small cracked mounds in the soil resulting from the swelling and expansion of fruiting bodies beneath the surface [13,16]. The truffles are then extracted by hand or with the use of digging sticks. They may also be partly exposed above ground if winds have blown away the sand [16]. In Botswana, Basarwa and Khoisan, communities are highly knowledgeable in finding truffles. Harvesting is from March to June [9].

PROCESSING AND HANDLING. Soil should be washed off the truffles after collecting. They can then be sliced and sun-dried for future use.

STORAGE AND VIABILITY. Spores from desert truffles can be sun-dried and stored for nine months in a cold room without losing viability (Figure 4).

Figure 4 – Drying truffles at the BCA glasshouse.
(Photo: A. Hudson, RBG Kew.)

PROPAGATION

Spores do not exhibit dormancy and no pre-treatment is needed. However, germination of spores has been shown to improve if the sporocarps are first dried in the sun.

TRADE

Kalahari truffles are sold in local markets, providing gatherers with an important source of income. Tsetseng and Kacgae communities sell truffles for income generation. Commercial harvesting of truffles is increasing [7,16]. There is a need to ensure local communities are the primary beneficiaries of commercial development (Mshigeni *et al.*, 2005, cited in [15]).

Authors

Kebadire K. Mogotsi, Bamphitlhi Tiroesele, Cecilia Amosso, Moctar Sacandé, Tiziana Ulian, Efisio Mattana and Ester Gaya.

References

[1] Ackerman, L. G., Van Wyk, P. J. & du Plessis, L. M. (1975). Some aspects of the composition of the Kalahari truffle or N'abba. *South African Food Review* 2 (5): 145–147.

[2] Ferdman, Y., Aviram, S., Roth-Bejerano, N., Trappe, J. M. & Kagan-Zur, V. (2005). Phylogenetic studies of *Terfezia pfeilii* and *Choiromyces echinulatus* (Pezizales) support new genera for southern African truffles: *Kalaharituber* and *Eremiomyces*. *Mycological Research* 109 (2): 237–245.

[3] Giovannetti, G., Roth-Bejerano, N., Zanini, E. & Kagan-Zur, V. (1994). Truffles and their cultivation. *Horticultural Reviews* 16: 71–107.

[4] IUCN (2017). *The IUCN Red List of Threatened Species.* Version 2017-3. http://www.iucnredlist.org/

[5] Kagan-Zur, V., Kuang, J., Tabak, S., Taylor, F. W. & Roth-Bejerano, N. (1999). Potential verification of a host plant for the desert truffle *Terfezia pfeilii* by molecular methods. *Mycological Research* 103 (10): 1270–1274.

[6] Kagan-Zur, V. & Roth-Bejerano, N. (2008). Desert truffles. *Fungi* 1: 32–37.

[7] Kagan-Zur, V., Roth-Bejerano, N. & Taylor, F. W. (1995). *Cultivation of Terfezia pfeilii – the Kalahari Desert Truffle. Final Report (March 1991 – February 1995).* The Institutes for Applied Research, Ben-Gurion University of the Negev, Beer-Sheva, Israel.

[8] Kasterine, A. & Hughes, K. (2012). *The North American Market for Natural Products: Prospects for Andean and African Products.* International Trade Centre (ITC), Geneva, Switzerland.

[9] Khonga, E. B., Modise, D. M., Mogotsi, K. K., Machacha, S. & Mpotokwane, S. (2005). Review of macrofungi and their potential as alternative crops in Botswana. In: *Proceedings of the Second Crop Science and Production Conference, CICE, Botswana College of Agriculture, Botswana, 6–8 September 2005 – Sustainable Strategies for Development of Commercial Crop Production in Botswana.*

[10] Minter, D. W. (2013). *Kalaharituber pfeilii.* IMI Descriptions of Fungi and Bacteria, No. 198, Sheet 1972.

[11] Roth-Bejerano, N., Li, Y. F. & Kagan-Zur, V. (2004). Homokaryotic and heterokaryotic hyphae in *Terfezia*. *Antonie van Leeuwenhoek* 85 (2): 165–168.

[12] Sawaya, W. N., Al-Shalhat, A., Al-Sogair, A. & Al-Mohammad, M. (1985). Chemical composition and nutritive value of truffles of Saudi Arabia. *Journal of Food Science* 50 (2): 450–453.

[13] Story, R. (1958). Some plants used by Bushmen in obtaining food and water. *Botanical Survey of South Africa Memoir* 30: 1–115.

[14] Taylor, F. W., Thamage, D. M., Baker, N., Roth-Bejerano, N. & Kagan-Zur, V. (1995). Notes on the Kalahari desert truffle, *Terfezia pfeilii*. *Mycological Research* 99 (7): 874–878.

[15] *The Global Fungal Red List* (2017). http://iucn.ekoo.se/iucn/species_view/369302

[16] Trappe, J. M., Claridge, A. W., Arora, D. & Smit, W. A. (2008). Desert truffles of the African Kalahari: ecology, ethnomycology, and taxonomy. *Economic Botany* 62 (3): 521–529.

Lippia javanica (Burm.f.) Spreng.

TAXONOMY AND NOMENCLATURE

FAMILY. Verbenaceae

SYNONYMS. *Lantana capensis* (Thunb.) Spreng., *L. galpiniana* H.Pearson, *L. indica* Moldenke, *L. scabra* Hochst., *L. whytei* Moldenke, *Verbena capensis* Thunb., *V. javanica* Burm.f.

VERNACULAR / COMMON NAMES. Fever tea, fever tree, lemon bush, wild sage, wild tea (English); bokhukhwane, mosukudu, musukudu (Tswana).

PLANT DESCRIPTION

LIFE FORM. Shrub.

PLANT. Erect, multi-stemmed, <2(–3) m tall, much-branched, often from the base (Figure 1). Stems subterete or somewhat angular (square in cross-section), green, soon turning brown, striate. **LEAVES.** Strongly aromatic (aroma lemon-like) when crushed, opposite-decussate, 2–10 x 1–3 cm, lanceolate, wrinkled, dull green above, grey green and pubescent below. **FLOWERS.** Small (c. 3 mm long), tubular, creamy white, in dense rounded flower heads [8] (Figure 1).

Figure 1 – *L. javanica* planted along a fence in the Tsetseng community garden and close-up of the flowers.
(Photos: C. Varet, The Ivory Foundation, Paris, and B. Wursten, Flora of Zimbabwe: https://www.zimbabweflora.co.zw/index.php.)

FRUITS AND SEEDS

Mericarps (hereafter seeds) brown on the convex face, white on the flat, commissural face, 1.5–1.75 x 1–1.25 mm, half ovoid or subspheroidal (Figure 2) [8].

FLOWERING AND FRUITING. Flowers mainly in summer and autumn in Southern Africa, but in some places flowers found all year round [7]. Seeds available from January to March.

Figure 2 – Schizocarps with the two attached mericarps, and detached mericarps showing the white colour on the flat commissural face. (Photo: P. Gómez-Barreiro, RBG Kew.)

DISTRIBUTION

Widespread, from tropical East Africa (Ethiopia, Kenya, Tanzania and Uganda) and Central Africa (Central African Republic and Democratic Republic of Congo) south to Botswana, South Africa, Malawi, Mozambique, Zambia and Zimbabwe [16].

HABITAT

Woodland and forest margins, wooded grassland, riverine vegetation, margins of swampy ground, disturbed and cultivated lands, at altitudes of 10–2,240 m a.s.l. [3].

USES

In traditional medicine, an infusion of the leaves is used to treat coughs, common cold, fever, bronchitis and malaria, and is a calmative. A lotion made from the leaves is applied topically for scabies, measles and lice [14], and to treat skin rashes, stings and bites. The essential oil shows moderate antibacterial activity and promising anti-inflammatory activity [12]. The volatile oil could have commercial value as a perfume and insect repellent, for example, for the control of bark beetles of the genus *Ips* [1,13].

KNOWN HAZARDS AND SAFETY

Iridoid glycosides and toxic triterpenoids (icterogenins) have been detected in some *Lippia* spp. Ingestion of icterogenins by livestock results in the animals becoming photosensitive, and is suspected of causing geeldikkop, a serious disease of sheep in Southern Africa [14,15]. When handling and applying plant materials in the field, for example as insect repellents, contact with the skin should be avoided. In the case of accidental contact, affected areas should be washed immediately with clean running water [1].

CONSERVATION STATUS

Assessed as Least Concern (LC) in South Africa according to IUCN Red List criteria [4], but has yet to be assessed in Botswana and to have its status confirmed in the IUCN Red List [5,10].

SEED CONSERVATION

HARVESTING. Seeds are collected when those of the entire infructescence are ready for dispersal. Care should be taken not to drop the tiny seeds.

PROCESSING AND HANDLING. Seeds are removed from the infructescence by gently rubbing them between the fingers. Seeds are separated from other plant material by sieving.

STORAGE AND VIABILITY. Seeds of this species are reported to be desiccation tolerant [9]. X-ray analysis carried out on seed lots stored at the MSB highlighted 20–60% empty seeds.

PROPAGATION

SEEDS

Dormancy and pre-treatments: Seeds of *Lippia* spp. are reported to be both non-dormant and physiologically dormant [2]. Seeds of this species are suspected to be non-dormant because applied treatments (gibberellic acid [GA_3] and potassium nitrate [KNO_3]) did not improve seed germination compared to untreated seeds. However, further studies are needed to exclude any type of physiological dormancy [6]. Base temperature for germination (Tb, °C) and the thermal constant for 50% germination (S, °Cd) of untreated seeds were 7.5°C and 84.3°Cd respectively [6].

Germination, sowing and planting: Seed lots stored at the MSB attained high germination percentages (80–100%) when incubated in the light at constant temperatures (20–30°C) or at alternating temperature regimes of 25/10°C and 35/20°C [9]. Good germination takes place if the seeds are sown in sand, although germination may take up to 1.5 months. Plants grow relatively fast, preferring sunny places and most soil types [7].

VEGETATIVE PROPAGATION

Can be propagated by apical and basal cuttings using (ideally) a pine bark medium, and treating the cuttings with 0.3% IBA (indole-3-butyric acid) rooting hormone. The rooted cuttings are ready to be transplanted after 15–20 days [11].

TRADE

Herbal teas derived from this species have been produced and packaged for sale in shops and pharmacies in Botswana and for export by Thusano Lefatsheng (an NGO) since the early 1990s. Informal trade also occurs in major villages, towns and cities.

Authors

Seoleseng O. Tshwenyane, Cecilia Amosso, Kebadire K. Mogotsi, Moctar Sacande, Tiziana Ulian, Steve Davis and Efisio Mattana.

References

[1] Anjarwalla, P., Belmain, S., Koech, G., Jamnadass, R. & Stevenson, P. C. (2015). *Lippia javanica* (Burm.f.) Spreng. *Pesticidal Plant Leaflet.* World Agroforestry Centre, Nairobi, Kenya and University of Greenwich, Natural Resources Institute, Chatham, UK.

[2] Baskin, C. C. & Baskin, J. M. (2014). *Seeds: Ecology, Biogeography, and Evolution of Dormancy and Germination*, 2nd edition. Academic Press, San Diego, USA.

[3] Coates Palgrave, K. & Coates Palgrave, M. (2002). *Trees of Southern Africa*, 3rd edition. Struik Publishers, Cape Town, South Africa.

[4] Foden, W. & Potter, L. (2005). *Lippia javanica* (Burm.f.) Spreng. In: *National Assessment: Red List of South African Plants.* Version 2015.1. http://redlist.sanbi.org/t

[5] IUCN (2017). *The IUCN Red List of Threatened Species.* Version 2017-3. http://www.iucnredlist.org/

[6] Mattana, E., Sacande, M., Sanogo, A. K., Lira, R., Gómez-Barreiro, P., Rogledi, M. & Ulian, T. (2017). Thermal requirements for seed germination of underutilised *Lippia* species. *South African Journal of Botany* 109: 223–230.

[7] Le Roux, L.-N. (2004). *Lippia javanica* (Burm.f.) Spreng. South African National Biodiversity Institute (SANBI), South Africa. http://pza.sanbi.org/

[8] Pope, G. V. & Martins, E. S. (eds) (2005). Verbenaceae. In: *Flora Zambesiaca*, Vol. 8, part 7. Flora Zambesiaca Managing Committee, Royal Botanic Gardens, Kew.

[9] Royal Botanic Gardens, Kew (2017). *Seed Information Database (SID).* Version 7.1. http://data.kew.org/sid/

[10] Setshogo, M. P. & Hargreaves, B. (2002). Botswana. In: *Southern African Plant Red Data Lists.* Edited by J. S. Golding. Southern African Botanical Diversity Network Report No. 14, SABONET, Pretoria, South Africa. pp. 16–20.

[11] Soundy, P., Mpati, K. W., du Toit, E. S., Mudau, F. N. & Araya, H. T. (2008). Influence of cutting position, medium, hormone and season on rooting of fever tea (*Lippia javanica* L.) stem cuttings. *Medical and Aromatic Plant Science and Biotechnology* 2 (2): 114–116.

[12] Van Wyk, B.-E. (2008). A broad review of commercially important Southern African medicinal plants. *Journal of Ethnophamacology* 119 (3): 342–355.

[13] Van Wyk, B.-E. & Gericke, N. (2000). *People's Plants: A Guide to Useful Plants of Southern Africa.* Briza Publications, Pretoria, South Africa.

[14] Van Wyk, B.-E., Van Oudtshoorn, B. & Gericke, N. (2009). *Medicinal Plants of South Africa*, 2nd edition. Briza Publications, Pretoria, South Africa.

[15] Watt, J. M. & Breyer-Brandwijk, M. G. (1962). *The Medicinal and Poisonous Plants of Southern and Eastern Africa*, 2nd edition. E. & S. Livingstone, Edinburgh & London.

[16] WCSP (2017). *World Checklist of Selected Plant Families.* Royal Botanic Gardens, Kew. http://wcsp.science.kew.org/

Myrothamnus flabellifolia Welw.

TAXONOMY AND NOMENCLATURE

FAMILY. Myrothamnaceae

SYNONYMS. None.

VERNACULAR / COMMON NAMES. Bush tea, resurrection bush, resurrection plant (English); galalatshwene, maritatshwene, monnaonkganang (Tswana); ikalimela (Shona).

PLANT DESCRIPTION

LIFE FORM. Shrub.

PLANT. Woody, dioecious, prostrate, ascending or erect, <90 cm (occasionally 2 m) [5]. Young branches 4-angled, soon becoming sub-spinulose and woody. **LEAVES.** Small, opposite, flat, usually 10–14 x 6–8 mm, or folded, blackening and folding fan-like when dry, expanding and greening again after rain [14]. **FLOWERS.** Inflorescences 2–3 cm long, on short lateral branches [2,5]. Male flowers consist of 3–6 stamens with reddish anthers that dehisce longitudinally; female flowers are zygomorphic and consist of 3 basally attached carpels with papillose/spatulate stigmas that are reddish purple and feather-like in appearance [7]. The vegetative tissue of the plant, in particular the leaves, can dehydrate to an air-dry state; the leaves and stem segments curl and change colour from green to dull brown (Figure 1), re-hydrating when water is available (poikilohydry) [10,12]. Leaves that may appear 'dead' revive quickly; hence the English vernacular name 'resurrection plant'.

Figure 1 – *M. flabellifolia* in hydrated and dehydrated states. (Photos: O. Morebodi, BUAN.)

FRUITS AND SEEDS

The fruits are 3-lobed coriaceous, dehiscent capsules, which are slightly larger than the carpels at anthesis. Seeds 0.3–0.5 mm long, ovoid; seed coats wrinkled (Figure 2) [7].

FLOWERING AND FRUITING. Flowers at the start of the rainy season, from September to December. Seeds are available from January to March.

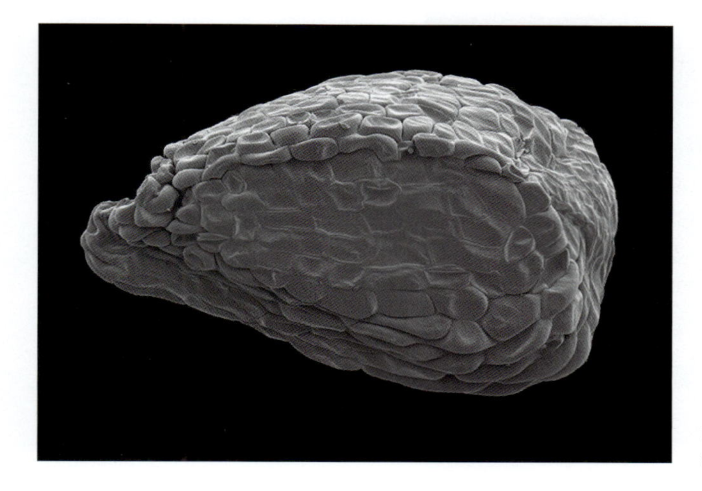

Figure 2 – *M. flabellifolia* seeds.
(Photo: W. Stuppy and R. Kesseler, RBG Kew.)

DISTRIBUTION

Botswana, South Africa, Zimbabwe, Mozambique, Malawi, Tanzania and Kenya, with isolated populations in Zambia and south-eastern Democratic Republic of Congo [1,5,6,16].

HABITAT

Rocky inselbergs and shallow soils with depths of c. 150 mm. A pioneer of bare rock slopes [13,16], at 365–1,200 m a.s.l., but can also occur up to 2,000 m a.s.l. [1,8].

USES

A very important species in traditional medicine, used to treat severe depression and a wide range of other ailments [7]. Smoke from burning the leaves is inhaled to treat chest complaints and asthma [15]; an aromatic salve is used to sterilise wounds; herbal teas and decoctions are prepared to treat coughs, influenza, mastitis, backache, kidney disorders, haemorrhoids and abdominal pains; the leaves are chewed to treat scurvy, halitosis and Vincent's gingivitis [13,15]. The essential oil distilled from the leaves has been demonstrated to have activity against a number of microbial pathogens [7]. Some properties of the essential oil and galloylquinic acids are the basis of a number of applications in traditional medicine [7]. Extracts of the leaves are used in some commercial skin creams. Herbivores and pests usually avoid this species. However, beetles have been observed chewing the leaves [7]. In Pilikwe, the community use a hammer mill to process the plant for medicines for income generation (Figure 3).

Figure 3 – Hammer mill machine used for processing the plant. (Photo: S. Ngwako, BUAN.)

KNOWN HAZARDS AND SAFETY

No known hazards.

CONSERVATION STATUS

Assessed as Data Deficient (DD) in South Africa according to IUCN Red List criteria [17], but has yet to be assessed in Botswana and to have its status confirmed in the IUCN Red List [4,11].

SEED CONSERVATION

HARVESTING. After checking to ensure the seeds have reached maturity, the whole infructescence should be cut before seed dispersal.

PROCESSING AND HANDLING. Seeds can be removed by gently rubbing the infructescence, the seeds can then be separated from other plant material by using a sieve.

STORAGE AND VIABILITY. X-ray analysis carried out on seed lots stored at the MSB highlighted high percentages of filled seeds (>95%). Seeds of this species are reported to be desiccation tolerant [3,9].

PROPAGATION

SEEDS

Dormancy and pre-treatments: High germination percentages have been reported for seeds incubated without any pre-treatments [9].

Germination, sowing and planting: Seeds are reported to germinate after 10–15 days when incubated at c. 25°C [3]. Experiments carried out on seed lots stored at the MSB highlighted high germination percentages (>85%) when seeds were incubated at constant temperatures of >20°C [9].

VEGETATIVE PROPAGATION

No protocols available.

TRADE

Traded for medicinal purposes in Botswana, but there are no official trade statistics.

Authors

Samodimo Ngwako, Kebadire K. Mogotsi, Moctar Sacande, Tiziana Ulian, Steve Davis and Efisio Mattana.

References

[1] Conservatoire et Jardin botaniques & South African National Biodiversity Institute (SANBI) (2016). *African Plant Database*. Version 3.4.0. http://www.ville-ge.ch/musinfo/bd/cjb/africa/

[2] Glen, H. F., Sherwin, H. & Condy, G. (1999). *Flowering Plants of Africa*. Vol. 56, Plate 2151. NBI Publications, Pretoria, South Africa.

[3] Goldsworthy, D. A. (1992). *Desiccation Tolerance in* Myrothamnus flabellifolia *Welw*. Master's thesis, University of Natal, Pietermaritzburg, South Africa.

[4] IUCN (2017). *The IUCN Red List of Threatened Species*. Version 2017-3. http://www.iucnredlist.org/

[5] Launert, E. (ed.) (1978). Myrothamnaceae. In: *Flora Zambesiaca*, Vol. 4. Flora Zambesiaca Managing Committee, Royal Botanic Gardens, Kew.

[6] Lisowski, S., Malaisse, F. & Symoens, J. J. (1970). Les Myrothamnaceae, nouvelle famille pour la flore phanérogamique du Congo-Kinshasa. *Bulletin de la Jardin Botanique Nationale de Belgique* 40 (3): 225–229.

[7] Moore, J. P., Lindsey, G. G., Ferrant, J. M. & Brandt, W. F. (2007). An overview of the biology of the desiccation-tolerant resurrection plant *Myrothamnus flabellifolia*. *Annals of Botany* 99 (2): 211–217.

[8] Puff, C. (1978). Zur Biologie von *Myrothamnus flabellifolius* Welw. (Myrothamnaceae). *Dinteria* 14: 1–20.

[9] Royal Botanic Gardens, Kew (2017). *Seed Information Database (SID)*. Version 7.1. http://data.kew.org/sid/

[10] Schneider, H., Wistuba, N., Wagner, H.-J., Thürmer, F. & Zimmermann, U. (2000). Water rise kinetics in refilling xylem after desiccation in a resurrection plant. *New Phytologist* 148 (2): 221–238.

[11] Setshogo, M. P. & Hargreaves, B. (2002). Botswana. In: *Southern African Plant Red Data Lists*. Edited by J. S. Golding. Southern African Botanical Diversity Network Report No. 14, SABONET, Pretoria, South Africa. pp. 16–20.

[12] Sherwin, H. W., Pammenter, N. W., February, E., Vander Willigen, C. & Farrant, J. M. (1998). Xylem hydraulic characteristics, water relations and wood anatomy of the resurrection plant *Myrothamnus flabellifolius* Welw. *Annals of Botany* 81 (4): 567–575.

[13] Van Wyk, B.-E., Van Oudtshoorn, B. & Gericke, N. (2009). *Medicinal Plants of South Africa*, 2nd edition. Briza Publications, Pretoria, South Africa.

[14] Watson, L. & Dallwitz, M. J. (1992 onwards). Myrothamnaceae Niedenzu. In: *The Families of Flowering Plants: Descriptions, Illustrations, Identification, and Information Retrieval*. Version 19 October 2016. http://delta-intkey.com

[15] Watt, J. M. & Breyer-Brandwijk, M. G. (1962). *The Medicinal and Poisonous Plants of Southern and Eastern Africa*, 2nd edition. E. & S. Livingstone, Edinburgh & London.

[16] Weimarck, H. (1936). *Myrothamnus flabellifolia* Welw., eine polymorphe Pflanzenart. *Botaniska Notiser* 1936: 451–462.

[17] Williams, V. L., Raimondo, D., Crouch, N. R., Cunningham, A. B., Scott-Shaw, C. R., Lötter, M. & Ngwenya, A. M. (2008). *Myrothamnus flabellifolius* Welw. In: *National Assessment: Red List of South African Plants*. Version 2015.1. http://redlist.sanbi.org/

Schinziophyton rautanenii (Schinz) Radcl.-Sm.

TAXONOMY AND NOMENCLATURE

FAMILY. Euphorbiaceae

SYNONYMS. *Ricinodendron rautanenii* Schinz, *R. viticoides* Mildbr., *Vitex lukafuensis* De Wild.

VERNACULAR / COMMON NAMES. Mangetti, manketti, mongongo, mugongo (English); mongongo, mugongo (Tswana); mongongo (Herero); omunkete (Ndonga, Namibia); omangette (Oshikwanyama, Namibia).

PLANT DESCRIPTION

LIFE FORM. Tree.

PLANT. Dioecious, <20 m tall, with a rounded or spreading crown (Figure 1); trunk <1 m in diameter; bark whitish, pale grey or pale brown; young branches smooth, later becoming reticulate and flaking. **LEAVES.** Digitately compound; leaflets (3)5–7, elliptical-ovate, oblanceolate, obtuse or acute, shortly acuminate, entire or sub-entire. **FLOWERS.** 10 mm in diameter, pale yellow to white, in slender loose sprays or cymose panicles. Male and female flowers on separate inflorescences; male inflorescences 10–22 cm long, 4–8 cm wide; male flower petals 6–7 x 2–3 mm; female inflorescences 5–6 cm long, 2–3 cm wide; female flower petals 9 x 4 mm [3,4,8].

Figure 1 – *S. rautanenii* tree. (Photo: C. Amosso, BUAN.)

FRUITS AND SEEDS

Fruits are ovoid-ellipsoid drupes <7 cm long and 5 cm in diameter, light green, covered with velvety hairs, turning yellow when ripe (Figure 2). The fruits contain 1–2 seeds [3] which are ellipsoid and compressed, 18–25 mm long and 15–20 mm wide, and have a very hard, thick endocarp (Figure 3).

FLOWERING AND FRUITING. Trees fruit after 15–25 years, and may live up to 100 years [14]. The trees are leafless from March/May to October/November; flowers from September to December; fruits available from March to May [7,9].

DISTRIBUTION

Angola, Botswana, Democratic Republic of Congo, Malawi, Mozambique, Namibia, South Africa, Tanzania and Zimbabwe [3,13].

Figure 2 – *S. rautanenii* immature and mature fruits and pyrenes. (Photo: K. K. Mogotsi, BUAN.)

Figure 3 – External views of the pyrenes and longitudinal section showing the thick endocarp and the seed. (Photos: P. Gómez-Barreiro, RBG Kew.)

HABITAT

Deciduous woodland and grassland with scattered trees, sometimes in pure stands, on deep sandy soil with 94–99% fine sand component, among sand dunes (sometimes on their crests), usually at low to medium altitudes (200–1,500 m a.s.l.) [8], with mean annual temperatures of c. 20°C. Mature plants tolerate light frost, but seedlings cannot tolerate temperatures below 7°C [4].

USES

Mongongo 'nuts' were, and locally still are, an important food source for hunter-gatherer communities. The fruit pulp is eaten raw or cooked, the flavour being similar to that of dates (*Phoenix dactylifera* L.) but not as sweet. A strong alcoholic distillate is made from the pulp which has a high sugar content. The fruit pulp has high carbohydrate content, 6–9% protein (by dry weight) and 85–100 mg calcium/100 g edible portion. The kernel oil is rich in linoleic and eleostearic acids and in gamma-tocopherol. The wood is traditionally used for making canoes, fishing floats, dart and drawing boards, crates and coffins [3,4]. The tree is highly valued culturally and is sacred in some places [7].

Figure 4 – Preparation for oil extraction at the Shaikarawe and Tsetseng communities.
(Photos: T. T. Mogotsi, BUAN; E. Mattana, RBG Kew.)

KNOWN HAZARDS AND SAFETY

No known hazards.

CONSERVATION STATUS

Assessed as Least Concern (LC) in South Africa according to IUCN Red List criteria [1], but has yet to be assessed in Botswana and to have its status confirmed in the IUCN Red List [5,11].

SEED CONSERVATION

HARVESTING. Harvesting begins in March at the end of the rainy season, when ripe fruits fall from the tree and are picked up from the ground [4]. Gathering continues until the end of the dry season (September–November) when half of the fruits have already lost their pulp due to insects [4,6].

PROCESSING AND HANDLING. The fruits are normally skinned after steaming with a small amount of water. They are then boiled in fresh water to separate the nuts. If the fruits are dry, the pulp can be separated from the pyrene with the help of a mortar and pestle. The pyrenes are then removed and washed with water. The seeds are extracted by cracking the hard endocarp with a small axe (Figure 4) [6].

STORAGE AND VIABILITY. X-ray analysis carried out on seed lots stored at the MSB highlighted a high percentage of filled seeds (c. 100%). If seeds are stored at a temperature of 10°C, they remain viable for up to two years [4]. However, seeds of this species are reported to be orthodox [10] and after appropriate drying can therefore be stored at –20°C to extend their longevity.

PROPAGATION

SEEDS

Dormancy and pre-treatments: 60% germination (based on the total number of viable seeds) has been achieved on untreated seeds [2]. 80% germination has been achieved when seeds extracted from the woody endocarp have been chipped and then sown in a solution of gibberellic acid [10] and >80% aqueous smoke solution [Sacande *et al.*, data unpublished], suggesting the presence of physiological dormancy.

Germination, sowing and planting: Very poor and slow germination is reported for seeds of this species [2,12]. Seeds treated as above can be incubated for germination at 26°C with 12 hours of light per day [10].

VEGETATIVE PROPAGATION

Schinziophyton rautanenii can be propagated by cuttings as carried out by the local community partners in the UPP. Truncheons (used for fencing) can also take root [7].

TRADE

Shaikarawe and Tsetseng communities sell the seed oil on a small scale in local markets.

Authors

Kebadire K. Mogotsi, Seoleseng O. Tshwenyane, Cecilia Amosso, Moctar Sacande, Tiziana Ulian and Efisio Mattana.

References

[1] Archer, R. H. & Victor, J. E. (2005). *Schinziophyton rautanenii* (Schinz) Radcl.-Sm. In: *National Assessment: Red List of South African Plants*. Version 2015.1. http://redlist.sanbi.org/

[2] Chimbelu, E. G. (1983). A germination and survival study of mugongo (*Ricinodendron rautanenii* Schinz) under laboratory and greenhouse conditions. *The Indian Forester* 109 (11): 804–809.

[3] Coates Palgrave, K. (1988). *Trees of Southern Africa*, 2nd edition. Struik Publishers, Cape Town, South Africa.

[4] Graz, F. P. (2007). *Schinziophyton rautanenii* (Schinz) Radcl.-Sm. In: *Plant Resources of Tropical Africa 14: Vegetables Oils*. Edited by H. A. M. van der Vossen & G. S. Mkamilo. Backhuys Publishers, Leiden, The Netherlands. pp. 150–153.

[5] IUCN (2017). *The IUCN Red List of Threatened Species.* Version 2017-3. http://www.iucnredlist.org/

[6] Letloa LLHRC (Lands, Livelihoods and Heritage Resource Centre) (2007). *The Khwe of the Okavango Panhandle: the Use of Veld Plants for Food and Medicine.* CTP Book Printers, Cape Town, South Africa.

[7] Orwa, C., Mutua, A., Kindt, R., Jamnadass, R. & Simons, A. (2009). *Agroforestree Database: A Tree Reference and Selection Guide*. Version 4.0. World Agroforestry Centre, Kenya. http://www.worldagroforestry.org/output/agroforestree-database

[8] Pope, G. V. (ed.) (1996). Euphorbiaceae. In: *Flora Zambesiaca*, Vol. 9, part 4. Flora Zambesiaca Managing Committee, Royal Botanic Gardens, Kew.

[9] Royal Botanic Gardens, Kew (1999–2015). *Survey of Economic Plants for Arid and Semi-Arid Lands (SEPASAL) Database*. http://apps.kew.org/sepasalweb/sepaweb

[10] Royal Botanic Gardens, Kew (2017). *Seed Information Database (SID).* Version 7.1. http://data.kew.org/sid/

[11] Setshogo, M. P. & Hargreaves, B. (2002). Botswana. In: *Southern African Plant Red Data Lists.* Edited by J. S. Golding. Southern African Botanical Diversity Network Report No. 14, SABONET, Pretoria, South Africa. pp. 16–20.

[12] Tietema, T., Merkesdal, E. & Schroten J. (1992). *Seed Germination of Indigenous Trees in Botswana.* ACTS Press, African Centre for Technology Studies, Nairobi, Kenya.

[13] WCSP (2017). *World Checklist of Selected Plant Families.* Royal Botanic Gardens, Kew. http://wcsp.science.kew.org/

[14] Wehmeyer, A. S., Lee, R. B. & Whiting, M. (1969). The nutrient composition and dietary importance of some vegetable foods eaten by the !Kung Bushman. *South African Medical Journal* 43 (50): 1529–1530.

Sclerocarya birrea subsp. *caffra* (Sond.) Kokwaro

TAXONOMY AND NOMENCLATURE

FAMILY. Anacardiaceae

SYNONYMS. *Commiphora acutidens* Engl., *Poupartia caffra* (Sond.) H.Perrier, *P. excelsa* Marchand, *Sclerocarya caffra* Sond., *S. schweinfurthiana* Schinz

VERNACULAR / COMMON NAMES. Cider tree (English); marula, morula (Tswana); omukongo (Herero, Namibia); morula (Kalanga); goaras (Khoekhoe, Namibia); omugongo (Ndonga, Namibia); eegongo (Oshikwanyama, Namibia); marula (Shona).

PLANT DESCRIPTION

LIFE FORM. Tree.

PLANT. Deciduous, mostly dioecious, thick-boled, laxly branched (Figure 1), <18 m tall, with a widely reticulate, greyish bark and spreading branches. **LEAVES.** Imparipinnate, with 7–13(17) leaflets, petiole and rachis 15–30 cm long, hemicylindric, with longitudinal channels above, glabrous; leaflets discolorous, ovate, oblong-elliptic or elliptic, acuminate, margin entire or sometimes dentate-serrate on suckers, asymmetric and slightly cuneate or rounded at the base. **FLOWERS.** Male flower sepals c. 2 x 2 mm, petals yellow to purplish pink (red in bud), 4–6 x 3–4 mm (Figure 1); female inflorescences subterminal, shorter than the male, and with fewer flowers; the axis and pedicels become thickened when fruiting [4].

Figure 1 – *S. birrea* tree and close-up of flowers and fruits. (Photos: O. Morebodi, BUAN, and B. Wursten, Flora of Zimbabwe: https://www.zimbabweflora.co.zw/index.php.)

FRUITS AND SEEDS

Fruits are fleshy drupes, yellow, 3–3.5 cm in diameter, with a very juicy mesocarp (Figures 1 and 2). The hard endocarp (stone) is thick and obovoid, 2–3 x 2.5 cm, with one to four cavities each containing a seed (Figure 3); seeds c. 1.5–2 x 0.5–1 cm. Each cavity (locule) is covered by a lid (operculum) that remains firmly attached until germination [4,6].

FLOWERING AND FRUITING. Flowers during the dry season when leafless (mainly August to September in South Africa; October to January in Namibia); fruits ripen February to June in Southern Africa, falling green, then turning yellow on the ground [2].

Figure 2 – Ripe and unripe fruits of *S. birrea.*
(Photo: O. Morebodi, BUAN.)

Figure 3 – External view of the endocarp (stone) with three loculi (each one bearing a seed) and the opercula removed and, on the right, one of the opercula.
(Photo: P. Gómez-Barreiro, RBG Kew.)

DISTRIBUTION

Throughout Southern Africa, from Angola to South Africa, and including Botswana, Malawi, Namibia, Zambia, Zimbabwe and Mozambique, north to tropical East Africa (Kenya, Tanzania); also native to Madagascar [2,4,13]. Throughout Botswana, except the deep sand areas of the Kgalagadi [9].

HABITAT

Wooded grassland, woodland and bushland, on rocky hills and sandy soils, at 500–1,800 m a.s.l., with mean annual rainfall of 200–1,100 mm and temperatures between 18–40°C [12].

USES

The fruits are edible and can be eaten fresh or made into a delicious jelly (Figure 4). They are also used to make popular alcoholic beverages (amarula liqueur and marula beer) which are available commercially [12]. The seeds can be eaten raw; they are rich in oil, protein and vitamin C. The seed oil is used to preserve meat, and for cooking and skin care [8]. The leaves are claimed to have hypoglycaemic effects and several studies have shown antidiabetic activity [14]. The bark is used in traditional medicine for treating dysentery and diarrhoea, and for preventing malaria [2]. The wood is used for fencing and making pestles, mortars, bowls, saddles and furniture [12]. The fruits are eaten by game animals after they have fallen from the tree [2]. The bark is eaten by elephants, which can strip the entire bark from a tree [9].

KNOWN HAZARDS AND SAFETY

No known hazards.

CONSERVATION STATUS

Sclerocarya birrea is often protected by local communities, and retained in cultivated areas, on account of its social and economic value. In parts of its range, including Namibia and South Africa, marula is afforded protection under forestry legislation [3,10]. *Sclerocarya birrea* subsp. *caffra* is assessed as Least Concern (LC) in South Africa according to IUCN Red List criteria [15], but has yet to be assessed in Botswana and to have its status confirmed in the IUCN Red List [5,11].

SEED CONSERVATION

HARVESTING. The fruits can be harvested from the ground after they have fallen from the tree [6].

PROCESSING AND HANDLING. Before depulping, the fruits should be soaked in water for 24 hours. The fleshy pulp can then be removed using a cement mixer. The fruits are first mixed with gravel (1 kg of gravel to 2 kg of fruits) and a large amount of water. The mixture is stirred in the mixer and when the stones are clean they are separated from the pulp and gravel, and then washed with water. After extraction, the stones are spread out on a mesh and dried in the sun for at least two days [6]. Alternatively, the seeds are separated by eating the fruit pulp, or by rubbing the fruit pulp over a wire mesh until clean seeds remain. The seeds can then be dried.

STORAGE AND VIABILITY. X-ray analysis carried out on seed lots stored at the MSB showed a high percentage of filled seeds (>80%), although c. 40% of empty seeds were detected in one seed lot. Seeds are orthodox and can withstand substantial water loss and low storage temperatures [6].

PROPAGATION

SEEDS

Dormancy and pre-treatments: The hard endocarp forms a physical barrier to germination; removal of the opercula significantly improves germination [6]. After-ripening has also been reported to improve germination [7]. Further studies are needed to determine the type of seed dormancy (i.e. physical, physiological or combinational) [1].

Germination, sowing and planting: Germination is epigeal. If opercula are removed, germination is fast, reaching 70% after one week and 85% after two weeks from sowing [6]. In the laboratory, the highest germination percentage was recorded under constant dark conditions at 25°C for

Figure 4 – Marula jam and oil in jars. (Photo: S. Ngwako, BUAN.)

stones that had their opercula removed and an after-ripening period of 12 months [7]. In the nursery, optimum results are achieved when seeds are sown on a well-drained mixture of soil, coarse sand and manure (in a ratio of 3:1:1 or in equal parts by volume). Seedlings should be transplanted when they are 20–30 cm high.

VEGETATIVE PROPAGATION

Can be propagated by cuttings and truncheons [9].

TRADE

A number of products derived from marula fruits are commercially marketed. The most prominent of these is amarula liqueur. Other marula products include jam, marula beer, oil and soap (Figure 4). In Botswana, the UPP community in Pilikwe plans to market the oil and to make products such as soap. There is informal trade in marula liqueur in rural communities in the Tswapong and Bobirwa districts of Botswana.

Authors

Samodimo Ngwako, Kebadire K. Mogotsi, Moctar Sacande, Tiziana Ulian and Efisio Mattana.

References

[1] Baskin, C. C. & Baskin, J. M. (2014). *Seeds: Ecology, Biogeography, and Evolution of Dormancy and Germination*, 2nd edition. Academic Press, San Diego, USA.

[2] Coates Palgrave, K. & Coates Palgrave, M. (2002). *Trees of Southern Africa*, 3rd edition. Struik Publishers, Cape Town, South Africa.

[3] Curtis, B. A. & Mannheimer, C. A. (2005). *Tree Atlas of Namibia*. National Botanical Research Institute of Namibia, Windhoek, Namibia.

[4] Exell, A. W., Fernandes, A. & Wild, H. (eds) (1966). Anacardiaceae. In: *Flora Zambesiaca*, Vol. 2, part 2. Flora Zambesiaca Managing Committee, Royal Botanic Gardens, Kew.

[5] IUCN (2017). *The IUCN Red List of Threatened Species.* Version 2017-3. http://www.iucnredlist.org/

[6] Jøker, D. & Erdey, D. (2003). *Sclerocarya birrea* (A.Rich.) Hochst. *Seed Leaflet* No. 72. Danida Forest Seed Centre, Humlebaek, Denmark.

[7] Moyo, M., Kulkarni, M. G., Finnie, J. F. & Van Staden, J. (2009). After-ripening, light conditions, and cold stratification influence germination of marula [*Sclerocarya birrea* (A.Rich.) Hochst. subsp. *caffra* (Sond.) Kokwaro] seeds. *HortScience* 44 (1): 119–124.

[8] Palmer, E. (1977). *A Field Guide to the Trees of Southern Africa*. Collins, London.

[9] Roodt, V. (1998). *Trees & Shrubs of the Okavango Delta, Medicinal Uses and Nutritional Value*. Shell Field Guide Series: Part 1. Shell Oil Botswana, Gaborone, Botswana.

[10] SA Forestry Online (2015). *Protected Trees in South Africa (2015). List of Protected Trees Under The National Forest Act, 1998 (Act No. 84 of 1998).* http://saforestryonline.co.za/indigenous/protected-trees-in-south-africa/

[11] Setshogo, M. P. & Hargreaves, B. (2002). Botswana. In: *Southern African Plant Red Data Lists.* Edited by J. S. Golding. Southern African Botanical Diversity Network Report No. 14, SABONET, Pretoria, South Africa. pp. 16–20.

[12] Van Wyk, B. & Van Wyk, P. (1997). *Field Guide to Trees of Southern Africa*. Struik Nature, Cape Town, South Africa.

[13] Van Wyk, B., Van Wyk, P. & Van Wyk, B.-E. (2000). *Photographic Guide to Trees of Southern Africa*. Struik Nature, Cape Town, South Africa.

[14] Van Wyk, B.-E., Van Oudtshoorn, B. & Gericke, N. (2009). *Medicinal Plants of South Africa*, 2nd edition. Briza Publications, Pretoria, South Africa.

[15] Williams, V. L., Raimondo, D., Crouch, N. R., Cunningham, A. B., Scott-Shaw, C. R., Lötter, M., Ngwenya, A. M. & Helm, C. (2008). *Sclerocarya birrea* (A.Rich.) Hochst. subsp. *caffra* (Sond.) Kokwaro. In: *National Assessment: Red List of South African Plants.* Version 2015.1. http://redlist.sanbi.org/

Stomatostemma monteiroae (Oliv.) N.E.Br.

TAXONOMY AND NOMENCLATURE

FAMILY. Apocynaceae

SYNONYMS. *Cryptolepis monteiroae* Oliv.

VERNACULAR / COMMON NAMES. Monteiro vine (English); mashedza, mosada, mosata (Tswana).

PLANT DESCRIPTION

LIFE FORM. Perennial climber (rarely a shrub).

PLANT. Vigorous liana-like, caudiciform, with twining stems to 10 m long, coiling around other plants for support (Figure 1), growing from a succulent tuber (5–20 cm in diameter); older stems woody; milky latex present. **LEAVES.** Opposite, lanceolate-elliptic to narrowly ovate, tapering at both ends, <10 cm long, 1–3 cm wide; hairless, dark green above, with conspicuous yellowish veins, paler below. **FLOWERS.** Mostly in showy, terminal inflorescences, strongly scented; corolla <2 cm long, creamy white to yellowish green, with maroon to purple spotting in tube and on throat; corona lobes 2 mm long, green, fitted between the corolla lobes [9] (Figure 1) .

Figure 1 – *S. monteiroae* in flower. (Photo: B. Wursten, Flora of Zimbabwe: https://www.zimbabweflora.co.zw/index.php.)

FRUITS AND SEEDS

Fruits are green, paired, cylindric follicles, each 6–9 x 1.5–2.5 cm, grooved or slightly winged. Seeds 8–10 x 3 mm (Figure 2), with a plume of silky hairs (coma) <27 mm long.

FLOWERING AND FRUITING. Flowers from end of October to March, peaking in January; fruits from January to July [9].

Figure 2 – External views of the cleaned seeds without plumes. (Photo: P. Gómez-Barreiro, RBG Kew.)

DISTRIBUTION

Widely distributed in central and eastern Southern Africa, including Botswana (where it is locally common in central areas), Mozambique, South Africa, Swaziland, Zambia and Zimbabwe [9].

HABITAT

Dry woodland, sandy plains, and rocky habitats of hills, mountains and cliffs [9], at altitudes of 20–1,300 m a.s.l. [1], growing with the support of associated trees, such as *Brachystegia spiciformis* Benth., *Colophospermum mopane* (Benth.) J.Léonard, *Commiphora* spp. and *Sclerocarya birrea* (A.Rich.) Hochst. [9].

USES

The fruits are edible and used in traditional medicine. They are sliced and sun-dried after being boiled in water for 30 minutes [5]. The flesh of the dried fruits has a meaty taste and is eaten by vegetarians as a meat substitute [4,5].

KNOWN HAZARDS AND SAFETY

No known hazards.

CONSERVATION STATUS

Assessed as Least Concern (LC) in South Africa according to IUCN Red List criteria [2], but has yet to be assessed in Botswana and to have its status confirmed in the IUCN Red List [3,8].

SEED CONSERVATION

HARVESTING. Fruits for consumption are harvested from the plant when still green [6]. To obtain seeds for propagation, fruits should be picked when they turn brown at maturity, before they break open and disperse the seeds.

PROCESSING AND HANDLING. Fruits are opened by hand and the seeds removed. The plume of silky hairs is then separated from the seeds.

STORAGE AND VIABILITY. No information available on long-term seed storage behaviour [7]. However, if kept at a temperature of 10°C seeds remain viable for up to two years.

PROPAGATION

SEEDS

Dormancy and pre-treatments: No information available on the requirements to break seed dormancy. For optimum germination, seeds are chipped with a scalpel or they can be soaked in water for two hours. However, high germination percentages have been achieved on seed lots stored at the MSB without any pre-treatments.

Germination, sowing and planting: High germination levels (>95%) were achieved on seed lots stored at the MSB under laboratory controlled conditions when seeds were incubated with 8 hours of light per day at constant temperatures of 20°C and 25°C. In the nursery, seeds can be sown in sandy or loamy soil. Germination is slow; the germination rate is c. 50%.

VEGETATIVE PROPAGATION

No protocols available.

TRADE

In Botswana, the community in Tswapong, through the Kgetsi-ya-tsie Trust in Lerala and Pilikwe villages, sell food and medicinal products made from *S. monteiroae*.

Authors

Samodimo Ngwako, Cecilia Amosso, Kebadire K. Mogotsi, Moctar Sacande, Tiziana Ulian, Steve Davis and Efisio Mattana.

References

[1] Conservatoire et Jardin botaniques & South African National Biodiversity Institute (SANBI) (2016). *African Plant Database*. Version 3.4.0. http://www.ville-ge.ch/musinfo/bd/cjb/africa/

[2] Foden, W. & Potter, L. (2005). *Stomatostemma monteiroae* (Oliv.) N.E.Br. In: *National Assessment: Red List of South African Plants.* Version 2015.1. http://redlist.sanbi.org/

[3] IUCN (2017). *The IUCN Red List of Threatened Species.* Version 2017-3. http://www.iucnredlist.org/

[4] Mogotsi, K. K. (2012). *Climate-Smart Agriculture and Role of Emerging Crops in Africa: A Case of Botswana*.http://doclinks2012residentialschool.myevent.com/clients/3/31/314effb673ed43804a4bfe2fd166864f/File/Climate%20Smart%20Agriculture%20and%20Role%20of%20Emerging%20Crops%20in%20Africa-%20Case%20of%20Botswana.pdf

[5] Mogotsi, K. K. & Ngwako, S. (2011). Towards afforestation and livelihoods improvement: MSB–UPP contribution in Botswana. In: *Forest Landscape Africa: MSB International Forestry Workshop (Forest 2011), 5–10 December 2011, Nairobi, Kenya*. Millennium Seed Bank Partnership, Royal Botanic Gardens, Kew and World Agroforestry Centre, Kenya.

[6] Motlhanka, D. M. T., Motlhanka, P. & Selebatso, T. (2008). Edible indigenous wild fruit plants of eastern Botswana. *International Journal of Poultry Science* 7 (5): 457–460.

[7] Royal Botanic Gardens, Kew (2017). *Seed Information Database (SID).* Version 7.1. http://data.kew.org/sid/

[8] Setshogo, M. P. & Hargreaves, B. (2002). Botswana. In: *Southern African Plant Red Data Lists*. Edited by J. S. Golding. Southern African Botanical Diversity Network Report No. 14, SABONET, Pretoria, South Africa. pp. 16–20.

[9] Venter, H. J. T. & Verhoeven, R. L. (1993). A taxonomic account of *Stomatostemma* (Periplocaceae). *South African Journal of Botany* 59 (1): 50–56.

Strychnos cocculoides Baker

TAXONOMY AND NOMENCLATURE

FAMILY. Loganiaceae

SYNONYMS. *Strychnos paralleloneura* Gilg & Busse, *S. schumanniana* Gilg, *S. suberosa* De Wild.

VERNACULAR / COMMON NAMES. Monkey orange, wild orange, yellow monkey-orange (English); mogorogoro, mogorogorwana, mogorogorwane, mohoruhoru, moruda, nhume (Tswana); omusu (Herero, Namibia); omuguni (Ndonga, Namibia); omauni (Oshikwanyama, Namibia); mungono (Shona).

PLANT DESCRIPTION

LIFE FORM. Shrub or small tree (Figure 1).

PLANT. Deciduous [9], or semi-deciduous [4], <6(–8) m high, branching low down; bark pale grey to pale brown, thick, corky, longitudinally ridged; branches pale or dark brown, ± fissured and corky, spiny (especially when young), and terminating in a straight spine. **LEAVES.** Pale to dark green above, paler beneath, coriaceous, very variable in shape and size, usually orbicular, ovate-elliptic, those on old fire-cut stumps smaller. **FLOWERS.** In dense compact heads, seemingly umbellate, terminal on the main branches or on short lateral shoots, sometimes flowering on one-year-old shoots; flowers pentamerous, sepals pale green, petals pale green, white or greenish yellow, sparsely pubescent or glabrous outside [9].

Figure 1 – *S. cocculoides* with unripe fruits. (Photo: K. K. Mogotsi, BUAN.)

FRUITS AND SEEDS

Fruits are large, hard, resembling an orange, globose, yellow or orange, often speckled with green, dark green when nearly mature, with an edible pulp and c. 10–100 seeds [9]. Seeds light brownish yellow, flattened, ± plano-convex, obliquely ovate or elliptic, usually irregularly curved, slightly longer than wide, c. 12–22 x 8–15 x 2.5–4 mm, slightly rough, pubescent [9].

FLOWERING AND FRUITING. Flowers mainly from September to December in Southern Africa (October to February in East Africa); fruits ripen April to August in Southern Africa (November to May in Namibia), taking up to a year to ripen [3,4,13].

DISTRIBUTION

Widespread in sub-Saharan Africa, from Gabon to Kenya, south to South Africa. In Botswana, it has a patchy distribution in the south-east, central, eastern and northern parts of the country [2,9,13].

HABITAT

Woodlands and mixed forests [9], preferring sites with deep sandy soil, or on rocky slopes and on acidic dark grey clays and red or yellow red loams, at 400–2,000 m a.s.l., with mean annual temperatures of 14–25°C and mean annual rainfall of 600–1,200 mm [13].

USES

The juicy fruit pulp is edible. It is refreshing with a delicate flavour when fresh [15]. The fruit is sometimes buried in the sand to allow the pulp to liquefy [15]. Note that the seeds are reported to contain toxic substances and should not be eaten, and unripe fruits are purgative [13,16]. In traditional medicine, the root is chewed to treat eczema; a root decoction is drunk as a cure for gonorrhoea, and pounded leaves are used to treat sores [11]. The fruit is mixed with honey or sugar to treat coughs [6]. The wood is whitish, tough, pliable, and used for making handles and other small items [15].

KNOWN HAZARDS AND SAFETY

All parts of the plant contain low concentrations of strychnine. The seeds and bark are said to be toxic [16]. The unripe fruits are purgative [13]. Excessive consumption of the ripe fruit can cause stomach pains and diarrhoea.

CONSERVATION STATUS

Sometimes protected and cultivated by local communities in parts of its range on account of its food and economic value, and also afforded protection under forestry legislation in Namibia [4]. Assessed as Least Concern (LC) in South Africa according to IUCN Red List criteria [5], but has yet to be assessed in Botswana and to have its status confirmed in the IUCN Red List [8,14].

SEED CONSERVATION

HARVESTING. Ripe fruits may be picked from the tree or collected from the ground. Only the ripe yellow fruits should be collected.

PROCESSING AND HANDLING. Seeds are easily removed from the pulp and then cleaned and sorted by mixing with sand and then sieving.

STORAGE AND VIABILITY. Seeds of this species are reported to be orthodox [12]. However, they are short-lived at room temperature (remaining viable for up to two months) [11].

PROPAGATION

SEEDS

Dormancy and pre-treatments: Seeds of this species are reported to be physiologically dormant [1]. In nursery conditions, germination may be improved by soaking the seeds in hot water and allowing the seeds to cool for 24–48 hours before sowing [7,12].

Germination, sowing and planting: Seed lots banked at the MSB reached high germination percentages when seeds were incubated under laboratory controlled conditions at a constant 30°C, with 8 hours of light per day, without any pre-treatments [12]. In the nursery, seeds sown at a depth of 20–30 mm in pots or seedbeds germinate after about a month [11]. After germination the seedlings should be kept in shade at first, before being exposed to full light after 14 days (Figure 2). Direct sowing of seeds in the field is possible, but this results in patchy germination and poor seedling survival.

VEGETATIVE PROPAGATION

Can be propagated from coppice shoots and root suckers [7,10]. Scions can also be grafted on to *S. spinosa* Lam. rootstocks resulting in faster than normal growth of *S. cocculoides* [10].

Figure 2 – Seedlings of *S. cocculoides* in BUAN. (Photo: E. Mattana, RBG Kew.)

TRADE

In a farmer survey in Southern Africa, *S. cocculoides* was ranked among the most popular indigenous fruit trees. Nevertheless, the species is not traded in most parts of the region except in Zimbabwe, which exports fruits to Botswana [10]. In Botswana, there is a small amount of informal trade, but no statistics are available. In Pilikwe, fruits are sold locally to generate income.

Authors

Samodimo Ngwako, Kebadire K. Mogotsi, Moctar Sacande, Tiziana Ulian, Steve Davis and Efisio Mattana.

References

[1] Baskin, C. C. & Baskin, J. M. (2014). *Seeds: Ecology, Biogeography, and Evolution of Dormancy and Germination*, 2nd edition. Academic Press, San Diego, USA.

[2] Bruce, E. A., Lewis, J., Hubbard, C. E. & Milne-Redhead, E. (1960). Loganiaceae (including Buddleiaceae). In: *Flora of Tropical East Africa.* Crown Agents for Oversea Governments and Administrations, London.

[3] Coates Palgrave, K. & Palgrave, M. (2002). *Trees of Southern Africa*, 3rd edition. Struik Publishers, Cape Town, South Africa.

[4] Curtis, B. A. & Mannheimer, C. A. (2005). *Tree Atlas of Namibia*. National Botanical Research Institute of Namibia, Windhoek, Namibia.

[5] Foden, W. & Potter, L. (2005). *Strychnos cocculoides* Baker. In: *National Assessment: Red List of South African Plants*. Version 2015.1. http://redlist.sanbi.org/

[6] Food and Agriculture Organization of the United Nations (FAO) (1983). *Food and Fruit-Bearing Forest Species. 1: Examples from East Africa*. FAO Forestry Paper 44/1. FAO, Rome, Italy.

[7] Hines, D. A. & Eckman, K. (1993). *Indigenous Multipurpose Trees of Tanzania: Uses and Economic Benefits for People*. FO:Misc/93/9 Working Paper. Food and Agriculture Organization of the United Nations (FAO), Rome, Italy.

[8] IUCN (2017). *The IUCN Red List of Threatened Species.* Version 2017-3. http://www.iucnredlist.org/

[9] Launert, E. (ed.) (1983). Loganiaceae. In: *Flora Zambesiaca*, Vol. 7, part 1. Flora Zambesiaca Managing Committee, Royal Botanic Gardens, Kew.

[10] Mkonda, A., Akinifesi, F. K., Maghembe, J. A., Swai, R., Kadzere, I., Kwesiga, F. R., Saka, J., Lungu, S. & Mhango, J. (2002). *Towards Domestication of 'Wild Orange'* Strychnos cocculoides *in Southern Africa: A Synthesis of Research and Development Efforts.* Paper presented at the Southern Africa Regional Agroforestry Conference, 20–24 May 2002, Warmbaths, South Africa. http://www.worldagroforestry.org/downloads/Publications/PDFS/PP04098.pdf

[11] Orwa, C., Mutua, A., Kindt, R., Jamnadass, R. & Simons, A. (2009). *Agroforestree Database: A Tree Reference and Selection Guide*. Version 4.0. World Agroforestry Centre, Kenya. http://www.worldagroforestry.org/output/agroforestree-database

[12] Royal Botanic Gardens, Kew (2017). *Seed Information Database (SID).* Version 7.1. http://data.kew.org/sid/

[13] SCUC (2006). *Monkey Orange* (Strychnos cocculoides)*: Field Manual for Extension Workers and Farmers.* Practical Manual No. 8. Southampton Centre for Underutilised Crops (SCUC), University of Southampton, UK.

[14] Setshogo, M. P. & Hargreaves, B. (2002). Botswana. In: *Southern African Plant Red Data Lists.* Edited by J. S. Golding. Southern African Botanical Diversity Network Report No. 14, SABONET, Pretoria, South Africa. pp. 16–20.

[15] Van Wyk, B. & Van Wyk, P. (1997). *Field Guide to Trees of Southern Africa.* Struik Nature, Cape Town, South Africa.

[16] Watt, J. M. & Breyer-Brandwijk, M. G. (1962). *The Medicinal and Poisonous Plants of Southern and Eastern Africa*, 2nd edition. E. & S. Livingstone, Edinburgh & London.

Strychnos pungens Soler.

TAXONOMY AND NOMENCLATURE

FAMILY. Loganiaceae

SYNONYMS. *Strychnos henriquesiana* Baker

VERNACULAR / COMMON NAMES. Monkey apple, spine-leaved monkey orange, wild orange (English); mogwagwa, motu, mutu, ntamba, witu (Tswana); omuhuruhuru (Herero, Namibia); tah (Khoisan, Jul'hoansi); omupwaka (Ndonga, Namibia); omiwapaka, omupuaka (Oshikwanyama, Namibia); ihlala (Shona).

PLANT DESCRIPTION

LIFE FORM. Tree or shrub.

PLANT. Deciduous, c. 3–7 m tall (Figure 1), bark grey or brown, smooth in smaller trees, becoming rough and shallowly reticulate, but not corky, with age; inner bark yellow; wood yellowish; branches with conspicuous lenticels, with short rigid lateral branchlets. **LEAVES.** Glabrous, shining, dark green above, sometimes less so and partly pubescent beneath; 3–8 cm long, 3–5 cm wide, elliptic, opposite, smooth but stiff, with a sharp spine-like tip; petiole short (1–4 mm), glabrous. **FLOWERS.** Greenish creamy white, <9 mm long, in inflorescences c. 2(–5) cm long, 1–2 cm wide, in axils of current leaves or on older branches [1,4].

Figure 1 – *S. pungens* tree. (Photo: C. Amosso, BUAN.)

FRUITS AND SEEDS

The fruits are round, large (12 cm in diameter), with a smooth, hard, woody shell. They are bluish green, turning yellow when ripe, and each contains 20–100 seeds [1,4,6]. The seeds are hard, obliquely ovate, elliptic and flattened, often irregularly curved, 20–24 mm long, 7–12 mm wide, pale brownish yellow or whitish, and covered in very short, erect yellowish hairs [1,4].

FLOWERING AND FRUITING. Flowers from September to October; fruits from March to August [9].

DISTRIBUTION

Angola, northern and south-eastern Botswana, Democratic Republic of Congo, Malawi, South Africa, Tanzania, Zambia and Zimbabwe [1,4].

HABITAT

Open woodland or bushland on well-drained alluvial or sandy soil, at low to medium altitudes (120–2,000 m a.s.l.) [1,4], with mean annual rainfall 250 mm or more. The species is salt tolerant [1].

USES

The fruits are edible when ripe, although they are less palatable than those of some related species. The shell of the dried fruit is used in craftwork and as a sounding board for musical instruments. The timber is used in carving and for making tools [10]. The bark and roots are used as a traditional medicine to treat a variety of illnesses [10]. A decoction of the roots is used to treat stomach ache and bronchitis. An infusion of crushed leaves is used as a lotion for sore eyes. The Khwe use the roots to heal sick children and to cleanse the body of a deceased spouse or child [5]. A root decoction is used to promote the appetite of children [5]. The seeds are eaten to treat snake bites. The presence of strychnine or strychnine-type alkaloids that stimulate the central nervous system may overcome the respiratory depression that causes death as a result of venomous snake bites [10]. See notes below on known hazards.

KNOWN HAZARDS AND SAFETY

Only the pulp of ripe fruits should be eaten. Eating too many ripe fruits can be purgative. Unripe fruits can cause headaches, dizziness and vomiting. As noted above, the seeds contain strychnine or strychnine-type alkaloids. They are intensely bitter tasting and cause diarrhoea when eaten in large quantities [10,11].

CONSERVATION STATUS

Assessed as Least Concern (LC) in South Africa according to IUCN Red List criteria [2], but has yet to be assessed in Botswana and to have its status confirmed in the IUCN Red List [3,8].

SEED CONSERVATION

HARVESTING. Fruits (Figure 2) are collected from the ground when they fall at maturity as the tree is usually too tall to allow the fruits to be hand-picked from the branches.

PROCESSING AND HANDLING. Seeds are easily removed from the pulp. After sorting the seeds, they can be cleaned by mixing with sand and then gently rubbing over a wire mesh.

STORAGE AND VIABILITY. X-ray analysis carried out on a seed lot stored at the MSB highlighted that 100% of seeds were filled. No information is available on the seed storage behaviour of this species. However, all six *Strychnos* spp. stored at the MSB are reported to be desiccation tolerant, suggesting that the seeds of *S. pungens* may also be orthodox [7]. In rural communities, seeds are stored in cold rooms, where they may remain viable for more than five years.

PROPAGATION

SEEDS

Dormancy and pre-treatments: Germination can be improved by scarifying the base of the seeds with sandpaper prior to sowing. This suggests the presence of non-deep physiological dormancy. However, further studies under controlled conditions need to be carried out to characterise seed dormancy in this species.

Germination, sowing and planting: For optimum results, seeds should be sown in sandy soil. Given the right conditions, percentage germination is c. 100%. Germination takes c. 30–45 days. No information is available on the requirements for optimum germination under laboratory controlled conditions.

VEGETATIVE PROPAGATION

No protocols available.

Figure 2 – Ripe fruits collected by students of the Pilikwe Primary School. (Photo: T. Ulian, RBG Kew.)

TRADE

Fruits are sold in local markets.

Authors

Kebadire K. Mogotsi, Cecilia Amosso, Seoleseng O. Tshwenyane, Moctar Sacande, Tiziana Ulian and Efisio Mattana.

References

[1] Coates Palgrave, K. (1988). *Trees of Southern Africa*, 2nd edition. Struik Publishers, Cape Town, South Africa.

[2] Foden, W. & Potter, L. (2005). *Strychnos pungens* Soler. In: *National Assessment: Red List of South African Plants*. Version 2015.1. http://redlist.sanbi.org/

[3] IUCN (2017). *The IUCN Red List of Threatened Species.* Version 2017-3. http://www.iucnredlist.org/

[4] Launert, E. (ed.) (1983). Loganiaceae. In: *Flora Zambesiaca*, Vol. 7, part 1. Flora Zambesiaca Managing Committee, Royal Botanic Gardens, Kew.

[5] Letloa LLHRC (Lands, Livelihoods and Heritage Resource Centre) (2007). *The Khwe of the Okavango Panhandle: the Use of Veld Plants for Food and Medicine.* CTP Book Printers, Cape Town, South Africa.

[6] Peters, C. R., O'Brien, E. M. & Drummond, R. B. (1992). *Edible Wild Plants of Sub-Saharan Africa*. Royal Botanic Gardens, Kew.

[7] Royal Botanic Gardens, Kew (2017). *Seed Information Database (SID).* Version 7.1. http://data.kew.org/sid/

[8] Setshogo, M. P. & Hargreaves, B. (2002). Botswana. In: *Southern African Plant Red Data Lists*. Edited by J. S. Golding. Southern African Botanical Diversity Network Report No. 14, SABONET, Pretoria, South Africa. pp. 16–20.

[9] Tietema, T., Merkesdal, E. & Schroten J. (1992). *Seed Germination of Indigenous Trees in Botswana*. ACTS Press, African Centre for Technology Studies, Nairobi, Kenya.

[10] Van Wyk, B.-E. & Gericke, N. (2000). *People's Plants: A Guide to Useful Plants of Southern Africa*. Briza Publications, Pretoria, South Africa.

[11] Watt, J. M. & Breyer-Brandwijk, M. G. (1962). *The Medicinal and Poisonous Plants of Southern and Eastern Africa*, 2nd edition. E. & S. Livingstone, Edinburgh & London.

Tylosema esculentum (Burch.) A.Schreib.

TAXONOMY AND NOMENCLATURE

FAMILY. Leguminosae – Cercidoideae

SYNONYMS. *Bauhinia bainesii* Schinz, *B. esculenta* Burch.

VERNACULAR / COMMON NAMES. Camel's foot, gemsbok bean, mangetti, marama bean, morama, moramma nut (English); marama, morama (Tswana); ombamui (Herero); gami (Khoekhoe); tsi, tsin (Khoisan); marumana (Tsonga).

PLANT DESCRIPTION

LIFE FORM. Perennial herb.

PLANT. Stems prostrate or trailing up to several metres, herbaceous or woody below (Figure 1), branches reddish to greyish hairy at first, then hairless. **LEAVES.** Petiole 1.5–3.5 cm long; lamina 3–7.5 x 4–10 cm, bilobed halfway or more, cordate at the base, lobes kidney-shaped; stipules 3–5 mm long, elliptic-oblong to obovate. **FLOWERS.** Racemes 4–12 cm long, c. 8–20-flowered, with short flower stalks; sepals 8–12 mm long, oblong-lanceolate, tapered distally, pubescent, the upper pair fused; petals yellow, the 4 larger ones 1.5–2.5 cm long, obovate above a long claw, slightly crinkled [15].

Figure 1 – Right: *T. esculentum* with flowers and closed leaves and developed pods under stressed conditions. Left: Plant with pods under non-stressed conditions. (Photos: E. Mattana, RBG Kew; C. Amosso, BUAN.)

FRUITS AND SEEDS

Pods red turning brown (Figure 1), c. 7 x 5.5 cm, ovate-oblong, hairless with one to six seeds [5]. Seeds brown to brownish black, 1.3–2 x 1.2–1.8 cm, ovoid to subspherical (Figure 2) [15].

FLOWERING AND FRUITING. Flowers from October to March [18]; fruits ripen from March to May [13].

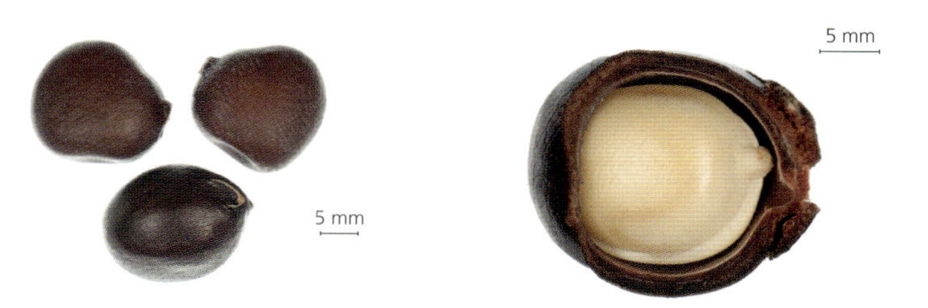

Figure 2 – External views and longitudinal section of the seeds. (Photo: P. Gómez-Barreiro, RBG Kew.)

DISTRIBUTION

Native to Botswana (where it is found in the Kalahari Desert), Namibia, South Africa (Limpopo, North West and Gauteng) and Zimbabwe [6,15].

HABITAT

Dry sandy places, flat grassy places, and also associated with dolomite in grassland or wooded grassland [6,15], at altitudes of 1,000–1,500 m a.s.l. [10].

USES

Widely used as an edible plant, the roasted seeds are rather like cashew nuts (Figure 3). The extracted oil is similar to almond oil and is suitable for domestic purposes, having a pleasant, nutty flavour, with a slightly bitter after-taste [8]. Beans are also boiled with maize meal or ground and pounded to a powder for making porridge or a cocoa-like beverage. Beans can be used as flour, butter, oil, milk and as meat analogues or a snack food [4]. Small tubers and young stems can also be roasted and eaten. Some village elders claim to use this species to cure or prevent illnesses such as diarrhoea, headaches and gynaecological problems [8]. Bean and tuber extracts have been used in traditional medicine for the general upkeep of health [1]. Tubers more than two years old are used as a source of water, as they contain up to about 90% water by weight [4].

Figure 3 – Preparation of morama milk at a workshop led by BCA staff and ICDT, Tsetseng. (Photo: P. Bakewell-Stone, RBG Kew.)

KNOWN HAZARDS AND SAFETY

No toxic substances have been detected in any of the seed components [19]. The tuber flesh, after water has been squeezed out of it, can cause vomiting if eaten [7].

CONSERVATION STATUS

Assessed as Least Concern (LC) in South Africa according to IUCN Red List criteria [2], but has yet to be assessed in Botswana and to have its status confirmed in the IUCN Red List [3,14].

SEED CONSERVATION

HARVESTING. Pods are collected from the plant but in most cases they dehisce or shatter when ripe and release the seeds, which are then collected from the ground [15]. The plant has never been cultivated in the Kalahari region, the pods are instead harvested from scattered natural populations [9,13].

PROCESSING AND HANDLING. After harvesting, seeds are extracted from the tough, woody pods mostly by using stones (Figure 3). Care is needed to avoid crushing the seeds inside.

STORAGE AND VIABILITY. X-ray analysis carried out at the MSB found that 100% of seeds were filled. Seeds are reported to be orthodox [12]. Dry storage at ambient room temperature is the most common storage method in local communities. The seeds are stored in breathable brown sacks away from ground level where there is sufficient air flow in the traditional huts. When stored in these conditions, seeds can remain viable for up to two years.

PROPAGATION

SEEDS

Dormancy and pre-treatments: Seed germination can exceed 60% when seeds are sown without any pre-treatments [16,17]. However, mechanical scarification (abrasion of the seed between two sheets of sandpaper), immersion in water at room temperature for 20 hours or in hot water (100°C) for 2–4 minutes, or dry heating for 5 minutes at 100–150°C, significantly improves final germination to >80% [16,17]. At the MSB, seeds are sterilised by immersion in saturated calcium hypochloride for 5 minutes and then scarified by filing before sowing [12].

Germination, sowing and planting: Seeds can germinate at 30°C in complete darkness [16,17]. High germination percentages (85%) can also be achieved for pre-treated seeds (see above) incubated at 26°C, with 12 hours of light per day [12]. Sandy soils improve seedling emergence while clay loam soils are less suitable [16,17]. Plants can be grown successfully in arid regions, at least under experimental conditions, and a healthy seed crop can be expected in about 4.5 years [11].

VEGETATIVE PROPAGATION

Preliminary investigation (under laboratory conditions) suggests that vegetative propagation is possible using young shoots ('sprouts') [18].

TRADE

In Botswana, the community in Tsetseng, through their community trust, have become leading innovators in creating marketable products from *T. esculentum* (Figure 4) [8]. No information available on international trade.

Figure 4 – Display of some morama products. (Photo: T. Ulian, RBG Kew.)

Authors

Kebadire K. Mogotsi, Moctar Sacande, Tiziana Ulian, Efisio Mattana, Cecilia Amosso and Albertinah T. Matsika.

References

[1] Chingwaru, W., Majinda, R. T., Yeboah, S. O., Jackson, J. C., Kapewangolo, P. T., Kandawa-Schulz, M. & Cencic, A. (2011). *Tylosema esculentum* (marama) tuber and bean extracts are strong antiviral agents against rotavirus infection. *Evidence-Based Complementary and Alternative Medicine* 2011: 284795.

[2] Foden, W. & Potter, L. (2005). *Tylosema esculentum* (Burch.) A.Schreib. In: *National Assessment: Red List of South African Plants*. Version 2015.1. http://redlist.sanbi.org/

[3] IUCN (2017). *The IUCN Red List of Threatened Species*. Version 2017-3. http://www.iucnredlist.org/

[4] Jackson, J. C., Duodu, K. G., Holse, M., Lima de Faria, M. D., Jordaan, D., Chingwaru, W., Hansen, A., Cencic, A., Kandawa-Schultz, M., Mpotokwane, S. M., Chimwamurombe, P., de Kock, H. L. & Minnaar, A. (2010). The morama bean (*Tylosema esculentum*): a potential crop for Southern Africa. *Advances in Food and Nutrition Research* 61: 187–246.

[5] Keegan, A. B. & Van Staden, J. (1981). Marama bean, *Tylosema esculentum*, a plant worthy of cultivation. *South African Journal of Science* 77 (9): 387.

[6] Lebrun, J.-P. & Stork, A. L. (2008). *Tropical African Flowering Plants, Ecology and Distribution*. Vol. 3 — Mimosaceae — Fabaceae (incl. *Derris*). Conservatoire et Jardin Botaniques de la Ville de Genève, Switzerland.

[7] Leger, S. (1997). *The Hidden Gifts of Nature: A Description of Today's Use of Plants in West Bushmanland (Namibia)*. German Development Service and Ministry of Environment and Tourism, Directorate of Forestry, Windhoek, Namibia.

[8] Mogotsi, K. & Ulian, T. (2013). Conserving indigenous food plants in Botswana – the case of morama bean. *Samara* 24: 4.

[9] Monaghan, B. G. & Halloran, G. M. (1996). RAPD variation within and between natural populations of morama [*Tylosema esculentum* (Burchell) Schreiber] in Southern Africa. *South African Journal of Botany* 62 (6): 287–291.

[10] Müseler, D. L. & Schönfeldt, H. C. (2006). The nutrient content of the marama bean (*Tylosema esculentum*), an underutilised legume from Southern Africa. *Agricola* 16 (2): 7–13.

[11] Powell, A. M. (1987). Marama bean (*Tylosema esculentum*, Fabaceae) seed crop in Texas. *Economic Botany* 41 (2): 216–220.

[12] Royal Botanic Gardens, Kew (2017). *Seed Information Database (SID)*. Version 7.1. http://data.kew.org/sid/

[13] Schippers, R. R. (2000). *African Indigenous Vegetables: An Overview of the Cultivated Species*. University of Greenwich/ACP-EU Technical Centre for Agricultural and Rural Cooperation, Chatham, UK.

[14] Setshogo, M. P. & Hargreaves, B. (2002). Botswana. In: *Southern African Plant Red Data Lists*. Edited by J. S. Golding. Southern African Botanical Diversity Network Report No. 14, SABONET, Pretoria, South Africa. pp. 16–20.

[15] Timberlake, J. R., Pope, G. V., Polhill, R. M. & Martins, E. S. (eds) (2007). Leguminosae (Caesalpinioideae). In: *Flora Zambesiaca*, Vol. 3, part 2. Flora Zambesiaca Managing Committee, Royal Botanic Gardens, Kew.

[16] Travlos, I. S., Economou, G. & Karamanos, A. J. (2007). Effects of heat and soil texture on seed germination and seedling emergence of marama bean, *Tylosema esculentum* (Burch.) A.Schreib. *Journal of Food, Agriculture & Environment* 5 (2): 153–156.

[17] Travlos, I. S., Economou, G. & Karamanos, A. J. (2007). Germination and emergence of the hard seed coated *Tylosema esculentum* (Burch) A.Schreib [*sic*] in response to different pre-sowing seed treatments. *Journal of Arid Environments* 68: 501–507.

[18] van der Maesen, L. J. G. (2006). *Tylosema esculentum* (Burch.) A.Schreib. In: *PROTA (Plant Resources of Tropical Africa/Ressources Végétales de l'Afrique Tropicale)*. Edited by M. Brink & G. Belay. Wageningen, The Netherlands. http://www.prota4u.org/search.asp

[19] Watt, J. M. & Breyer-Brandwijk, M. G. (1962). *The Medicinal and Poisonous Plants of Southern and Eastern Africa*, 2nd edition. E. & S. Livingstone, Edinburgh & London.

Ximenia caffra Sond.

TAXONOMY AND NOMENCLATURE

FAMILY. Olacaceae

SYNONYMS. *Ximenia caffra* Sond. var. *caffra*

VERNACULAR / COMMON NAMES. Large sour plum (English); meretologa, moretologa, moretologa wa kgomo (Tswana); omumbeke, ozoninga (Herero); nswanja (Kalanga); oshipeke (Ndonga, Namibia); oipeke oimbyu (Oshikwanyama, Namibia); itsengeni, musanza, mutengeni, mutengeno, mutenguru, nhengeni, tsvanzva (Shona).

PLANT DESCRIPTION

LIFE FORM. Shrub (usually) or tree.

PLANT. Semi-deciduous, <6 m tall (Figure 1), armed with spines; young branches and leaves densely rusty, tomentose to glabrous [7]; bark, grey to dark grey, smooth at first, rough and fissured later [4,12]. **LEAVES.** Coriaceous, with elliptic or oblong-elliptic lamina, 2.5–8 x 1.2–4 cm, obtuse to retuse at apex. **FLOWERS.** Solitary or fasciculate in the axils of leaves, small, tubular, sweet-scented, petals 6–12 x 1.5–2.5 mm [7], creamy white to greenish, sometimes pink to red bearded around the throat of the corolla [12].

Figure 1 – *X. caffra* in fruit. (Photo: C. Amosso, BUAN.)

FRUITS AND SEEDS

Fruits are smooth skinned, oval, c. 2.5 x 3.5 cm, greenish when young, soft orange to bright red when ripe, with juicy pulp and one woody seed [11] which is smooth, mottled, yellowish brown to red, hard coated, ellipsoid, c 2.5 cm long, 1 cm wide, sometimes bordered [12]. The hard outer part of the seed has a white oily nut inside (Figure 2).

FLOWERING AND FRUITING: Flowers from September to October during the dry season towards the onset of rains. Fruits ripen from November to February during the rainy season, but some ripen later during early winter (July) [4,12].

Figure 2 – External views and longitudinal and cross-sections of the seeds. (Photos: P. Gómez-Barreiro, RBG Kew.)

DISTRIBUTION

Tropical Africa (Democratic Republic of Congo) to Southern Africa (including Botswana, Mozambique, Zambia and Zimbabwe) and north-eastern South Africa; also in East Africa (Kenya, Tanzania and Uganda) [2,4,7,8].

HABITAT

Dry wooded bushland and wooded grassland, on plains, along dry river courses, rocky hillsides and on termite mounds, on a variety of soils, including sand, sandy loams to sandy clay loams, from sea level up to 2,000 m a.s.l. [5,7,12]. *Ximenia caffra* withstands moderate frost and is drought resistant when mature [3].

USES

Fruits are eaten raw when fresh or left to dry, although they taste bitter or sour [15]. They are also suitable for making jam [12]. A drink is prepared by soaking the fruits overnight, then hand-pressing, sieving and sweetening with sugar. The fresh juice or dried fruit pulp can be added to porridge to add protein and taste. Seeds contain an edible oil, which the Basarwa people of the Kalahari Desert also use to soften leather, to treat bows and bow strings to prevent breakage, and to rub on to chapped hands and feet [3,4,12]. In traditional medicine, the roots are used to treat abscesses, dysentery, diarrhoea, fever, severe stomach aches or colic, malaria, schistosomiasis (bilharzia), syphilis, hookworm, coughs, chest pains and generalised body pain [12,15]. The juice from rotten fruits is applied to the pustules of chickenpox. The rationale behind this is that fungal growth on decaying fruits produces antibiotics such as penicillin [16]. Some communities swallow the seeds to aid digestion or as a laxative. The wood is hard and fine-grained, and is used to make tool handles, spoons and for general construction [12]. The leaves are browsed by game and livestock [15].

KNOWN HAZARDS AND SAFETY

The skin of the fruit contains hydrocyanic (prussic) acid, which could cause poisoning if consumed in excess [18].

CONSERVATION STATUS

Assessed as Least Concern (LC) in South Africa according to IUCN Red List criteria [8], but has yet to be assessed in Botswana and to have its status confirmed in the IUCN Red List [9,14].

SEED CONSERVATION

HARVESTING. Ripe fruits are picked by hand from the plant or when they have fallen to the ground.

PROCESSING AND HANDLING. Seeds can be easily extracted and cleaned by washing the fruits in water to separate them from the fleshy pulp and then allowed to dry (Figure 3).

STORAGE AND VIABILITY. X-ray analysis carried out on seed lots stored at the MSB highlighted <20% of empty seeds. No information is available on the response of the seeds of this species to storage. Two congeneric taxa stored at the MSB, namely *X. americana* L. and *X. americana* var. *microphylla* Welw., are both reported to be desiccation tolerant, suggesting that the seeds of *X. caffra* are also likely to be orthodox [13]. However, low viability of seeds of *X. americana* and *X. americana* var. *microphylla* was detected after storage at the MSB for three and two months, respectively. Seeds of *X. caffra* lose their viability within three months when stored at room temperature [10]. Further studies are needed to determine the storage physiology of *X. caffra* seeds.

PROPAGATION

SEEDS

Dormancy and pre-treatments: Fresh seeds germinate easily. Germination is improved if seeds are soaked in water for 24 hours and exogenous gibberellic acid is applied [11].

Germination, sowing and planting: Fresh seeds can be sown in a 5:1 sand-compost mixture. If kept moist, seeds germinate after 14–30 days [1,17]. Seedlings should be transplanted into nursery bags when they reach the 2-leaf stage. After one season, they can be planted into open ground. Growth rate is moderate (up to c. 50 cm per year) [17].

VEGETATIVE PROPAGATION

Can be propagated by root suckers [10,12].

Figure 3 – Seed processing in the field. (Photos: C. Amosso, BUAN.)

TRADE

There is informal trade in products made from *X. caffra* in both urban and rural areas of Botswana. There is reported to be a high demand for medicinal products in Botswana [6].

Authors

Bamphitlhi Tiroesele, Cecilia Amosso, Kebadire K. Mogotsi, Moctar Sacande, Tiziana Ulian, Steve Davis and Efisio Mattana.

References

[1] Baloyi, J. K. & Reynolds, Y. (2004). *Ximenia caffra* Sond. South African National Biodiversity Institute (SANBI), South Africa. http://pza.sanbi.org/

[2] Beentje, H. J. (1994). *Kenya Trees, Shrubs and Lianas*. National Museums of Kenya, Nairobi, Kenya.

[3] Chivandi, E., Davidson, B. C. & Erlwanger, K. H. (2012). The red sour plum (*Ximenia caffra*) seed: a potential non-conventional energy and protein source for livestock feeds. *Int. J. Agric. Biol.* 14: 540–544.

[4] Coates Palgrave, K. (1988). *Trees of Southern Africa*, 2nd edition. Struik Publishers, Cape Town, South Africa.

[5] Curtis, B. A. & Mannheimer, C. A. (2005). *Tree Atlas of Namibia*. National Botanical Research Institute of Namibia, Windhoek, Namibia.

[6] Diederichs, N. (ed.) (2006). *Commercialising Medicinal Plants: A Southern African Guide*. Sun Press, Stellenbosch, South Africa.

[7] Exell, A. W., Fernandes, A. & Wild, H. (eds) (1963). Olacaceae. In: *Flora Zambesiaca*, Vol. 2, part 1. Flora Zambesiaca Managing Committee, Royal Botanic Gardens, Kew.

[8] Foden, W. & Potter, L. (2005). *Ximenia caffra* Sond. var. *caffra*. In: *National Assessment: Red List of South African Plants*. Version 2015.1. http://redlist.sanbi.org/

[9] IUCN (2017). *The IUCN Red List of Threatened Species*. Version 2017-3. http://www.iucnredlist.org/

[10] Katende, A. B., Birnie, A. & Tengnäs, B. (2000). *Useful Trees and Shrubs for Uganda: Identification, Propagation and Management for Agricultural and Pastoral Communities*. Regional Land Management Unit, RELMA/SIDA, Nairobi, Kenya.

[11] Mng'omba, S. A., du Toit, E. S. & Akinnifesi, F. K. (2007). Germination characteristics of tree seeds: spotlight on Southern African tree species. *Tree and Forestry Science and Biotechnology* 1 (1): 8 pp.

[12] Orwa, C., Mutua, A., Kindt, R., Jamnadass, R. & Simons, A. (2009). *Agroforestree Database: A Tree Reference and Selection Guide*. Version 4.0. World Agroforestry Centre, Kenya. http://www.worldagroforestry.org/output/agroforestree-database

[13] Royal Botanic Gardens, Kew (2017). *Seed Information Database (SID).* Version 7.1. http://data.kew.org/sid/

[14] Setshogo, M. P. & Hargreaves, B. (2002). Botswana. In: *Southern African Plant Red Data Lists*. Edited by J. S. Golding. Southern African Botanical Diversity Network Report No. 14, SABONET, Pretoria, South Africa. pp. 16–20.

[15] Uiras, M. M. (2001). A taxonomic study of the genus *Ximenia* in Namibia. *Agricola* 12 (22): 111–115.

[16] Van Wyk, B.-E. & Gericke, N. (2000). *People's Plants: A Guide to Useful Plants of Southern Africa*. Briza Publications, Pretoria, South Africa.

[17] Venter, F. & Venter, J.-A. (2016). *Making The Most of Indigenous Trees*, 3rd edition. Briza Publications, Pretoria, South Africa.

[18] Von Koenen, E. (2001). *Medicinal, Poisonous, and Edible Plants in Namibia*. Edition Namibia, Vol. 4. Klaus Hess Publishers, Windhoek, Namibia.

Community nursery managed by a women's group in Tharaka.
(Photo: T. Ulian, RBG Kew.)

KENYA

Albizia coriaria Welw. ex Oliv.

TAXONOMY AND NOMENCLATURE

FAMILY. Leguminosae – Caesalpinioideae

SYNONYMS. None.

VERNACULAR / COMMON NAMES. Woman's tongue tree (English); mukurue (Kikuyu); omubele (Luhya); kumupeli, kumuyebeye (Luhya, Bukusu dialect); ober (Luo); etekwa (Teso).

PLANT DESCRIPTION

LIFE FORM. Tree.

PLANT. Deciduous, 6–36 m tall, crown spreading, flat (Figure 1); bark grey black, rough, flaking; young branchlets puberulous or shortly pubescent, later glabrescent. **LEAVES.** Compound, rachis thinly puberulous or shortly pubescent, pinnae in (2)3–6 pairs, leaflets in (4)6–11(12) pairs, oblong to elliptic or ovate-oblong, 13–33 x 5–14(17) mm, rounded at apex, subglabrous except for a few hairs on midrib beneath. **FLOWERS.** Subsessile, or with pedicels <2 mm long, calyx puberulous outside with a few shortly stipitate glands, corolla white or whitish, puberulous outside [2].

Figure 1 – General habit. (Photo: V. Otieno, KEFRI.)

FRUITS AND SEEDS

Unripe fruits bright green, turning pale yellow; pods glossy purple brown when ripe, <16 cm long, narrowed at both ends, and containing 3–5 seeds [6]. Seeds are brown and flattened, 9–12 x 8–9 mm [2], with a funicular aril (Figure 2).

FLOWERING AND FRUITING. In Kenya, flowering occurs from March to May; fruiting in October to December in Bungoma [6], Siaya and Busia counties.

Figure 2 – External views of the seeds showing the areole surrounded by the pleurogram and the funiculus, and longitudinal section showing the plumule, embryonic axis and cotyledon. (Photos: P. Gómez-Barreiro, RBG Kew.)

DISTRIBUTION

From West Africa to Sudan and south to Angola. In Kenya, found only in the west, in the Lake Victoria region [6] (i.e. in Siaya, Bungoma, Busia, Kisumu and Homabay counties).

HABITAT

Riverine and lakeside forests to open or wooded grassland, on a variety of soils (including gravel), 1,140–1,700 m a.s.l. A pioneer species [8], and often left standing in crop fields [6].

USES

In traditional medicine, a bark decoction is used for treating menhorrhagia, threatened abortion and postpartum haemorrhage. A bark decoction is drunk, or bark ash is licked, for whooping cough, tonsillitis, pneumonia, anthrax, kidney ailments, and anaemia and heart problems. The roots are used for treating venereal diseases, and steamed to treat sore eyes. In traditional veterinary medicine, a root decoction is used for gastrointestinal problems, measles, and foot and mouth disease in cattle [5]. The wood is used for firewood, charcoal, timber, furniture, carvings, poles and in boat building. *Albizia coriaria* fixes nitrogen and is used in agroforestry. It is a good shade tree, and is often planted as an ornamental. The foliage is browsed by livestock, and the flowers attract bees which then produce a good honey [6,8].

KNOWN HAZARDS AND SAFETY

No known hazards.

CONSERVATION STATUS

A widely distributed species assessed as Not Threatened (nt) on a global scale according to previous IUCN criteria [3], but assessed as Near Threatened (NT) in Uganda on account of harvesting for charcoal and timber [7]. The status of this species has yet to be confirmed in the IUCN Red List [4].

SEED CONSERVATION

HARVESTING. Mature purple brown pods are collected from the crown by spreading a net or canvas under the tree and climbing the tree to hand-pick the pods or shaking the branches to dislodge the pods.

PROCESSING AND HANDLING. Pods are dried in the sun and then threshed lightly to minimise mechanical damage to the seeds. Seeds can be extracted manually by hand and cleaned by hand-sorting, winnowing or sieving.

STORAGE AND VIABILITY. Seeds of this species are reported to be orthodox [9] and after appropriate drying can therefore be stored at sub-zero temperatures for many years without significant loss of viability. Dried seeds can also be stored in airtight containers (e.g. kilner jars and aluminium packets) in a cool dry place for short- to medium-term storage at room conditions.

PROPAGATION

SEEDS

Dormancy and pre-treatments: Seeds of this species are reported to be non-dormant [1]. No pre-treatments are therefore required before sowing [6], although chipping by using a scalpel, knife or nail clipper can improve germination.

Germination, sowing and planting: Germination tests carried out on seed lots stored at the MSB under laboratory controlled conditions highlighted germination percentages ≥75% for scarified seeds incubated with 8 hours of light per day at constant temperatures of ≥20°C [9]. In the nursery, pre-treated seeds germinate readily after 7–21 days, reaching final germination percentages of 60–80%. Pre-treated seeds can be sown in a seedbed or in polythene tubes. Seedlings are ready for pricking out three weeks after germination, and are ready for transplanting in the field after five months. For field establishment and intercropping, seedlings should be planted 15–20 m apart. *Albizia coriaria* is slow growing; recommended management practices include lopping and pollarding [6,8].

VEGETATIVE PROPAGATION

Can be propagated from wildings [8].

TRADE

Carvings made from the wood are sold in local markets.

Authors

William Omondi, Victor Otieno, David Meroka, Steve Davis and Efisio Mattana.

References

[1] Baskin, C. C. & Baskin, J. M. (2014). *Seeds: Ecology, Biogeography, and Evolution of Dormancy and Germination*, 2nd edition. Academic Press, San Diego, USA.

[2] Brenan, J. P. M. (1959). Leguminosae Subfamily Caesalpinioideae. In: *Flora of Tropical East Africa*. Edited by C. E. Hubbard & E. Milne-Redhead. Crown Agents for Oversea Governments and Administrations, London.

[3] ILDIS (2006–2013). *World Database of Legumes*. International Legume Database and Information Service. http://www.legumes-online.net/ildis/aweb/database.htm

[4] IUCN (2017). *The IUCN Red List of Threatened Species.* Version 2017-3. http://www.iucnredlist.org/

[5] Kokwaro, J. O. (2009). *Medicinal Plants of East Africa*, 3rd edition. University of Nairobi Press, Nairobi, Kenya.

[6] Maundu, P. & Tengnäs, B. (eds) (2005). *Useful Trees and Shrubs for Kenya.* Technical Handbook No. 35, World Agroforestry Centre, Nairobi, Kenya.

[7] National Red List (2017). *Albizia coriaria*. National Red List Network. http://www.nationalredlist.org/

[8] Orwa, C., Mutua, A., Kindt, R., Jamnadass, R. & Simons, A. (2009). *Agroforestree Database: A Tree Reference and Selection Guide*. Version 4.0. World Agroforestry Centre, Kenya. http://www.worldagroforestry.org/output/agroforestree-database

[9] Royal Botanic Gardens, Kew (2017). *Seed Information Database (SID).* Version 7.1. http://data.kew.org/sid/

Annona senegalensis Pers.

TAXONOMY AND NOMENCLATURE

FAMILY. Annonaceae

SYNONYMS. *Annona arenaria* Thonn., *A. chrysophylla* Bojer, *A. porpetac* Boivin ex Baill.

VERNACULAR / COMMON NAMES. Wild custard apple, wild soursop (English); chekwa, mkonokono, mtomoko, mtomoko-mwitu, mutopetope mwitu (Swahili); malamuti, mlamote (Boni); mbokwe (Digo); mutakuma (Giriama); kitomoko, makulo, mutomoko wa kitheka (Kamba); omokera (Kisii); kumufwora, muvulu (Luhya); nyabolo, obolo, obolobolo (Luo).

Figure 1 – Branch and unripe fruit.
(Photos: H. Migiro, KEFRI.)

PLANT DESCRIPTION

LIFE FORM. Shrub or small tree.

PLANT. Semi-deciduous, 1.5–10 m tall, bark grey brown, often rough and corrugated, young stems rust-coloured, velvety to greyish tomentose. **LEAVES.** Oblong to ovate or elliptic, c. 5–18 x 2.5–12 cm, apex obtuse or apiculate to rounded or slightly emarginate, base cuneate or truncate to cordate, papery or coriaceous, glabrous or sparsely pubescent above, midrib hairy; glabrescent, rust-coloured and velvety beneath (Figure 1). **FLOWERS.** Solitary or in groups of 2–4, <3 cm in diameter, usually fragrant, pedicels <2 cm long, sepals ovate triangular, petals 8–12 mm long, greenish outside, creamy to yellowish or pinkish within, sometimes blotched with purple or crimson blotched at base inside [10].

FRUITS AND SEEDS

Fruits 2.5–5 x 2.5–4 cm, ovoid or globose, smooth, with divisions; green when unripe (Figure 1), turning orange yellow and smelling like pineapple when ripe [5,10]. Seeds (Figure 2) numerous, cylindric, oblong, orange brown [10].

FLOWERING AND FRUITING. In Kenya, trees flower in late February to mid-March. Fruits mature in July and August in Makueni, Kitui and Machakos [5].

Figure 2 – External (left) and internal (right) views of the seeds showing the ruminate endosperm, the vascular bundle and the small embryo at the micropylar end. (Photos: P. Gómez-Barreiro, RBG Kew.)

DISTRIBUTION

Native from West and Central Africa east to Ethiopia and Kenya, and south to Lesotho, South Africa and Swaziland. Also native in Madagascar [10].

HABITAT

Wooded or bushed grassland, riverine woodland, savanna, secondary (fire-induced) bushland, and on the coast in evergreen forest, from sea level to 2,400 m a.s.l. on various soils, including sandy loam on coral rocks [2,5,10].

USES

In traditional medicine, a decoction of the roots is taken as a remedy for chest colds, stomach ache, diarrhoea, vomiting, impotence, sterility and flatulence. The fruits are edible, and are also used to treat diarrhoea, dysentery and vomiting. The leaves are applied to snake bites, or the leaf juice is swallowed. The gum is applied to cuts and wounds [4]. The wood is used for poles, tool handles and low-quality firewood [7]. The leaves are browsed by livestock. The bark is used for fibre and produces a yellow or brown dye. The tree is planted as an ornamental and as a windbreak.

KNOWN HAZARDS AND SAFETY

A study on rats to investigate the safety profile of *A. senegalensis* indicated that root bark extracts are safe at low doses, but high dosage levels caused some liver damage in rats [6].

CONSERVATION STATUS

Assessed as Least Concern (LC) in South Africa according to IUCN Red List criteria [8], but this widespread species has yet to be assessed in Kenya and to have its status confirmed in the IUCN Red List [3].

SEED CONSERVATION

HARVESTING. Mature dull green or yellowish green fruits can be picked by hand from the crown of trees. Fruits ripen 2–3 days after harvesting.

PROCESSING AND HANDLING. Seeds can be extracted from the fruits by hand or by macerating ripe fruits.

STORAGE AND VIABILITY. No information available on seed storage behaviour. However, other *Annona* spp. are reported to have desiccation-tolerant seeds, suggesting that the seeds of *A. senegalensis* are likely to be orthodox [9] and therefore suitable for long-term conservation at sub-zero temperatures. When seeds are stored in room conditions, they are susceptible to insect damage and lose viability within six months [5].

PROPAGATION

SEEDS

Dormancy and pre-treatments: No information is available on seed dormancy. Seeds of other *Annona* spp. are reported to have morphophysiological (MPD) dormancy [1]. The seed morphology of *A. senegalensis* indicates that the small embryo (Figure 2) should grow inside the seed before radicle emergence, suggesting the presence of morphological dormancy (MD) in this species. However, further studies should be carried out to confirm MD and to determine whether a physiological component of dormancy (PD) is also present. Seeds require no pre-treatments for direct sowing [5].

Germination, sowing and planting: Further studies are needed to identify the optimum incubation temperatures and light requirements for germination under laboratory controlled conditions. In the nursery, germination of untreated seeds is sporadic [5]. Seeds can be sown directly in the field in prepared and refilled pits at a depth of 7 cm. The pits should be watered as necessary.

VEGETATIVE PROPAGATION

Wildings and root suckers from wounded roots [5].

TRADE

The fruits are sold in local markets.

Authors

William Omondi, Victor Otieno, David Meroka, Steve Davis and Efisio Mattana.

References

[1] Baskin, C. C. & Baskin, J. M. (2014). *Seeds: Ecology, Biogeography, and Evolution of Dormancy and Germination*, 2nd edition. Academic Press, San Diego, USA.

[2] Beentje, H. J. (1994). *Kenya Trees, Shrubs and Lianas*. National Museums of Kenya, Nairobi, Kenya.

[3] IUCN (2017). *The IUCN Red List of Threatened Species.* Version 2017-3. http://www.iucnredlist.org/

[4] Kokwaro, J. O. (2009). *Medicinal Plants of East Africa*, 3rd edition. University of Nairobi Press, Nairobi, Kenya.

[5] Maundu, P. & Tengnäs, B. (eds) (2005). *Useful Trees and Shrubs for Kenya*. Technical Handbook No. 35, World Agroforestry Centre, Nairobi, Kenya.

[6] Okoye, T. C., Akah, P. A., Ezike, A. C., Okoye, M. O., Onyeto, C. A., Ndukwu, F., Ohaegbulam, E. & Ikele, L. (2012). Evaluation of the acute and sub acute toxicity of *Annona senegalensis* root bark extracts. *Asian Pacific Journal of Tropical Medicine* 5 (4): 277–282.

[7] Orwa, C., Mutua, A., Kindt, R., Jamnadass, R. & Simons, A. (2009). *Agroforestree Database: A Tree Reference and Selection Guide*. Version 4.0. World Agroforestry Centre, Kenya. http://www.worldagroforestry.org/output/agroforestree-database

[8] Raimondo, D., von Staden, L., Foden, W., Victor, J. E., Helme, N. A., Turner, R. C., Kamundi, D. A. & Manyama, P. A. (2009). *Red List of South African Plants. Strelitzia* 25, South African National Biodiversity Institute, Pretoria, South Africa.

[9] Royal Botanic Gardens, Kew (2017). *Seed Information Database (SID).* Version 7.1. http://data.kew.org/sid/

[10] Verdcourt, B. (1971). Annonaceae. In: *Flora of Tropical East Africa*. Edited by E. Milne-Redhead & R. M. Polhill. Crown Agents for Oversea Governments and Administrations, London.

Balanites aegyptiaca (L.) Delile

TAXONOMY AND NOMENCLATURE

FAMILY. Zygophyllaceae

SYNONYMS. *Agialid abyssinica* Tiegh., *A. aegyptiaca* (L.) Kuntze, *A. arabica* Tiegh., *Balanites arabica* (Tiegh.) Blatt., *B. fischeri* Mildbr. & Schltr., *B. latifolia* (Tiegh.) Chiov., *B. suckertii* Chiov., *Canthium zizyphoides* Mildbr. & Schltr., *Ximenia aegyptiaca* L.

VERNACULAR / COMMON NAMES. Desert date (English); mjunju (Swahili); baddan (Boran); mulului, ndului (Kamba); ngoswet (Kipsigis); otho (Luo); olngoswa (Maasai); ngoswa (Marakwet); mububua (Mbeere); tuyunwo (Pokot); lowvai (Samburu); ngonswo (Tugen); eroronyit (Turkana).

PLANT DESCRIPTION

LIFE FORM. Tree or rarely a shrub.

PLANT. Evergreen, sometimes deciduous, <12(–15) m tall (Figure 1), usually spiny (spines <8 cm long; Figure 1), trunk usually straight, 60 cm in diameter, often fluted; branches spreading irregularly or pendulous; bark hard, becoming rough, corky and deeply fissured, dark grey; branchlets and spines greyish green becoming light brown. **LEAVES.** Subsessile or with petiole <3.5 cm long; 0.5–1 mm long, stipules triangular and caducous; leaflets very variable in shape, narrowly spatulate or elliptic to broadly ovate or obovate, sometimes very broadly spatulate, 2.5–6 x 1.5–4 cm (Figure 1).

FLOWERS. Pentamerous, sweet-scented, green to pale yellow or creamy white, in inflorescences of up to 20 (or more), flowers in cymose fascicles at spineless or spiniferous nodes, or closely arranged on shoots of short internodes [2,10].

Figure 1 – *B. aegyptiaca* tree and details of the leaves and spines. (Photos: NMK and B. Wursten, Flora of Zimbabwe: https://www.zimbabweflora.co.zw/index.php.)

FRUITS AND SEEDS

Fruits are ellipsoid, green, turning brown to yellowish when ripe; brittle outside, yellowish brown, fibrous and oily inside. Seeds (Figure 2) are hard woody stones (pyrenes) [10].

FLOWERING AND FRUITING. In the equatorial zone, flowering and fruiting occurs over a prolonged season with a peak at the height of the dry season [12].

Figure 2 – Seeds of *B. aegyptiaca.* (Photo: P. Gómez-Barreiro, RBG Kew.)

DISTRIBUTION

Native throughout much of Africa, from Mauritania to Somalia and Egypt southwards to Zambia and Zimbabwe; also native in the Arabian Peninsula and Jordan valley [8,10]. In Kenya, it is found in the arid, semi-arid and subhumid regions of Northern, Southern, Coast and Nyanza provinces [1]. The natural distribution of this species is obscured by a long history of cultivation and semi-domestication and subsequent naturalisation.

HABITAT

Dry bushland, bushed grassland, wooded grassland or woodland, also riverine [2]. Intolerant of shade so prefers open woodland or savanna. Mainly found on flat, alluvial sites with deep sandy loam, up to 1,000 m a.s.l. in areas with a mean annual temperature of 20–30°C and mean annual rainfall of 250–400 mm [12], but can be found from 350–2,100 m a.s.l., especially where the water table is seasonally high [8].

USES

The fruit is edible, and yields an edible oil used for cooking [2]. The fruit has a sweet taste, but is rather bitter near the stone. The tree has many uses in traditional medicine. For example, the fruits have been used to treat liver and spleen diseases [5], gum from the wood is used to treat chest complaints, and an infusion of the roots is used as an emetic in the treatment of abdominal pains and malaria [2]. An emulsion of the fruit and seeds, or from the bark, has molluscicidal properties, and is lethal to the snail hosts of schistosomiasis [3,6]. The bark has anti-inflammatory, anti-nociceptive and antioxidant properties [13]. Fresh and dried leaves, young shoots and fruit are eaten by livestock. The wood is used as firewood; it produces considerable heat and very little smoke making it particularly suitable for indoor use. It also produces high-quality charcoal. The nutshell has been suggested to be suitable for industrial activated charcoal. A strong fibre obtained from the bark is used for making baskets and ropes. The wood is durable and used for making yokes, wooden spoons, pestles, mortars, tool handles, stools and combs, particularly among the Kamba in Kenya. The branches are cut and coiled to make livestock enclosures. *Balanites aegyptiaca* is often used as a live fence and boundary marker, and has potential for use in shelterbelts [6].

KNOWN HAZARDS AND SAFETY

Despite its edibility, the fruit has been regarded as emetic and purgative [14] and the seeds are mildly purgative [7]. The bark juice is also poisonous [7]. The spines on the stems are sharp so care should be taken when harvesting the fruits [12].

CONSERVATION STATUS

This widespread and abundant species is not considered to be either rare or threatened, but its status has yet to be confirmed in the IUCN Red List [4].

SEED CONSERVATION

HARVESTING. Fruits should be harvested immediately they change colour from green to yellow [1]. They can be collected by spreading a tarpaulin under the tree and shaking the branches to dislodge the fruits. Care should be taken to avoid the spines on branches when climbing the tree [12].

PROCESSING AND HANDLING. The fruit pulp must be removed as soon as possible to avoid fermentation. If extraction is not possible in the field, the fruits should be kept dry and spread in a thin layer during temporary field storage. The fruit pulp can be removed after soaking the fruits in water and then letting them dry [12].

STORAGE AND VIABILITY. Seeds are reported to be orthodox [9] and after appropriate drying can therefore be stored at sub-zero temperatures for long-term conservation. In addition, mature and properly dried seeds can be stored in airtight containers for several years at 3°C [1].

PROPAGATION

SEEDS

Dormancy and pre-treatments: Several pre-treatments, such as different types of scarification and gibberellic acid treatments, were applied on seeds of this species to examine whether they possess physical (PY), physiological (PD) or combinational dormancy (PY+PD) [11]. Neither scarification nor gibberellic acid treatments resulted in significantly higher germination than the control, confirming that seeds of this species do not exhibit dormancy [11]. No pre-treatments are therefore necessary before sowing in the nursery [1].

Germination, sowing and planting: High germination percentages (c. 80%) are reported under laboratory controlled conditions for scarified seeds incubated in sand in an alternating temperature regime of 20/35 °C in the light [9]. Similar results (60–80% germination according to provenance) were obtained for intact seeds sown at 25°C on filter paper [11].

VEGETATIVE PROPAGATION

Can be propagated from stem cuttings [8] and root suckers [6].

TRADE

No data available.

Authors

Veronicah Ngumbau, Patrick Muthoka, Tiziana Ulian, Steve Davis and Efisio Mattana.

References

[1] Albrecht, J. (1993). *Tree Seed Handbook of Kenya*. GTZ Forestry Seed Centre, Muguga, Kenya.

[2] Beentje, H. J. (1994). *Kenya Trees, Shrubs and Lianas.* National Museums of Kenya, Nairobi, Kenya.

[3] Booth, F. E. M. & Wickens, G. E. (1988). *Non-Timber Uses of Selected Arid Zone Trees and Shrubs in Africa.* Food and Agriculture Organization of the United Nations (FAO), Rome, Italy.

[4] IUCN (2017). *The IUCN Red List of Threatened Species.* Version 2017-3. http://www.iucnredlist.org/

[5] Maundu, P. M., Ngugi, G. W. & Kabuye, C. H. S. (1999). *Traditional Food Plants of Kenya.* Kenya Resource Centre for Indigenous Knowledge (KENRIK), National Museums of Kenya, Nairobi, Kenya.

[6] Orwa, C., Mutua, A., Kindt, R., Jamnadass, R. & Simons, A. (2009). *Agroforestree Database: A Tree Reference and Selection Guide*. Version 4.0. World Agroforestry Centre, Kenya. http://www.worldagroforestry.org/output/agroforestree-database

[7] Quattrocchi, U. (2012). *CRC World Dictionary of Medicinal and Poisonous Plants: Common Names, Scientific Names, Eponyms, Synonyms, and Etymology, Vol. I: A–B.* CRC Press, Boca Raton, Florida, USA.

[8] Retallick, S. J. & Mbah, J. M. (1992). Vegetative propagation of *Balanites aegyptiaca* (L.) Del. *Commonwealth Forestry Review* 71 (1): 52–56.

[9] Royal Botanic Gardens, Kew (2017). *Seed Information Database (SID).* Version 7.1. http://data.kew.org/sid/

[10] Sands, M. J. S. (2003). Balanitaceae. In: *Flora of Tropical East Africa*. Edited by H. J. Beentje & S. A. Ghazanfar. A. A. Balkema, Lisse, The Netherlands.

[11] Schelin, M., Tigabu, M., Eriksson, I., Sawadogo, L. & Odén, P. C. (2003). Effects of scarification, gibberellic acid and dry heat treatments on the germination of *Balanites aegyptiaca* seeds from the Sudanian savanna in Burkina Faso. *Seed Science and Technology* 31 (3): 605–617.

[12] Schmidt, L. & Jøker, D. (2000). *Balanites aegyptiaca* (L.) Delile. *Seed Leaflet* No. 21. Danida Forest Seed Centre, Humlebaek, Denmark.

[13] Van Wyk, B.-E., Van Oudtshoorn, B. & Gericke, N. (2009). *Medicinal Plants of South Africa*, 2nd edition. Briza Publications, Pretoria, South Africa.

[14] Watt, J. M. & Breyer-Brandwijk, M. G. (1962). *The Medicinal and Poisonous Plants of Southern and Eastern Africa*, 2nd edition. E. & S. Livingstone, Edinburgh & London.

Berchemia discolor (Klotzsch) Hemsl.

TAXONOMY AND NOMENCLATURE

FAMILY. Rhamnaceae

SYNONYMS. *Adolia discolor* (Klotzsch) Kuntze, *Araliorhamnus punctulata* H.Perrier, *A. vaginata* H.Perrier, *Phyllogeiton discolor* (Klotzsch) Herzog, *Scutia discolor* Klotzsch

VERNACULAR / COMMON NAMES. Bird plum, dog plum, mountain date (English); mkulu, mnago (Swahili); jajab (Boran); mkulu (Giriama); kisaaya, kisanawa (Kamba); muthuama (Marakwet); mu-thwana (Mbeere); muchukwa (Pokot, Tugen); santaita (Samburu); deen (Somali); mzwana (Taita); muthwana (Tharaka); emeyen (Turkana).

PLANT DESCRIPTION

LIFE FORM. Shrub or tree.

PLANT. Evergreen or deciduous, <10(–25) m tall (Figure 1); wood very hard, heavy; bark reticulately fissured, dark grey brown, shedding in sheets; young branches with conspicuous lenticels. **LEAVES.** Entire, alternate at base of shoots, opposite or nearly so distally, elliptic to ovate-oblong, acute, (2)3–5(9) x (1.5)2–3.5(6) cm, base cuneate or usually rounded, green, glabrous or minutely pubescent near midrib above, slightly paler, microvesiculate and glabrous to densely pubescent beneath (Figure 1). **FLOWERS.** Yellow green, usually in clusters of 2–6, sometimes solitary, sepals 2–3 mm long, petals <3 mm long [3,5,7].

Figure 1 – General habit of *B. discolor* and details of leaves and unripe fruits. (Photos: V. Otieno, KEFRI and B. Wursten, Flora of Zimbabwe: https://www.zimbabweflora.co.zw/index.php.)

FRUITS AND SEEDS

Fruits are fleshy, yellow egg-shaped drupes, c. 2 cm long and 1 cm in diameter, containing two seeds. Fruits are 2-locular with two seeds enclosed in the endocarp [9]. The stone (endocarp) is bony and thickish; the seed coat is thin and adherent to the endocarp (Figure 2). The endocarp does not completely enclose the seeds. Soft tissue occurs at the fruit stalk end of the stone, which is separated by septa from the embryos (Figure 2) [16].

FLOWERING AND FRUITING. Flowering and fruiting vary according to climatic conditions; fruits usually develop between January and May in the Kitui, Kibwezi and Taita regions of Kenya [1].

Figure 2 – Above: Longitudinal external view of the stone, showing the brown endocarp and the white soft tissue region, and cross-section showing the thick endocarp and the two locules with the seeds. Below: Longitudinal sections of the stone showing the two loculi and the soft region at the top, and a seed with small testa and excised embryo. (Photos: P. Gómez-Barreiro, RBG Kew.)

DISTRIBUTION

Eastern Africa, from Sudan and Ethiopia south to the Northern Provinces of South Africa, and west to Angola and Namibia. Also found in the Arabian Peninsula [7].

HABITAT

Widespread in thicket, semi-desert grassland and wooded grassland. Although less abundantly, it grows tallest in stream valleys and riverine forests [7]. It is mainly found in lower rainfall areas below 1,000 m [5], but can occur up to 1,600 m a.s.l. in Kenya [3], where it mainly occurs in arid and semi-arid areas of the north-eastern coast, Kitui, Machakos, Baringo and Turkana [1].

USES

The resinous wood is used in construction and for making tools, poles and furniture. The fruit is edible and can be eaten raw, or fermented to make alcoholic beverages [13]. A tea is made from the leaves. A good-quality honey is made from the pollen [12]. In traditional medicine, a bark infusion is used by the Turkana to treat liver ailments [3], and the fruits are used in Kenya for treating inflammation of the throat and tonsils [8]. The leaves and young branches are browsed by livestock [4]. A brown or black dye is made from the bark and root bark [13]; resin can be obtained from the heartwood [1]. The tree is also planted as an ornamental [10].

KNOWN HAZARDS AND SAFETY

No known hazards.

CONSERVATION STATUS

This widespread species is not considered to be either rare or threatened. Assessed as Least Concern (LC) in South Africa according to IUCN Red List criteria [11], but has yet to be assessed in Kenya and to have its status confirmed in the IUCN Red List [6].

SEED CONSERVATION

HARVESTING. Fruits are collected by picking them from the crown [1].

PROCESSING AND HANDLING. After collection, the fruits are squeezed to remove the endocarp from the pulp. Endocarps and pulp are then further separated by floating the mixture in water [1].

STORAGE AND VIABILITY. Seeds of this species are reported as likely to be orthodox [14]. Viability can be maintained for several years in hermetic storage at 3°C with 7–13% moisture content [1].

PROPAGATION

SEEDS

Dormancy and pre-treatments: Seeds of this species are reported to have combinational (i.e. physical and physiological) dormancy [2]. However, physical dormancy has been excluded for seeds of the congeneric *Berchemia scandens* (Hill) K.Koch, the seed morphology of which is very similar to that of *B. discolor*. In particular, the soft tissue region at the fruit stalk end of the stone has been identified as the primary area of water entrance during imbibition [16]. Scarification has been reported to improve seed germination of *B. discolor* [15] and this could indicate the presence of non-deep physiological dormancy [2]. More studies are needed to confirm the imbibition

Figure 3 – Seedlings propagated in the nursery. (Photo: T. Ulian, RBG Kew.)

pattern in *B. discolor*, to confirm the lack of physical dormancy and to identify pre-treatments that could break physiological dormancy.

Germination, sowing and planting: In the nursery, germination is good but sporadic. When seeds are pre-treated by nicking and soaking in water, germination reaches 30% after six weeks and 65% after 10 weeks from sowing [9]. Untreated seeds are reported to germinate in 50–60 days with an expected germination percentage of 50% [1]. No information is available on seed germination under laboratory controlled conditions.

VEGETATIVE PROPAGATION

Can be propagated from coppice shoots and root suckers [10].

TRADE

No data available.

Authors

Veronicah Ngumbau, Patrick Muthoka, Steve Davis and Efisio Mattana.

References

[1] Albrecht, J. (1993). *Tree Seed Handbook of Kenya*. GTZ Forestry Seed Centre, Muguga, Kenya.

[2] Baskin, C. C. & Baskin, J. M. (2014). *Seeds: Ecology, Biogeography, and Evolution of Dormancy and Germination*, 2nd edition. Academic Press, San Diego, USA.

[3] Beentje, H. J. (1994). *Kenya Trees, Shrubs and Lianas.* National Museums of Kenya, Nairobi, Kenya.

[4] Curtis, B. A. & Mannheimer, C. A. (2005). *Tree Atlas of Namibia*. National Botanical Research Institute of Namibia, Windhoek, Namibia.

[5] Exell, A. W., Fernandes, A. & Wild, H. (eds) (1966). Rhamnaceae. In: *Flora Zambesiaca*, Vol. 2, part 2. Flora Zambesiaca Managing Committee, Royal Botanic Gardens, Kew.

[6] IUCN (2017). *The IUCN Red List of Threatened Species.* Version 2017-3. http://www.iucnredlist.org/

[7] Johnston, M. C. (1972). Rhamnaceae. In: *Flora of Tropical East Africa*. Edited by E. Milne-Redhead & R. M. Polhill. Crown Agents for Oversea Governments and Administrations, London.

[8] Maundu, P. M., Ngugi, G. W. & Kabuye, C. H. S. (1999). *Traditional Food Plants of Kenya.* Kenya Resource Centre for Indigenous Knowledge (KENRIK), National Museums of Kenya, Nairobi, Kenya.

[9] Msanga, H. P. (1998). *Seed Germination of Indigenous Trees in Tanzania: Including Notes on Seed Processing and Storage, and Plant Uses.* Canadian Forest Service, Northern Forestry Centre, Edmonton, Canada.

[10] Orwa, C., Mutua, A., Kindt, R., Jamnadass, R. & Simons, A. (2009). *Agroforestree Database: A Tree Reference and Selection Guide*. Version 4.0. World Agroforestry Centre, Kenya. http://www.worldagroforestry.org/output/agroforestree-database

[11] Raimondo, D., von Staden, L., Foden, W., Victor, J. E., Helme, N. A., Turner, R. C., Kamundi, D. A. & Manyama, P. A. (2009). *Red List of South African Plants. Strelitzia* 25, South African National Biodiversity Institute, Pretoria, South Africa.

[12] Roodt, V. (1998). *Trees & Shrubs of the Okavango Delta, Medicinal Uses and Nutritional Value*. Shell Field Guide Series: Part 1. Shell Oil Botswana, Gaborone, Botswana.

[13] Royal Botanic Gardens, Kew (1999–2015). *Survey of Economic Plants for Arid and Semi-Arid Lands (SEPASAL) Database*. http://apps.kew.org/sepasalweb/sepaweb

[14] Royal Botanic Gardens, Kew (2017). *Seed Information Database (SID).* Version 7.1. http://data.kew.org/sid/

[15] Tibugari, H., Kaundura, F. & Mandumbu, R. (2013). *Berchemia discolour* [*sic*] response to different scarification methods. *Journal of Agricultural Technology* 9 (7): 1927–1935.

[16] Walck, J. L., Shea Cofer, M., Gehan Jayasuriya, K. M. G., Thilina, R., Fernando, M. & Hidayati, S. N. (2012). A temperate rhamnaceous species with a non-enclosing stone and without physical dormancy. *Seed Science Research* 22: 269–278.

Blighia unijugata Baker

TAXONOMY AND NOMENCLATURE

FAMILY. Sapindaceae

SYNONYMS. *Phialodiscus unijugatus* (Baker) Radlk.

VERNACULAR / COMMON NAMES. Triangle tops tree (English); mkivule, mwakamwatu (Swahili); mpwakapwaka (Digo); muikoni (Kikuyu); shiarambatsa (Luhya); achak, bilo, ochol, ochond achak (Luo); muthiama (Meru); mubo, mubonyeni (Pokomo).

PLANT DESCRIPTION

LIFE FORM. Tree.

PLANT. Evergreen, dioecious, 6–9(–30) m tall, crown dense, rounded (Figure 1), bark smooth, but often with horizontal ridges and warts, grey to dark green; young twigs finely orange brown hairy, becoming smooth. **LEAVES.** Reddish when young, drying bright green or brownish; alternate, compound with up to 5 pairs of leaflets which are opposite, elliptic to oblong, 6–22 x 1.5–8 cm, upper pair largest, apex tapering, margin often wavy. **FLOWERS.** Inflorescences in axils of current leaves, 5–10 cm long, petals c. 1.5 mm, flowers singly inserted in cymose groups, whitish or yellowish, sweet-scented [3].

Figure 1 – Tree with ripe fruits. (Photo: J. Stevens, African Plants - A Photo Guide. www.africanplants.senckenberg.de.)

FRUITS AND SEEDS

Fruits (Figure 2) are reddish triangular capsules 2–3 x 2.5–3 cm, pointed at apex, usually 3-seeded; pink endocarp with yellow margins. Seeds (Figure 2) are ovoid, 1–2 cm long, glossy dark brown to black, with a bright yellow cup-shaped aril <1 cm long at the base [3].

Figure 2 – External views of fruits, showing the reddish triangular three-seeded capsules and the seeds with yellow arils. (Photo: J. Stevens, African Plants - A Photo Guide. www.africanplants.senckenberg.de.)

FLOWERING AND FRUITING. In Kenya, the flowering period extends from April to June; fruits mature in October and November.

DISTRIBUTION

Native to tropical Africa, from Guinea Bissau and Democratic Republic of Congo east to Ethiopia and Kenya, and south to Angola, Zimbabwe, Mozambique and extending into the subtropics in South Africa [3].

HABITAT

Moist evergreen forest, semi-deciduous forest, dry areas of riverine forest, and in wooded grassland, sometimes persisting in cultivated areas [3]. Grown in central and western parts of Kenya.

USES

In traditional medicine, an infusion of the roots, and ash from roasted stems, are used to treat fever [5]. Nausea and vomiting are treated with the fruit and medicinal ointments made from the seed oil. A fragrant cosmetic lotion has been made from soaking the pleasantly scented flowers in water. The leaves are cooked and eaten as a vegetable [6]. The wood is used for light construction, furniture, agricultural implements and carving, firewood and charcoal production. The flowers provide nectar and pollen for honey bees. *Blighia unijugata* is often planted as a shade tree in villages and in coffee plantations [9].

KNOWN HAZARDS AND SAFETY

No known hazards.

CONSERVATION STATUS

Assessed as Least Concern (LC) in South Africa according to IUCN Red List criteria [10], and as Lower Risk-Near Threatened (LR-nt) in Zambia [2] and Zimbabwe [7], but has yet to be assessed in Kenya and to have its status confirmed in the IUCN Red List [4].

SEED CONSERVATION

HARVESTING. Mature red fruits are collected from the crown by hand-picking from low branches, or by spreading a net or canvas on the ground and then climbing the tree and shaking the branches to dislodge the fruits.

PROCESSING AND HANDLING. Fruits are dried in the sun and then threshed lightly to minimise damage to the seeds. Alternatively, small amounts of seeds can be extracted by hand. Seeds can be cleaned by hand-sorting and detaching the yellow arils from the seeds.
STORAGE AND VIABILITY. Seeds are orthodox [11] and after appropriate drying can therefore be stored at sub-zero temperatures for long-term conservation or in airtight containers in a cool dry place for short- to medium-term storage.

PROPAGATION

SEEDS

Dormancy and pre-treatments: Seeds of this species are reported to be non-dormant [1]. No pre-treatments are required [8].
Germination, sowing and planting: No information is available on optimum incubation temperature or on light requirements for germination under laboratory controlled conditions. Further studies are needed to characterise seed germination fully. However, the seeds germinate readily without any pre-treatments, so this species can be propagated by seed in the nursery [8].

VEGETATIVE PROPAGATION

Stem cuttings and wildings can be used [8], the former root readily in sand [6]. This species grows rapidly and can be managed by coppicing and pollarding [8].

TRADE

The wood is used locally and occasionally traded on the international market [9].

Authors

William Omondi, Victor Otieno, David Meroka, Steve Davis and Efisio Mattana.

References

[1] Baskin, C. C. & Baskin, J. M. (2014). *Seeds: Ecology, Biogeography, and Evolution of Dormancy and Germination*, 2nd edition. Academic Press, San Diego, USA.

[2] Bingham, M. G. & Smith, P. P. (2002). Zambia. In: *Southern African Plant Red Data Lists*. Edited by J. S. Golding. Southern African Botanical Diversity Network Report No. 14, SABONET, Pretoria, South Africa. pp. 135–156.

[3] Davies, F. G. & Verdcourt, B. (1998). Sapindaceae. In: *Flora of Tropical East Africa*. Edited by H. J. Beentje & C. M. Whitehouse. A.A. Balkema, Rotterdam, The Netherlands.

[4] IUCN (2017). *The IUCN Red List of Threatened Species.* Version 2017-3. http://www.iucnredlist.org/

[5] Kokwaro, J. O. (2009). *Medicinal Plants of East Africa*, 3rd edition. University of Nairobi Press, Nairobi, Kenya.

[6] Le Roux, L.-N. (2016). *Blighia unijugata* Baker. South African National Biodiversity Institute (SANBI), South Africa. http://pza.sanbi.org/

[7] Mapaura, A. & Timberlake, J. R. (2002). Zimbabwe. In: *Southern African Plant Red Data Lists*. Edited by J. S. Golding. Southern African Botanical Diversity Network Report No. 14, SABONET, Pretoria, South Africa. pp. 157–182.

[8] Maundu, P. & Tengnäs, B. (eds) (2005). *Useful Trees and Shrubs for Kenya*. Technical Handbook No. 35, World Agroforestry Centre, Nairobi, Kenya.

[9] Obeng, E. A. (2010). *Blighia unijugata* Baker. In: *PROTA (Plant Resources of Tropical Africa/Ressources Vegetales de l'Afrique Tropicale)*. Edited by R. H. M. J. Lemmens, D. Louppe & A. A. Oteng-Amoako. Wageningen, The Netherlands. http://www.prota4u.org

[10] Raimondo, D., von Staden, L., Foden, W., Victor, J. E., Helme, N. A., Turner, R. C., Kamundi, D. A. & Manyama, P. A. (2009). *Red List of South African Plants. Strelitzia* 25, South African National Biodiversity Institute, Pretoria, South Africa.

[11] Royal Botanic Gardens, Kew (2017). *Seed Information Database (SID).* Version 7.1. http://data.kew.org/sid/

Carissa spinarum L.

TAXONOMY AND NOMENCLATURE

FAMILY. Apocynaceae

SYNONYMS. *Carissa diffusa* Roxb., *C. edulis* (Forssk.) Vahl, *C. ovata* R.Br., *C. scabra* R.Br., *C. villosa* Roxb., *Jasminonerium madagascariense* (Thouars) Kuntze, *J. sechellense* (Baker) Kuntze, and many more (see [14]).

VERNACULAR / COMMON NAMES. Mtandamboo (Swahili); mulimuli (Boni); dagams (Boran, Giriama); mukawa (Embu); dagamsa (Giriama); kikawa, mutote, ngaawa (Kamba); mukawa (Kikuyu); legetetwet, legetiet (Kipsigis); omonyangateti (Kisii); munyoke (Kuria); kumurwa, oburwa, shikata (Luhya); ochuoga (Luo); ilamuriak, olamuriaki (Maasai); legatetwo (Marakwet, Tugen); kamuria (Meru); legetetuet, legetetwa (Nandi); lokotetwo (Pokot); lamuriai, lamuriei (Samburu); adishawei (Somali); kirumba (Taita); emuriei (Teso); ekamurai (Turkana).

PLANT DESCRIPTION

LIFE FORM. Shrub or climber.

PLANT. Evergreen, <5(–20 m) tall (Figure 1), bark grey, spines simple (rarely forked), <7 cm long; exudes a milky latex. **LEAVES.** Petiolate, petiole c. 0.5 mm long, simple, opposite, leathery, dark green, ovate, elliptic or almost orbicular, apex obtuse or acute, base rounded. **FLOWERS.** In terminal clusters, fragrant, white inside, pink to red outside, corolla tube <2 cm, lobes 4–9 mm, sepals ovate [2,9].

Figure 1 – Habit of *C. spinarum*. (Photo: M. Meso, KEFRI.)

FRUITS AND SEEDS

Fruits are ovoid to almost spherical berries, c. 1 cm in diameter, red black, ripening to purplish black (Figure 2) [6,9], containing 2–4 flat seeds (Figure 3) [9].

FLOWERING AND FRUITING. Flowers in all seasons with peaks in the rainy seasons or soon after. Fruits from April to July and from November to December in Bungoma. In Nairobi, flowers appear in May [6].

Figure 2 – Unripe and ripe fruits. (Photo: M. Meso, KEFRI.)

Figure 3 – External and internal views of the seeds showing the endosperm and the well-developed spatulate embryo. (Photos: P. Gómez-Barreiro, RBG Kew.)

DISTRIBUTION

Widespread in tropical Africa from Senegal to Somalia, south to Botswana, Mozambique and extending into subtropical South Africa [9]; found in most areas of Kenya, but rare on the coast [6]. Also native in the Arabian Peninsula, the Indian Ocean islands, India, and throughout most of tropical Asia to Australia [14].

HABITAT

Forest (including riverine forest) margins, bushland, thickets and bushed grassland, on rocky hillsides, and on clay soils (especially on black cotton soils in valley bottoms), up to 2,500 m a.s.l. [6,9].

USES

Carissa spinarum has many uses [11]. The fruit (ripe and unripe) and flowers are edible, the fruit being relished by children and adults. The roots are used for flavouring soups and stews. In traditional medicine, the roots are used for treating a wide range of conditions, including indigestion, constipation, abdominal pains, chest pains, malaria, poliomyelitis, arthritis, rheumatism, gonorrhoea, yellow fever, epilepsy and kidney problems, and as a cathartic [5]. The fruit is used in traditional veterinary medicine as an anthelmintic [13]. A dye is obtained from the fruit. The plant is browsed by livestock, and the flowers attract bees for honey production. *Carissa spinarum* is planted as an ornamental and live fence [7].

KNOWN HAZARDS AND SAFETY

The crushed roots emit a strong smell of methyl salicylate (as in oil of wintergreen) and if rubbed onto the fingers produce a prickly sensation [3]. The plant is armed with rigid spines, so care should be taken when handling.

CONSERVATION STATUS

This widely distributed species is assessed as Least Concern (LC) in South Africa [10] and Sri Lanka [8] according to IUCN Red List criteria. Although widespread in Kenya [2], populations elsewhere are considered to be threatened, e.g. Critically Endangered (CR) in Réunion; Endangered (EN) in Egypt and Seychelles; Vulnerable (VU) in Mayotte; and Rare (R) in Orissa (India) [8]. The global status has yet to be confirmed in the IUCN Red List [4].

SEED CONSERVATION

HARVESTING. Mature purple black fruits can be collected directly from the crown by hand-picking or by shaking the branches to dislodge the fruits..

PROCESSING AND HANDLING. Ripe fruits can be squeezed by hand or rubbed gently on a fine wire mesh or screen under running tap water to remove the pulp. Seeds can then be dried in the shade and cleaned by hand-sorting.

STORAGE AND VIABILITY. Seeds are orthodox [12] and after appropriate drying can therefore be stored at sub-zero temperatures for long-term conservation. Seeds lose viability fairly quickly when stored in airtight containers in ambient conditions [6].

PROPAGATION

SEEDS

Dormancy and pre-treatments: Seeds do not exhibit dormancy [1], so no pre-treatments are required.

Germination, sowing and planting: Germination experiments carried out under laboratory controlled conditions on seed lots stored at the MSB highlighted high germination percentages (≥90%) for untreated seeds, incubated in the light at constant temperatures of ≥15°C [12]. In the nursery, fresh seeds germinate well and can be sown in a seedbed and potted later into polythene tubes. The trees are slow-growing and respond well to pruning.

VEGETATIVE PROPAGATION

Can be propagated from wildings, stem cuttings and root suckers [6].

TRADE

Fruits are occasionally sold in local markets [6].

Authors

William Omondi, Victor Otieno, David Meroka, Tiziana Ulian, Steve Davis and Efisio Mattana.

References

[1] Baskin, C. C. & Baskin, J. M. (2014). *Seeds: Ecology, Biogeography, and Evolution of Dormancy and Germination*, 2nd edition. Academic Press, San Diego, USA.

[2] Beentje, H. J. (1994). *Kenya Trees, Shrubs and Lianas*. National Museums of Kenya, Nairobi, Kenya.

[3] Burkill, H. M. (1985). *The Useful Plants of West Tropical Africa, Vol. 1: Families A–D*. 2nd edition. Royal Botanic Gardens, Kew.

[4] IUCN (2017). *The IUCN Red List of Threatened Species.* Version 2017-3. http://www.iucnredlist.org/

[5] Kokwaro, J. O. (2009). *Medicinal Plants of East Africa*, 3rd edition. University of Nairobi Press, Nairobi, Kenya.

[6] Maundu, P. & Tengnäs, B. (eds) (2005). *Useful Trees and Shrubs for Kenya*. Technical Handbook No. 35, World Agroforestry Centre, Nairobi, Kenya.

[7] Mutshinyalo, T. & Malatji, R. (2012). *Carissa edulis* Vahl. South African National Biodiversity Institute (SANBI), South Africa. http://pza.sanbi.org/

[8] National Red List (2017). *Carissa spinarum.* National Red List Network. http://www.nationalredlist.org/ (See also National Red List entry for *Carissa edulis*.)

[9] Omino, E. A. (2002). Apocynaceae (Part 1). In: *Flora of Tropical East Africa*. Edited by H. J. Beentje & S. A. Ghazanfar. A.A. Balkema, Rotterdam, The Netherlands.

[10] Raimondo, D., von Staden, L., Foden, W., Victor, J. E., Helme, N. A., Turner, R. C., Kamundi, D. A. & Manyama, P. A. (2009). *Red List of South African Plants. Strelitzia* 25, South African National Biodiversity Institute, Pretoria, South Africa.

[11] Royal Botanic Gardens, Kew (1999–2015). *Survey of Economic Plants for Arid and Semi-Arid Lands (SEPASAL) Database*. http://apps.kew.org/sepasalweb/sepaweb

[12] Royal Botanic Gardens, Kew (2017). *Seed Information Database (SID).* Version 7.1. http://data.kew.org/sid/

[13] Van Wyk, B. & Van Wyk, P. (1997). *Field Guide to Trees of Southern Africa*. Struik Nature, Cape Town, South Africa.

[14] WCSP (2017). *World Checklist of Selected Plant Families.* Royal Botanic Gardens, Kew. http://wcsp.science.kew.org/

Centrapalus pauciflorus (Willd.) H.Rob.

TAXONOMY AND NOMENCLATURE

FAMILY. Compositae

SYNONYMS. *Cacalia pauciflora* Kuntze, *Centrapalus galamensis* Cass., *Conyza pauciflora* Willd., *Vernonia afromontana* R.E.Fr., *V. coelestina* Schrad. ex DC., *V. filisquama* M.G.Gilbert, *V. galamensis* (Cass.) Less., *V. petitiana* A.Rich., *V. senegalensis* Desf., *V. zernyi* Gilli

VERNACULAR / COMMON NAMES. Iron weed, vernonia (English).

PLANT DESCRIPTION

LIFE FORM. Herb.

PLANT. Annual, sometimes perennial, <5 m tall, usually smaller, erect or sometimes straggling, sometimes much-branched (Figure 1); stems ribbed, finely to coarsely hairy. **LEAVES.** Sessile, sometimes rather crowded, elliptic or linear to lanceolate, oblanceolate, 3–25 x 0.2–5 cm, base cuneate to attenuate, margins minutely denticulate. **FLOWERS.** Capitula solitary, or few to numerous in terminal compound corymbiform cymes (Figure 2). Corolla pink to violet, blue or pale greenish blue (Figure 1) [2].

Figure 1 – Plant in flower.
(Photo: KALRO.)

FRUITS AND SEEDS

Cypselas (Figure 2) are blackish, narrowly obovoid, c. 3.5–8 mm long, 10-ribbed, densely ascending-pubescent; outer pappus of very narrow scales, 0.5–2 mm long, inner pappus tawny, dark brown or grey, 5.5–11 mm long [2].

FLOWERING AND FRUITING. Flowering is induced by short days. Growth is indeterminate. Some plants may reach only 20 cm and form a single capitulum, whereas others are more vigorous and develop into much-branched shrubs >2.5 m tall with many capitula. In an experiment on cultivated plants, flowering occurred after 117 days from sowing and seeds matured after 161–261 days, suggesting weak sensitivity to daylength [1].

Figure 2 – Plant with ripening cypselas. (Photo: D. O. Nyamongo, KALRO.)

DISTRIBUTION

Occurs naturally from Cape Verde and Senegal east to Eritrea and throughout East Africa, south to Mozambique [1,2].

HABITAT

Semi-arid tropics, in dry bushland; more often ruderal and as a weed of cultivation, and in montane forests in undisturbed areas up to 2,500 m a.s.l. [1].

USES

Seeds contain c. 40% of a triglyceride oil, rich in vernolic acid. The oil ('vernonia oil') can be used in the chemical, pharmaceutical and agro-industrial industries. It is being tested as a component of low-volatile organic solvent paint [6]. The residue after oil extraction is suitable as animal feed [5]. In Kenya, the plant is used in traditional medicine to treat stomach pain [1].

KNOWN HAZARDS AND SAFETY

No known hazards.

CONSERVATION STATUS

This widespread species has yet to be assessed according to IUCN Red List criteria in Kenya and its status confirmed in the IUCN Red List [3].

SEED CONSERVATION

HARVESTING. Harvesting of flower heads takes place when the involucres surrounding the seeds are dry and spread out to release the fully mature seeds. At this stage, seeds are 90% black in colour and firm [1].

PROCESSING AND HANDLING. After harvesting, the seeds can be separated from the capitula and the pappus removed by hand, a labour-intensive operation [1].

STORAGE AND VIABILITY. Seeds are reported to be desiccation tolerant [7].

PROPAGATION

SEEDS

Dormancy and pre-treatments: Seeds of this species are reported to have non-deep physiological dormancy, which can be overcome by exposure to dry-after-ripening, cold stratification and gibberellic acid (GA_3) [4].

Germination, sowing and planting: Experiments carried out on seed lots stored at the MSB highlighted high germination percentages (>75%) when seeds were incubated at constant temperatures of 20°C and 25°C, and at an alternating temperature regime of 35/20°C [7].

VEGETATIVE PROPAGATION

No protocols available.

TRADE

Commercial production for 'vernonia oil' has started in Ethiopia. Large-scale commercial production is still developing [1]. No information available for Kenya.

Authors

Desterio O. Nyamongo and Efisio Mattana.

References

[1] Baye, T. M. & Oyen, L. P. A. (2007). *Vernonia galamensis* (Cass.) Less. In: *PROTA (Plant Resources of Tropical Africa/Ressources Végétales de l'Afrique Tropicale)*. Edited by H. A. M. van der Vossen & G. S. Mkamilo. Wageningen, The Netherlands. http://www.prota4u.org

[2] Beentje, H. J. & Smith, S. A. L. (eds) (2000). Compositae (Part 1). In: *Flora of Tropical East Africa*. A.A. Balkema, Rotterdam, The Netherlands.

[3] IUCN (2017). *The IUCN Red List of Threatened Species.* Version 2017-3. http://www.iucnredlist.org/

[4] Nyamongo, D. O., Nyabundi, J. & Daws, M. I. (2009). Germination and dormancy breaking requirements for *Vernonia galamensis* (Asteraceae). *Seed Science and Technology* 37: 1–9.

[5] Ologunde, M. O., Ayorinde, F. O. & Shepard, R. L. (1990). Chemical evaluation of defatted *Vernonia galamensis* meal. *Journal of the American Oil Chemists' Society* 67 (2): 92–94.

[6] Roseberg, R. J. (1997). Herbicide tolerance by vernonia grown in the temperate zone. *Industrial Crops and Products* 6 (2): 89–96.

[7] Royal Botanic Gardens, Kew (2017). *Seed Information Database (SID).* Version 7.1. http://data.kew.org/sid/

Combretum collinum Fresen.

TAXONOMY AND NOMENCLATURE

FAMILY. Combretaceae

SYNONYMS. *Combretum bongense* Engl. & Diels

VERNACULAR / COMMON NAMES. Variable bush-willow, variable combretum (English); murithi, mururuka (Embu); itithi, mutithi (Kamba); sheraha (Luhya); adugno, kech rachar, odugno, odugu, ohoro (Luo); asenuet (Nandi, Sebei); ekimeng' (Turkana).

PLANT DESCRIPTION

LIFE FORM. Tree or sometimes a shrub (Figure 1).

PLANT. Semi-deciduous, <12 m tall, crown flat or rounded; bark smooth and grey when young, later reddish brown or pale yellow, fissured. **LEAVES.** Opposite, alternate or verticillate, glossy above, coriaceous or subcoriaceous, 'metallic' in appearance, 22 x 8 cm, elliptic, obovate or ovate; upper surface drying reddish olive or yellowish brown, lower surface green, buff or silvery, glabrous to densely tomentose. **FLOWERS.** Inflorescences are simple spikes or panicles <10 cm long, axillary or supra-axillary, flowers yellow, cream or white, fragrant; petals transversely elliptic to obovate or subcircular, 1.5–2.5 mm long [2,12].

Figure 1 – Overall habit and fruiting branch.
(Photos: H. Migiro, KEFRI and H. Pickering, Flora of Zimbabwe: https://www.zimbabweflora.co.zw/index.php.)

FRUITS AND SEEDS

Four-winged, single-seeded indehiscent fruits, 2.5–5 x 2–4 cm (Figure 2). Seeds are black, ovoid, cerebriform with a folded embryo (Figure 2).

FLOWERING AND FRUITING. In West Africa, flowering occurs during the second half of the dry season or the beginning of the rainy season. In Southern Africa, flowering occurs from August to October, and fruiting occurs from January to August, with some old fruits remaining on the tree for most of the year [3].

Figure 2 – Left: External views of the fruit showing the wings, and cross-section showing the seeds. Right: Longitudinal and cross-sections of the seeds showing the folded embryo.
(Photos: P. Gómez-Barreiro, RBG Kew.)

DISTRIBUTION

Widespread in tropical and subtropical Africa [12]. In Kenya, mostly in the central, southern and western regions [2].

HABITAT

Savanna, woodlands, deciduous and evergreen thickets, termite mounds, and *Combretum*-wooded grassland where it is often a (sub-)dominant tree [6], up to 2,200 m a.s.l. [2].

USES

Combretum collinum has a large number of uses, especially in traditional medicine. The roots are boiled and the decoction used to treat dysentery and snake bites. The roots are also chewed to treat common cold. The heartwood is fairly hard, but not very durable. It is used for wagons, canoes and tool handles. The wood makes good charcoal, and the flowers produce good nectar for honey [7,10].

KNOWN HAZARDS AND SAFETY

No known hazards.

CONSERVATION STATUS

This widespread species is not considered to be either rare or threatened. The subspecies in South Africa are assessed as Least Concern (LC) according to IUCN Red List criteria [9]. The status of the species has yet to be confirmed in the IUCN Red List [5].

SEED CONSERVATION

HARVESTING. Fruits are collected by spreading a tarpaulin under the tree and then shaking the branches until the fruits are dislodged. Alternatively, fruits can be collected by climbing the tree and picking by hand [8].

PROCESSING AND HANDLING. The 4-winged indehiscent fruit (Figure 2) is both the dispersal and storage unit. Seeds can be extracted from the fruit (Figure 2) but their removal is not easy. Soaking the fruit in cold water before opening facilitates the process [7].

STORAGE AND VIABILITY. X-ray analysis carried out on seed lots stored at the MSB highlighted a high percentage (c. 80%) of filled seeds (Figure 3). Seeds of this species are reported to be desiccation tolerant [11] and after appropriate drying can therefore be stored at sub-zero temperatures for long-term conservation. In addition, seeds retain viability for several months if kept dry and stored at temperatures up to 20°C [7].

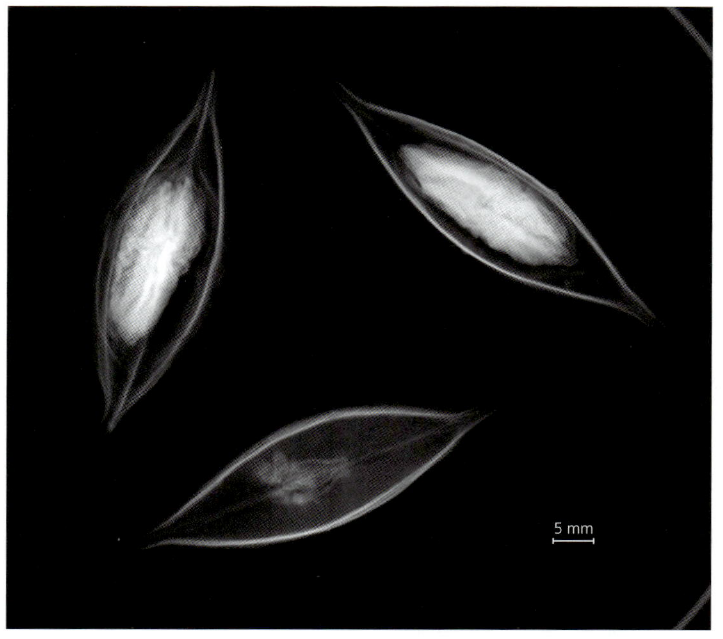

Figure 3 – X-ray images showing two filled fruits with sound seeds (top) and one empty fruit (below). (Photo: P. Gómez-Barreiro, RBG Kew.)

PROPAGATION

SEEDS

Dormancy and pre-treatments: Seeds of this species are reported to be non-dormant [1,4]. However, removal or chipping of the fruits, as well as soaking, have been reported to improve seed germination [7,11].

Germination, sowing and planting: High germination percentages (100%) have been achieved under laboratory controlled conditions for seeds incubated at 26°C with a photoperiod of 12 hours per day after the seeds have been removed from the fruit [11]. In the nursery, seedlings germinate after 8–20(–35) days; germination rates vary from 0–90%. Seedlings are vulnerable to fungal attack in humid conditions. Seedlings are usually transplanted into growing containers when they reach the two-leaf stage [7].

VEGETATIVE PROPAGATION

No protocols available.

TRADE

Parts of this species are sold at local markets for medicinal use [7].

Authors

Veronicah Ngumbau, Patrick Muthoka and Efisio Mattana.

References

[1] Baskin, C. C. & Baskin, J. M. (2014). *Seeds: Ecology, Biogeography, and Evolution of Dormancy and Germination*, 2nd edition. Academic Press, San Diego, USA.

[2] Beentje, H. J. (1994). *Kenya Trees, Shrubs and Lianas.* National Museums of Kenya, Nairobi, Kenya.

[3] Coates Palgrave, K. & Coates Palgrave, M. (2002). *Trees of Southern Africa*, 3rd edition. Struik Publishers, Cape Town, South Africa.

[4] Gashaw, M. & Michelsen, A. (2002). Influence of heat shock on seed germination of plants from regularly burnt savanna woodlands and grasslands in Ethiopia. *Plant Ecology* 159 (1): 83–93.

[5] IUCN (2017). *The IUCN Red List of Threatened Species.* Version 2017-3. http://www.iucnredlist.org/

[6] Launert, E. (ed.) (1978). Combretaceae. In: *Flora Zambesiaca*, Vol. 4. Flora Zambesiaca Managing Committee, Royal Botanic Gardens, Kew.

[7] Maroyi, A. (2013). *Combretum collinum* Fresen. In: *PROTA (Plant Resources of Tropical Africa 11 (2): Medicinal Plants 2.* Edited by G. H. Schmelzer & A. Gurib-Fakim. PROTA Foundation/Backhuys Publishers/ CTA, Wageningen, The Netherlands. pp. 66–69.

[8] Orwa, C., Mutua, A., Kindt, R., Jamnadass, R. & Simons, A. (2009). *Agroforestree Database: A Tree Reference and Selection Guide*. Version 4.0. World Agroforestry Centre, Kenya. http://www.worldagroforestry.org/output/agroforestree-database

[9] Raimondo, D., von Staden, L., Foden, W., Victor, J. E., Helme, N. A., Turner, R. C., Kamundi, D. A. & Manyama, P. A. (2009). *Red List of South African Plants. Strelitzia* 25, South African National Biodiversity Institute, Pretoria, South Africa.

[10] Royal Botanic Gardens, Kew (1999–2015). *Survey of Economic Plants for Arid and Semi-Arid Lands (SEPASAL) Database*. http://apps.kew.org/sepasalweb/sepaweb

[11] Royal Botanic Gardens, Kew (2017). *Seed Information Database (SID).* Version 7.1. http://data.kew.org/sid/

[12] Wickens, G. E. (1973). Combretaceae. In: *Flora of Tropical East Africa*. Edited by R. M. Polhill. Crown Agents for Oversea Governments and Administrations, London.

Croton macrostachyus Hochst. ex Delile

TAXONOMY AND NOMENCLATURE

FAMILY. Euphorbiaceae

SYNONYMS. *Croton acuminatus* R.Br., *C. butaguensis* De Wild., *C. guerzesiensis* Beille ex A.Chev., *Oxydectes macrostachya* (Hochst. ex Delile) Kuntze

VERNACULAR / COMMON NAMES. Forest fever tree, woodland croton (English); mtumbatu (Swahili); mukanisa (Boran); muthulu (Kamba); mutunduwa, njora (Kikuyu); tebesuet (Kipsigis); omosocho (Kisii); musine (Luhya); kumukunusia (Luhya, Bukusu dialect); ngong'ngo' (Luo); orkeparlu (Maasai); taboswa (Marakwet); mkigara (Taita).

PLANT DESCRIPTION

LIFE FORM. Tree or shrub.

PLANT. Dioecious, sometimes monoecious, 6–12(–25) m tall (Figure 1); bark light or dark grey, finely or deeply longitudinally fissured [11]. **LEAVES.** Large, <15 cm long, on long (to 11 cm) pedicels, ovate, base (sub)cordate, apex acuminate, crowded at end of branchlets, densely fulvous above, paler and softly hairy below, veins prominent, glands 2–8 (sometimes visible at leaf base) [8,11]; drought-deciduous [13], the leaves turn orange before falling [3]. **FLOWERS.** Yellowish, sweet-scented, to 3.5 mm long, in terminal racemes 15–32 x 3 cm wide (Figure 1); racemes with either all male or all female flowers, or mixed [3].

Figure 1 – General habit and flowers of *C. macrostachyus*. (Photos: D. O. Nyamongo, KALRO.)

FRUITS AND SEEDS

Fruits are subtrilobed, rarely 4-lobed, often more particularly so at the apex. Seeds (Figure 2) are ellipsoid 7 mm long, 4 mm wide, longitudinally wrinkled, grey, somewhat shiny, with a large waxy caruncle [11].

Figure 2 – External views of the seeds showing the waxy caruncle, and longitudinal section showing the embryo.
(Photos: P. Gómez-Barreiro, RBG Kew.)

FLOWERING AND FRUITING. Flowering occurs from January to October [3], with a peak between March and June in western Kenya, and from May to July in central Kenya. Fruit development takes 4–5 months [8].

DISTRIBUTION

Tropical Africa, from Guinea east to Ethiopia and south to Angola, Zambia, Malawi and Mozambique [11]. In Kenya, mainly in the west and centre [3].

HABITAT

Moist or dry evergreen upland forest, riverine forest or woodland, wooded grassland and bushland [3]. A pioneer species also common in secondary forest and forest edges, at 200–2,300 m a.s.l. [11].

USES

In traditional medicine, a decoction of the leaves and the ash from burnt leaves are used for treating coughs. The stem and root bark and the leaf sap are taken as a vermifuge and purgative [7]. A root decoction is used as an anthelmintic for tapeworm and as a purgative. Leaf juice is applied to wounds to hasten blood clotting, and the juice from boiled roots is drunk for malaria or venereal diseases. The bark is used as a remedy for skin rashes [6]. Stems and branches are used for firewood, poles and tool handles. The tree provides bee forage and leaf mulch, and is used for soil conservation [9]. It is commonly planted to provide shade and is suitable for intercropping [10].

KNOWN HAZARDS AND SAFETY

The bark contains crotin, which is purgative [14]. The seed oil is also a very powerful purgative [6,7].

CONSERVATION STATUS

This widespread species is not considered to be either rare or threatened, but its status has yet to be confirmed in the IUCN Red List [5].

SEED CONSERVATION

HARVESTING. Mature grey fruits can be collected by climbing the tree and hand-picking, or a canvas is spread on the ground and the branches shaken to dislodge the fruits. Timing of collection is very important because the fruit capsules open when mature, dispersing the seeds. Capsule opening is accompanied by a loud 'tack', which can be heard in the forest [1].
PROCESSING AND HANDLING. Capsules can be sun-dried to release the seeds, turning regularly to avoid overheating. Seeds are then cleaned by hand-sorting.
STORAGE AND VIABILITY. Seeds of this species are reported to be likely desiccation tolerant [12]. However, no germination occurred after storage for two years [4]. More recent studies on seeds collected in various regions of Ethiopia revealed an intermediate behaviour [13].

PROPAGATION

SEEDS

Dormancy and pre-treatments: Seeds of this species are reported to be physiologically dormant [2]. Experience in the nursery suggests that no pre-treatments are required before sowing [1]. Experiments under laboratory controlled conditions showed a positive effect of smoke-water solution on germination [13], as does chipping with a scalpel [12], confirming the presence of physiological dormancy.
Germination, sowing and planting: Further studies are needed to identify the optimum temperature and light regimes for germination under controlled laboratory conditions. However, >60% of treated seeds germinated when incubated at temperatures of 22–25°C [12,13]. Contradictory results were found regarding light requirements [2,4,13], highlighting the need for further research. In the nursery, seeds germinate in 45–60 days reaching 40% germination [1].

VEGETATIVE PROPAGATION

Can be propagated from wildings [10].

TRADE

The timber has little importance in the international market; most trade is for local use.

Authors

Desterio O. Nyamongo, Steve Davis and Efisio Mattana.

References

[1] Albrecht, J. (1993). *Tree Seed Handbook of Kenya*. GTZ Forestry Seed Centre, Muguga, Kenya.

[2] Baskin, C. C. & Baskin, J. M. (2014). *Seeds: Ecology, Biogeography, and Evolution of Dormancy and Germination*, 2nd edition. Academic Press, San Diego, USA.

[3] Beentje, H. J. (1994). *Kenya Trees, Shrubs and Lianas*. National Museums of Kenya, Nairobi, Kenya.

[4] Bussmann, R. W. & Lange, S. (2000). Germination of important East African mountain forest trees. *Journal of East African Natural History* 89 (1): 101–111.

[5] IUCN (2017). *The IUCN Red List of Threatened Species*. Version 2017-3. http://www.iucnredlist.org/

[6] Kokwaro, J. O. (2009). *Medicinal Plants of East Africa*, 3rd edition. University of Nairobi Press, Nairobi, Kenya.

[7] Mairura, F. S. (2007). *Croton macrostachyus* Hochst. ex Delile. In: *PROTA (Plant Resources of Tropical Africa/Ressources Végétales de l'Afrique Tropicale)*. Edited by G. H. Schmelzer & A. Gurib-Fakim. Wageningen, The Netherlands. http://www.prota4u.org

[8] Maundu, P. & Tengnäs, B. (eds) (2005). *Useful Trees and Shrubs for Kenya*. Technical Handbook No. 35, World Agroforestry Centre, Nairobi, Kenya.

[9] Msanga, H. P. (1998). *Seed Germination of Indigenous Trees in Tanzania: Including Notes on Seed Processing and Storage, and Plant Uses*. Canadian Forest Service, Northern Forestry Centre, Edmonton, Canada.

[10] Orwa, C., Mutua, A., Kindt, R., Jamnadass, R. & Simons, A. (2009). *Agroforestree Database: A Tree Reference and Selection Guide*. Version 4.0. World Agroforestry Centre, Kenya. http://www.worldagroforestry.org/output/agroforestree-database

[11] Radcliffe-Smith, A. (1987). Euphorbiaceae (Part 1). In: *Flora of Tropical East Africa*. Edited by R. M. Polhill. A.A. Balkema, Rotterdam, The Netherlands.

[12] Royal Botanic Gardens, Kew (2017). *Seed Information Database (SID)*. Version 7.1. http://data.kew.org/sid/

[13] Wakjira, K. & Negash, L. (2013). Germination responses of *Croton macrostachyus* (Euphorbiaceae) to various physico-chemical pretreatment conditions. *South African Journal of Botany* 87: 76–83.

[14] Watt, J. M. & Breyer-Brandwijk, M. G. (1962). *The Medicinal and Poisonous Plants of Southern and Eastern Africa*, 2nd edition. E. & S. Livingstone, Edinburgh & London.

Croton megalocarpus Hutch.

TAXONOMY AND NOMENCLATURE

FAMILY. Euphorbiaceae

SYNONYMS. *Croton elliotianus* Pax

VERNACULAR / COMMON NAMES. Croton (English); msenefu (Swahili); nyapo (Boran); nyaepo (Embu); muyama (Giriama); muthulu, nthulu (Kamba); mukinduri (Kikuyu, Meru); musine (Luhya); omolguruet (Maasai); masineitet (Nandi); marakuet (Samburu); mukigara (Taita); ortuet (Tugen).

PLANT DESCRIPTION

LIFE FORM. Tree.

PLANT. Deciduous, 15–25(–36 m) tall, bole clear up to 20 m, crown spreading, bark pale grey brown, closely longitudinally fissured; monoecious, occasionally dioecious. **LEAVES.** Dull green above, silvery below, ovate or elliptic, base subcordate, rounded or cuneate, 5–15 x 2–8 cm [3,11] (Figure 1). **FLOWERS.** Yellowish, <5 mm long [3], unisexual or hermaphrodite, short-lived but conspicuous, in pendulous spikes <25 cm long with only a few female flowers at the base [8].

Figure 1 – Details of leaves, inflorescences and fruits (Photo: P. Ballings, Flora of Zimbabwe: https://www.zimbabweflora.co.zw/index.php.)

FRUITS AND SEEDS

Fruits are ellipsoid-ovoid to subglobose, not lobed, 3–4.5 cm long, 2.5–3(–4) cm in diameter, loculicidal from the apex, lepidote. Endocarps are woody, 3–5.5 mm thick, with three double rows of pits outside corresponding to the septa. Seeds (Figure 2) are ellipsoid-ovoid or oblong-ellipsoid, somewhat ridged and wrinkled, white when fresh, grey brown when dry, with a minute caruncle [8].

FLOWERING AND FRUITING: Flowers throughout most of the year [3]. In Nyeri, Kiambu and Kericho counties of Kenya, flowering occurs between February and May; in Bungoma flowering occurs in September; and in Kakamega flowering occurs from October to January. Buds open after heavy rains [8]. Fruit development takes five months from pollination [1].

Figure 2 – Seeds.
(Photo: D.O. Nyamongo, KALRO.)

DISTRIBUTION

Tropical Africa, from Democratic Republic of Congo east to Somalia, and from East Africa (including Kenya) south to Malawi, Zambia and Mozambique [11,13]. In Kenya, *Croton megalocarpus* is widespread in the south, west and north [3], and is particularly common in coastal dry forests, around Nairobi, and on mountains in northern Kenya [8].

HABITAT

Dry upland evergreen or semi-deciduous forest, occasionally dominant; also in moist upland forest, dense woodland, riverine and scattered tree grassland [3], at 700–2,400 m a.s.l. [11]. A pioneer of forest clearings [10].

USES

In traditional medicine, a decoction of the bark is taken as a vermifuge and to treat stomach ache, whooping cough, pneumonia, and fever, including malarial fever. The wood is used for construction, flooring, railway sleepers and agricultural implements, and as fuelwood [1,6,8,9]. The flowers are foraged by bees, producing honey with a strong flavour. The tree provides shade and its leaves are used as a mulch (e.g. in coffee plantations). *C. megalocarpus* is also planted as a hedge and boundary marker [10], and is important in agroforestry in East Africa. The seed oil is being tested for biofuel production and large investments are being made to establish large-scale plantations in East Africa [7].

KNOWN HAZARDS AND SAFETY

No known hazards.

CONSERVATION STATUS

This widespread species is not considered to be either rare or threatened, but its status has yet to be confirmed in the IUCN Red List [5].

SEED CONSERVATION

HARVESTING. Mature greyish brown fruits can be collected from the crown by climbing the tree and hand-picking the seeds. Alternatively, a canvas can be spread on the ground and the branches shaken to dislodge the fruits. Strong shaking is not recommended as this can result in immature seeds dropping [1].

PROCESSING AND HANDLING. Seeds can be extracted by slightly cracking the dry fruit using appropriate tools. The seeds are cleaned by hand-sorting to remove other plant material. Seeds can be sun-dried to 5–9% moisture content [1] by spreading them thinly on a canvas and turning regularly to avoid overheating.

STORAGE AND VIABILITY. Seeds of this species are likely to be orthodox [12], although further studies are needed to confirm the physiological response to desiccation. Mature, properly dried seeds can be stored in airtight containers at 3°C in a coldstore for at least a year [1].

PROPAGATION

SEEDS

Dormancy and pre-treatments: Seeds of this species are reported to be non-dormant or physiologically dormant [2]. Experience in the nursery suggests that no pre-treatments are required before sowing [9]. Experiments under laboratory controlled conditions showed that the removal of the hard seed coat improved germination [4], suggesting the presence of non-deep physiological dormancy.

Figure 3 – Seedlings in the nursery. (Photo: V. Otieno, KEFRI.)

Germination, sowing and planting: No information available on the optimum conditions for germination under laboratory controlled conditions. When untreated seeds were incubated at 20°C (both in light and in darkness), 45% germinated in the first 20 days, reaching an overall germination rate of 76% in the light compared to only 46% in darkness [4]. In the nursery (Figure 3), seeds germinated within 35–45 days. The expected germination rate of mature and healthy seed lots is c. 95% [1].

VEGETATIVE PROPAGATION

Can be propagated by grafting or cuttings [10] and from wildings.

TRADE

The timber (commonly called musine) is sometimes traded internationally, but most trade is for local use.

Authors

Desterio O. Nyamongo, Steve Davis and Efisio Mattana.

References

[1] Albrecht, J. (1993). *Tree Seed Handbook of Kenya*. GTZ Forestry Seed Centre, Muguga, Kenya.

[2] Baskin, C. C. & Baskin, J. M. (2014). *Seeds: Ecology, Biogeography, and Evolution of Dormancy and Germination*, 2nd edition. Academic Press, San Diego, USA.

[3] Beentje, H. J. (1994). *Kenya Trees, Shrubs and Lianas*. National Museums of Kenya, Nairobi, Kenya.

[4] Bussmann, R. W. & Lange. S. (2000). Germination of important East African mountain forest trees. *Journal of East African Natural History* 89 (1): 101–111.

[5] IUCN (2017). *The IUCN Red List of Threatened Species.* Version 2017-3. http://www.iucnredlist.org/

[6] Kokwaro, J. O. (2009). *Medicinal Plants of East Africa*, 3rd edition. University of Nairobi Press, Nairobi, Kenya.

[7] Maroyi, A. (2010). *Croton megalocarpus* Hutch. In: *PROTA (Plant Resources of Tropical Africa/ Ressources Végétales de l'Afrique Tropicale)*. Edited by R. H. M. J. Lemmens, D. Louppe & A. A. Oteng-Amoako. Wageningen, The Netherlands. http://www.prota4u.org

[8] Maundu, P. & Tengnäs, B. (eds) (2005). *Useful Trees and Shrubs for Kenya*. Technical Handbook No. 35, World Agroforestry Centre, Nairobi, Kenya.

[9] Msanga, H. P. (1998). *Seed Germination of Indigenous Trees in Tanzania: Including Notes on Seed Processing and Storage, and Plant Uses*. Canadian Forest Service, Northern Forestry Centre, Edmonton, Canada.

[10] Orwa, C., Mutua, A., Kindt, R., Jamnadass, R. & Simons, A. (2009). *Agroforestree Database: A Tree Reference and Selection Guide*. Version 4.0. World Agroforestry Centre, Kenya. http://www.worldagroforestry.org/output/agroforestree-database

[11] Radcliffe-Smith, A. (1987). Euphorbiaceae (Part 1). In: *Flora of Tropical East Africa*. Edited by R. M. Polhill. A.A. Balkema, Rotterdam, The Netherlands.

[12] Royal Botanic Gardens, Kew (2017). *Seed Information Database (SID).* Version 7.1. http://data.kew.org/sid/

[13] WCSP (2017). *World Checklist of Selected Plant Families.* Royal Botanic Gardens, Kew. http://wcsp.science.kew.org/

Dovyalis macrocalyx (Oliv.) Warb.

TAXONOMY AND NOMENCLATURE

FAMILY. Salicaceae

SYNONYMS. *Dovyalis adolfi-friderici* Mildbr. & Gilg, *D. antunesii* Gilg, *D. chirindensis* Engl., *D. glandulosissima* Gilg, *D. luckii* R.E.Fr., *D. mildbraedii* Gilg, *D. retusa* Robyns & Lawalrée, *D. salicifolia* Gilg

VERNACULAR / COMMON NAMES. Kei apple (English); munyee, munyhee (Giriama); cheptabirbiriet (Kipsigis); shinapateria (Luhya); busongolomunwa, kumusongolamunwa (Luhya, Bukusu dialect); akutho, nyamtotia (Luo); enkoshopini, olaimurunyai (Masaai); kaptowinet (Nandi); chuchwenion (Pokot); tabirbirwo (Tugen).

PLANT DESCRIPTION

LIFE FORM. Shrub or tree.

PLANT. Dioecious, 3–8 m tall, often multi-stemmed, young branches slender, arching (Figure 1), bark smooth, grey; branches grey brown, often dotted with breathing pores (lenticels), and bearing straight axillary spines, 1–6 cm long. **LEAVES.** Simple, elliptic or ovate, base cuneate or rounded, apex obtuse or subacute, 4–9 cm long, pale green, thin (Figure 1). **FLOWERS.** Yellow green, petals absent; male flowers hairy, in clusters of 1–4, stamens 20; female flowers shortly stalked, solitary, calyx lobes 6–10, thin, sticky and densely covered with hairs [2,6].

Figure 1 – Stems and leaves of *D. macrocalyx*. (Photo: V. Otieno, KEFRI.)

FRUITS AND SEEDS

The plum-like fruits are orange red, fleshy, 2 cm long and 1 cm wide, hanging on stalks 8 mm long. The fruits can be distinguished from other *Dovyalis* spp. by the presence of enlarged, hairy, red calyx lobes (Figure 2) [2,6]. The seeds are whitish brown, c. 8 x 6 mm [3].

FLOWERING AND FRUITING: Flowers from January to March; fruits from March to June in Bungoma County. Elsewhere in Kenya, flowering is reported to occur from August to October and in December [2]. Fruits ripen in November and December in Nairobi County [6].

Figure 2 – Ripening fruits of *D. macrocalyx*. (Photo: V. Otieno, KEFRI.)

DISTRIBUTION

Widespread from Central Africa to Sudan and East Africa, south to Southern Africa [9]. In Kenya, *Dovyalis macrocalyx* is mainly found in the west and coastal districts, around Nairobi, along the Mara River, and in Uasin Gishu, Bungoma, West Pokot and Kisumu Counties [6].

HABITAT

Rainforest and dry evergreen forest, riverine forest, bushland and wooded grassland, up to 2,600 m a.s.l. [9].

USES

The fruits are edible, and the roots and bark are used in traditional medicine. A root decoction is taken for headache and to prevent nightmares. An infusion of the bark is taken for venereal diseases [5]. The wood is used for building poles and tool handles [8]. *D. macrocalyx* is planted as an ornamental and live fence [6].

KNOWN HAZARDS AND SAFETY

The spines along the stems are sharp, so care should be taken when handling the plant.

CONSERVATION STATUS

This widespread species has yet to be assessed according to IUCN Red List criteria in Kenya and its status has yet to be confirmed in the IUCN Red List [4].

SEED CONSERVATION

HARVESTING. Mature orange red fruits can be collected from the crown by hand-picking or by shaking the branches to dislodge the fruits.

PROCESSING AND HANDLING. Fruits can be soaked in water for 2–3 days. The water is then drained off and the fruits squeezed by hand or rubbed gently on a fine wire mesh or screen under running tap water to remove the fruit pulp. Seeds can then be dried in the shade and cleaned by hand-sorting [6].

STORAGE AND VIABILITY. No information available on seed storage behaviour. However, three congeneric species are reported to be likely desiccation tolerant [7].

PROPAGATION

SEEDS

Dormancy and pre-treatments: No information is available on seed dormancy. However, the congeneric *D. caffra* (Hook.f. & Harv.) Sim is reported to be non-dormant [1]. No pre-treatments are required before sowing seeds of *D. macrocalyx* [6].

Germination, sowing and planting: No information is available on optimum incubation temperature and light requirements for germination under laboratory controlled conditions. Further studies are needed to characterise seed germination fully. In the nursery, seeds germinate within 9–21 days depending on the medium, and the expected germination rate for mature and healthy seeds is 50–70%. Seeds are sown in a seedbed and seedlings transferred to polythene tubes after 3–4 weeks. Alternatively, seeds can be sown directly in a prepared site. Seedlings are grown in the nursery for up to four months before field planting. For live fencing, seedlings are planted 10–20 cm apart. *D. macrocalyx* is relatively fast growing when established and coppices well, but growth is slow initially. Regular pruning is necessary to make a good live fence.

VEGETATIVE PROPAGATION

No protocols available.

TRADE

The fruits are sold in local markets. No data are available on international trade.

Authors

William Omondi, Victor Otieno, David Meroka and Efisio Mattana.

References

[1] Baskin, C. C. & Baskin, J. M. (2014). *Seeds: Ecology, Biogeography, and Evolution of Dormancy and Germination*, 2nd edition. Academic Press, San Diego, USA.

[2] Beentje, H. J. (1994). *Kenya Trees, Shrubs and Lianas*. National Museums of Kenya, Nairobi, Kenya.

[3] Exell, A. W. & Wild, H. (eds) (1960). Flacourtiaceae. In: *Flora Zambesiaca*, Vol. 1, part 1. Flora Zambesiaca Managing Committee, Royal Botanic Gardens, Kew.

[4] IUCN (2017). *The IUCN Red List of Threatened Species.* Version 2017-3. http://www.iucnredlist.org/

[5] Kokwaro, J. O. (2009). *Medicinal Plants of East Africa*, 3rd edition. University of Nairobi Press, Nairobi, Kenya.

[6] Maundu, P. & Tengnäs, B. (eds) (2005). *Useful Trees and Shrubs for Kenya*. Technical Handbook No. 35, World Agroforestry Centre, Nairobi, Kenya.

[7] Royal Botanic Gardens, Kew (2017). *Seed Information Database (SID).* Version 7.1. http://data.kew.org/sid/

[8] Ruffo, C. K., Birnie, A. & Tengnäs, B. (2002). *Edible Wild Plants of Tanzania*. RELMA Technical Handbook Series No. 27. Regional Land Management Unit (RELMA) and Swedish International Development Agency (SIDA), Nairobi, Kenya.

[9] Sleumer, H. (1975). Flacourtiaceae. In: *Flora of Tropical East Africa*. Edited by R. M. Polhill. Crown Agents for Oversea Governments and Administrations, London.

Markhamia lutea (Benth.) K.Schum.

TAXONOMY AND NOMENCLATURE

FAMILY. Bignoniaceae

SYNONYMS. *Dolichandrone hildebrandtii* Baker, *D. lutea* (Benth.) Benth. ex B.D.Jacks., *D. platycalyx* Baker, *Markhamia hildebrandtii* (Baker) Sprague, *M. platycalyx* (Baker) Sprague, *Spathodea lutea* Benth.

VERNACULAR / COMMON NAMES. Bell bean tree, Nile tulip tree (English); mgambo (Swahili); muu (Embu, Kikuyu, Meru); kyoo (Kamba); lusiola (Luhya); kumusoola (Luhya, Bukusu dialect); siala (Luo); mung'uani (Meru); mobet (Nandi); sogdu (Somali); ekokwait (Turkana).

PLANT DESCRIPTION

LIFE FORM. Tree.

PLANT. Evergreen, upright, (6)10–15(30) m tall, crown narrow, sometimes irregular as trunk may be divided at base (Figure 1), bark grey black or reddish brown, smooth or rough, finely fissured, bole fluted in old trees. **LEAVES.** Compound, pinnate, <35 cm long, with 7–11 leaflets, cuneate to rounded at base of leaflets, often in bunches, thin and wavy, each leaflet to 10 cm, wider at the tip. **FLOWERS.** Bright yellow, in axillary or terminal panicle, trumpet-shaped, orange red stripe in the throat, buds furry, splitting on one side [3,4,8].

Figure 1 – General habit and fruits.
(Photos: KEFRI.)

FRUITS AND SEEDS

Fruits are thin, dehiscent capsules (Figure 1), <75 cm long, hanging in spiralling clusters [8]. Mature seeds are yellow white, flat-winged (Figure 2).

FLOWERING AND FRUITING. Flowers in August and September; fruits in February and March in western Kenya. Around Mt Elgon, flowering takes place in December and January, and fruiting in July and August [1,13]. Elsewhere, flowers are reported in January, March to April, July to August, and October to November [3].

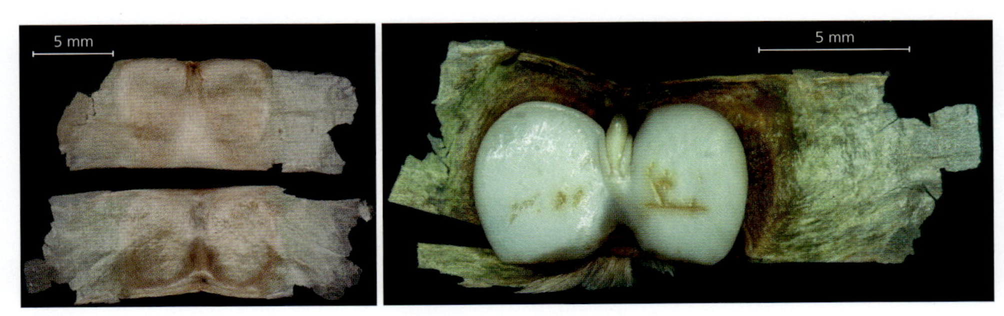

Figure 2 – External and internal views of the seeds showing the embryonic axis and the cotyledons. (Photos: P. Gómez-Barreiro, RBG Kew.)

DISTRIBUTION

Tropical Africa, from Ghana east to South Sudan and East Africa [4,14]. In Kenya, it is common around the Lake Basin, in Western Province, around Nairobi, and in highland areas of Embu and Nyeri [8].

HABITAT

Wooded grassland, submontane and riverine forest, and evergreen rainforest [4], 700–1,900 m a.s.l. [3], on red loam soils or acidic heavy clay soils, but does not tolerate waterlogging [8].

USES

The somewhat termite-resistant wood is used for making furniture, poles, tool handles, walking sticks, boats and banana props, as well as charcoal and firewood for domestic use and tobacco curing. In traditional medicine, young shoots and leaves are used for treating throat diseases and conjunctivitis. The bark is chewed for toothache. A bark decoction is drunk for syphilis. An infusion of fresh leaves is used for treating snake bites [6]. The flowers are a source of nectar for honey bees. *Markhamia lutea* is an important agroforestry tree and is often planted in indigenous agroforestry systems in western Kenya [7,11]. It is used as a shade tree for crops such as banana, beans and maize, and also planted in hedges along farm boundaries. The leaves provide a good mulch and the tree has a deep root system making it suitable for soil conservation. It is also planted as an ornamental for its showy flowers [7,8].

KNOWN HAZARDS AND SAFETY

No known hazards.

CONSERVATION STATUS

This widespread species has yet to be assessed according to IUCN Red List criteria in Kenya, and its status has yet to be confirmed in the IUCN Red List [5].

SEED CONSERVATION

HARVESTING. Timing of seed collection is important as the pods open when ripe and seeds are dispersed by the wind. Fruits should therefore be collected just prior to seed dispersal when the capsules turn greyish. Fruit colour alone is unreliable, so a cutting test should be carried out on the seeds to verify that they are not unripe [1]. Fruits can be collected from the crown by spreading a net or canvas under the tree and climbing to hand-pick the capsules, or by shaking branches to dislodge the fruits.

PROCESSING AND HANDLING. Mature capsules are sun-dried until they split to release the seeds [8]. Capsules that do not split can be either threshed or opened by hand. Seeds are cleaned by hand-sorting.

STORAGE AND VIABILITY. Seeds of this species are reported to be orthodox [12]. Viability can be maintained for several years in hermetic storage at 3°C with 5–10% moisture content [12]. At room temperature, seeds can be stored for a maximum of two years [9].

PROPAGATION

SEEDS

Dormancy and pre-treatments: Seeds of this species are reported to be non-dormant [2]. No pre-treatments are required before sowing [1,8].

Germination, sowing and planting: Under optimal conditions, mature seeds germinate within 20–30 days with a germination rate of 30–60% [1]. Seeds are sown in a seedbed or in polythene tubes. Seedlings are ready for potting when true leaves appear or after about four weeks. Seedlings are grown in the nursery until planted out in the field (Figure 3) at the onset of the long rains [10]. It is essential to remove weeds before planting. For woodlots, a spacing of c. 3 x 3 m is recommended. For scattered planting in cropland, trees should be planted 10–15 m apart. The species is fast-growing and coppices readily.

Figure 3 – Seedling planted in an experimental woodlot in Siaya. (Photo: A. Hudson, RBG Kew.)

VEGETATIVE PROPAGATION

Can be propagated by stem cuttings or by using longer cuttings (up to 1.5 m long and 6 cm diameter) from coppice shoots [7]. Wildings can also be used [11].

TRADE

The wood is sold in local markets and is not traded internationally [7].

Authors

William Omondi, Victor Otieno, David Meroka, Tiziana Ulian and Efisio Mattana.

References

[1] Albrecht, J. (1993). *Tree Seed Handbook of Kenya*. GTZ Forestry Seed Centre, Muguga, Kenya.

[2] Baskin, C. C. & Baskin, J. M. (2014). *Seeds: Ecology, Biogeography, and Evolution of Dormancy and Germination*, 2nd edition. Academic Press, San Diego, USA.

[3] Beentje, H. J. (1994). *Kenya Trees, Shrubs and Lianas*. National Museums of Kenya, Nairobi, Kenya.

[4] Bidgood, S., Verdcourt, B. & Vollesen, K. (2006). Bignoniaceae. In: *Flora of Tropical East Africa*. Edited by H. J. Beentje & S. A. Ghazanfar. Royal Botanic Gardens, Kew.

[5] IUCN (2017). *The IUCN Red List of Threatened Species.* Version 2017-3. http://www.iucnredlist.org/

[6] Kokwaro, J. O. (2009). *Medicinal Plants of East Africa*, 3rd edition. University of Nairobi Press, Nairobi, Kenya.

[7] Maroyi, A. (2012). *Markhamia lutea* (Benth.) K.Schum. In: *PROTA (Plant Resources of Tropical Africa/Ressources Végétales de l'Afrique Tropicale).* Edited by R. H. M. J. Lemmens, D. Louppe & A. A. Oteng-Amoako. Wageningen, The Netherlands. http://www.prota4u.org

[8] Maundu, P. & Tengnäs, B. (eds) (2005). *Useful Trees and Shrubs for Kenya*. Technical Handbook No. 35, World Agroforestry Centre, Nairobi, Kenya.

[9] Msanga, H. P. (1998). *Seed Germination of Indigenous Trees in Tanzania: Including Notes on Seed Processing and Storage, and Plant Uses*. Canadian Forest Service, Northern Forestry Centre, Edmonton, Canada.

[10] Omondi, W., Otieno, V., Nyamongo, D. O., Mattana, E., Hudson, A. & Ulian, T. (2015). *Useful Plants for Reforestation Activities in Kenya: Linking Environmental Challenges to the Well-Being of Local Rural Communities*. Conference Paper, XIV World Forestry Congress, Durban, South Africa, 7–11 September 2015. http://foris.fao.org/wfc2015/api/file/5542a04ae52d79267e89a2ee/contents/6dadc4b1-1a65-4b24-8a31-9be962d0f1ca.pdf

[11] Orwa, C., Mutua, A., Kindt, R., Jamnadass, R. & Simons, A. (2009). *Agroforestree Database: A Tree Reference and Selection Guide*. Version 4.0. World Agroforestry Centre, Kenya. http://www.worldagroforestry.org/output/agroforestree-database

[12] Royal Botanic Gardens, Kew (2017). *Seed Information Database (SID).* Version 7.1. http://data.kew.org/sid/

[13] Schmidt, L. H. & Mbora, A. (2008). *Markhamia lutea* (Benth) [*sic*] K.Schum. *Seed Leaflet* No. 140. Forest & Landscape Denmark, Hørsholm, Denmark.

[14] WCSP (2017). *World Checklist of Selected Plant Families.* Royal Botanic Gardens, Kew. http://wcsp.science.kew.org/

Melia volkensii Gürke

TAXONOMY AND NOMENCLATURE

FAMILY. Meliaceae

SYNONYMS. None.

VERNACULAR / COMMON NAMES. Bamba (Boran); kirumbuta (Digo); mukau (Kamba, Kikuyu); maramarui (Samburu); baba (Somali); kirumbutu (Taita); mkowe (Taita, Taveta dialect); mukau (Tharaka).

PLANT DESCRIPTION

LIFE FORM. Tree.

PLANT. Deciduous, <15(–20) m tall, crown often with browse-line produced by giraffe; bark with pronounced vertical fissures, grey. **LEAVES.** <35 cm long, compound, less divided than in *Melia azedarach* L., with leaflets 2–5 x 0.5–2.5 cm, nearly always entire (rarely a few leaflets have a single basal lobe or a few shallow teeth). **FLOWERS.** White, petals 5–7 mm, inflorescence congested, <12 cm long, axillary and on older branchlets [15] (Figure 1).

Figure 1 – Flowers and fruits of *M. volkensii*. (Photo: V. Otieno, KEFRI.)

FRUITS AND SEEDS

The fruits are 3–4 cm long, ovoid drupes, yellow at maturity later turning pale grey due to cork formation. Each fruit contains a pyrene with a very thick, bony, and terete endocarp, with a star-like, 5-lobed apical depression and a rose-like, 5-lobed basal depression with only two fertile locules (Figure 2). Seeds are oval, black at maturity, c. 2 cm long and 0.5 cm wide, with a caruncle at one end [4,9].

FLOWERING AND FRUITING. Flowering and fruiting are aseasonal, taking place two or three times per year. Fruits can be at very different stages of maturity even on the same branch [4]. In Kenya, and particularly in the Kitui and Makueni areas, flowering occurs in October and fruits ripen from March to August [9].

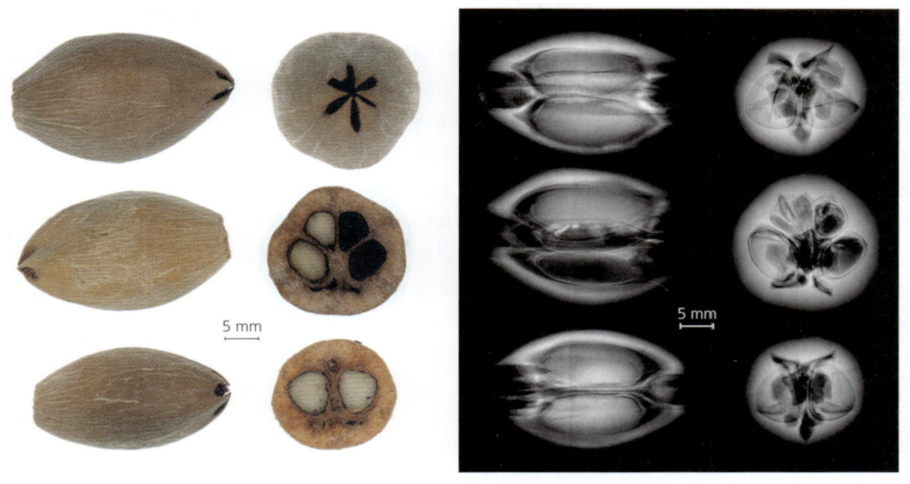

Figure 2 – External views, cross-sections and X-ray images of the pyrenes, showing the thick woody endocarp, the 5-lobed apical depression, and the empty and fertile locules with seeds. (Photos: P. Gómez-Barreiro, RBG Kew.)

DISTRIBUTION

Semi-arid areas of Kenya, Ethiopia, Somalia and Tanzania [15].

HABITAT

Deciduous bushland or bush woodland, often an emergent in *Acacia-Commiphora* bushland and thicket, at 350–1,675 m a.s.l. [15], in areas with a mean annual rainfall of 300–800 mm [1,4] on sandy, clay or shallow stony soils with good drainage [4].

USES

The leaves provide good quality fodder for both cattle and goats; goats also eat the fleshy drupes when they fall to the ground. Farmers lop branches for leaf fodder; dry branches are used for firewood. The timber is valued locally for door and window frames, door shutters, rafters, poles and furniture. In the Makueni region, *M. volkensii* is one of the principal species used for making log hives because the wood is easily worked and shaped. The flowers provide excellent bee forage for the production of honey. Leaf extracts have been traditionally used as a pesticide against ticks and fleas. Farmers value this species for soil conservation and improvement; leaf fall during the later stages of crop development increases soil fertility and, hence, crop yields [8,13].

KNOWN HAZARDS AND SAFETY

No known hazards.

CONSERVATION STATUS

This species is not considered to be either rare or threatened, but its status has yet to be confirmed in the IUCN Red List [3].

SEED CONSERVATION

HARVESTING. Ripe yellow fruits can be collected by spreading a net, canvas or other appropriate material on the ground, and shaking the branches to dislodge the fruits. Alternatively, the tree can be climbed and the fruits hand-picked [6]. During development, cork is deposited on the fruit surface, changing its colour and making it difficult to distinguish between ripe and unripe fruits [4]. So fruit colour alone should not be used to determine maturity.

PROCESSING AND HANDLING. Before sowing, seeds can be extracted from the hard endocarp (see Figure 2) by cracking the pyrene using either a knife and a hammer, or using a seed extractor developed by Kenya Forestry Research Institute (KEFRI) [7]. However, for conservation purposes, pyrenes should be kept intact and the seeds stored unextracted to avoid significant loss of viability [6].

STORAGE AND VIABILITY. Although seeds of this species are reported as likely desiccation tolerant (orthodox) [14], a low initial viability and a high level of desiccation sensitivity have been reported [12]. However, mature and properly dried stones can be stored for several years in airtight containers at a temperature of 3°C without significant loss of viability [1].

PROPAGATION

SEEDS

Dormancy and pre-treatments: Seed is the main method of propagation [11] (Figure 3). However, seed dormancy is a limiting factor [4]. After extraction from the pyrenes, seeds should be nicked and soaked in cold water for 24 hours, and then the seed coat should be longitudinally slit [11]. Although seeds are permeable to water, scarification has been reported to promote germination, highlighting a physiological component to seed dormancy [2]. The seeds are sensitive to fungal attack, so it is important to maintain a sterile environment during the scarification process [4].

Germination, sowing and planting: After pre-treatment, seeds are sown in a sterilised germination medium in high temperature and humidity. Seeds should be incubated at 27–37°C to avoid damage at higher and lower temperatures, 32°C being the optimum temperature for germination [10]. In practice, such conditions are maintained using non-mist propagators which provide greenhouse-like conditions [11]. Seed germination starts in 3–4 days and is completed in 14 days [10]. *Melia volkensii* is prone to high mortality after pricking out and during seedling establishment in the field [11].

Figure 3 – Seedlings propagated in the nursery and young tree planted in the field. (Photos: T. Ulian, RBG Kew.)

VEGETATIVE PROPAGATION

Planting wildings directly into the field is a common practice among farmers. Wildings can also be potted and grown in the nursery prior to planting in the field. Propagation by stem or root cuttings is possible with limited success [5,11].

TRADE

Various products derived from *M. volkensii* are traded locally in Kenya. These include timber, logs, sawdust and offcuts [8].

Authors

Veronicah Ngumbau, Patrick Muthoka, Tiziana Ulian and Efisio Mattana.

References

[1] Albrecht, J. (1993). *Tree Seed Handbook of Kenya*. GTZ Forestry Seed Centre, Muguga, Kenya.

[2] Baskin, C. C. & Baskin, J. M. (2014). *Seeds: Ecology, Biogeography, and Evolution of Dormancy and Germination*, 2nd edition. Academic Press, San Diego, USA.

[3] IUCN (2017). *The IUCN Red List of Threatened Species.* Version 2017-3. http://www.iucnredlist.org/

[4] Jøker, D. (2003). *Melia volkensii* Guerke. *Seed Leaflet* No. 71. Danida Forest Seed Centre, Humlebaek, Denmark.

[5] Kidundo, M. (1997). *Melia volkensii* – propagating the tree of knowledge. *Agroforestry Today* 9: 21–22. (Special Issue: *Improved Trees for Agroforestry.*)

[6] Kimondo, J. M. & Kiamba, K. (2005). An overview of natural distribution, propagation and management of *Melia volkensii*. In: *Research and Technology Development of Mukau (*Melia volkensii *Gürke).* Proceedings of the First National Workshop held at Kitui Regional Research Center, 16–19 November 2004. Edited by B. M. Kamondo, J. M. Kimondo, J. M. Mulatya & G. M. Muturi. Kenya Forestry Research Institute (KEFRI), Nairobi, Kenya. pp. 7–11.

[7] Lugadiru, J. (2005). *Melia volkensii* seed extractor. In: *Research and Technology Development of Mukau* (Melia volkensii *Gürke).* Proceedings of the First National Workshop held at Kitui Regional Research Center, 16–19 November 2004. Edited by B. M. Kamondo, J. M. Kimondo, J. M. Mulatya & G. M. Muturi. Kenya Forestry Research Institute (KEFRI), Nairobi, Kenya. pp. 25–27.

[8] Luvanda, A. M., Musyoki, J., Cheboiwo, J., Wekesa, L. & Ozawa, M. (2015). *An Assessment of the Socio-Economic Importance of* Melia volkensii *Based Enterprises in Kenya*. Kenya Forestry Research Institute (KEFRI), Nairobi, Kenya.

[9] Maundu, P. & Tengnäs, B. (eds) (2005). *Useful Trees and Shrubs for Kenya*. Technical Handbook No. 35, World Agroforestry Centre, Nairobi, Kenya.

[10] Milimo, P. B. & Hellum, A. K. (1990). The influence of temperature on germination of *Melia volkensii* seeds. In: *Tropical Tree Seed Research. Proceedings of an International Workshop held at the Forestry Training Centre, Gympie, Queensland, Australia, 21–24 August 1989*. Edited by J. W. Turnbull. ACIAR Proceedings Series No. 28, Australian Centre for International Agricultural Research, Canberra, Australia. pp. 29–32.

[11] Muok, B., Mwamburi, A., Kyalo, E. & Auka, S. (2010). *Growing* Melia volkensii: *A Guide for Farmers and Tree Growers in the Drylands*. KEFRI Information Bulletin No. 3, Kenya Forestry Research Institute (KEFRI), Nairobi, Kenya.

[12] Omondi, W. (2004). Desiccation sensitivity of seeds of four tree species of economic importance in Kenya. In: *Comparative Storage Biology of Tropical Tree Seeds*. Edited by M. Sacandé, D. Jøker, M. E. Dulloo & K. A. Thomsen. International Plant Genetic Resources Institute, Rome, Italy. pp. 75–86.

[13] Orwa, C., Mutua, A., Kindt, R., Jamnadass, R. & Simons, A. (2009). *Agroforestree Database: A Tree Reference and Selection Guide*. Version 4.0. World Agroforestry Centre, Kenya. http://www.worldagroforestry.org/output/agroforestree-database

[14] Royal Botanic Gardens, Kew (2017). *Seed Information Database (SID).* Version 7.1. http://data.kew.org/sid/

[15] Styles, B. T. & White, F. (1991). Meliaceae. In: *Flora of Tropical East Africa*. Edited by R. M. Polhill. A.A. Balkema, Rotterdam, The Netherlands.

Ocimum gratissimum L.

TAXONOMY AND NOMENCLATURE

FAMILY. Lamiaceae

SYNONYMS. *Geniosporum discolor* Baker, *Ocimum arborescens* Bojer ex Benth. *O. dalabaense* A.Chev., *O. febrifugum* Lindl., *O. guineense* Schumach. & Thonn., *O. heptodon* P.Beauv., *O. holosericeum* J.F.Gmel., *O. petiolare* Lam., *O. robustum* B.Heyne ex Hook.f., *O. sericeum* Medik., *O. suave* Willd., *O. trichodon* Baker ex Gürke, *O. urticifolium* Roth, *O. viride* Willd., *O. viridiflorum* Roth, *O. zeylanicum* Medik.

VERNACULAR / COMMON NAMES. African basil, tea bush, wild basil (English); manjabbi (Boran); vumba manga (Digo); murihani (Giriama); mukandu (Kamba), mkandu (Kikuyu); sivai (Kipsigis); esurancha (Kisii); bwar (Luo); sunoni (Maasai); chesimia, yoiyoiya (Marakwet); makandakandu (Meru); uvumbani (Pokomo); chemwoken (Pokot); lemurran (Samburu); mrumbawassi (Taita); loguru (Turkana).

PLANT DESCRIPTION

LIFE FORM. Woody herb or shrub.

PLANT. Aromatic, very variable, 0.3–2.5 m tall, stems erect, rounded-quadrangular, much-branched, often striate, woody at base, epidermis often peeling in strips; glabrous or with scattered hairs below, becoming pubescent at nodes and at inflorescence axis. **LEAVES.** Elliptic or ovate, 1.5–15 x 1–8.5 cm, serrate, apex obtuse, acute or acuminate, base cuneate or attenuate (Figure 1); upper surface glabrous or pubescent, lower surface usually paler, pubescent or tomentose or with scattered hairs on veins. **FLOWERS.** Inflorescences lax or dense, bracts erect, ovate, forming small green terminal coma; calyx horizontal ± downward-pointing or strongly reflexed against inflorescence axis; corolla greenish, dull white, pale purplish or pale yellow, 3–4 mm long [7].

FRUITS AND SEEDS

Nutlets brown (Figure 1), ± spherical, 1.5 mm in diameter. Minutely tuberculate, producing a small amount of mucilage when wet [7].

FLOWERING AND FRUITING. Flowers throughout the year [1].

DISTRIBUTION

Widely distributed throughout the Old World tropics and subtropics, including West and East Africa, and southwards to KwaZulu-Natal, South Africa. Also native in Madagascar, and throughout most of tropical Asia. Introduced to the Caribbean region and South America [7,12].

HABITAT

In Kenya, mostly found in disturbed areas, such as forest margins, secondary bushland, ruderal sites, grassland and riverine sites, from sea level to 2,400 m a.s.l. [1]. Also in moist savanna, scarp forest, dense bushland, coastal bush and thicket in South Africa [5].

Figure 1 – Infructescences and leaves. (Photo: KALRO.)

USES

The plant has many uses. In traditional medicine, the leaves are used to treat colic pains, upper respiratory tract infections, headache, diarrhoea, fever, eye problems, skin diseases and pneumonia [2]. The leaves are used for flavouring tea [4]; they are also used dried or fresh as an insect repellent and as an air freshener. The plant has antifungal and antimalarial activity. The plant contains essential oils, such as eugenol and thymol (depending on the chemotype), saponins and alkaloids. The essential oils are used in the perfume industry [2].

KNOWN HAZARDS AND SAFETY

Further research is needed on the potential acute and sub-chronic toxicity effects of the essential oil [8].

CONSERVATION STATUS

This widespread and common species is assessed according to IUCN Red List criteria as Least Concern (LC) in tropical East Africa [7], South Africa [9] and Sri Lanka [5], but its status has yet to be confirmed in the IUCN Red List [3].

SEED CONSERVATION

HARVESTING. Seeds can be harvested by stripping the ripe infructescences by hand.

PROCESSING AND HANDLING. Seeds can be separated by rubbing the infructescences between the hands or against a rough surface and then winnowing or seed blowing.

STORAGE AND VIABILITY. Seeds of this species are reported to be desiccation tolerant [10] and after appropriate drying can therefore be stored at sub-zero temperatures for long-term conservation.

PROPAGATION

SEEDS

Dormancy and pre-treatments: No information available on seed dormancy, highlighting the need for further research.

Germination, sowing and planting: Germination experiments carried out on seed lots stored at the MSB revealed high germination percentages (>80%) when seeds were incubated in the light at constant temperatures of 16–26°C [10].

VEGETATIVE PROPAGATION

Can be propagated from cuttings [6].

TRADE

Extracts and dusts are sold in local markets in small quantities in Kenya. The essential oil is a substitute for clove oil and thyme oil and is traded internationally to a small extent [11].

Authors

Desterio O. Nyamongo, Steve Davis and Efisio Mattana.

References

[1] Beentje, H. J. (1994). *Kenya Trees, Shrubs and Lianas*. National Museums of Kenya, Nairobi, Kenya.

[2] Ehiagbonare, J. E. (2007). Macropropagation of *Ocimum gratissimum* L [*sic*]: a multipurpose medicinal plant in Nigeria. *African Journal of Biotechnology* 6 (1): 13–14.

[3] IUCN (2017). *The IUCN Red List of Threatened Species.* Version 2017-3. http://www.iucnredlist.org/

[4] Maundu, P. M., Ngugi, G. W. & Kabuye, C. H. S. (1999). *Traditional Food Plants of Kenya.* Kenya Resource Centre for Indigenous Knowledge (KENRIK), National Museums of Kenya, Nairobi, Kenya.

[5] National Red List (2017). *Ocimum gratissimum*. National Red List Network. http://www.nationalredlist.org/

[6] Orwa, C., Mutua, A., Kindt, R., Jamnadass, R. & Simons, A. (2009). *Agroforestree Database*: *A Tree Reference and Selection Guide*. Version 4.0. World Agroforestry Centre, Kenya. http://www.worldagroforestry.org/output/agroforestree-database

[7] Paton, A. J., Bramley, G., Ryding, O., Polhill, R. M., Harvey, Y. B., Iwarson, M., Willis, F., Phillipson, P. B., Balkwill, K., Lukhoba, C. W., Otiend, D. F. & Harley, R. M. (2009). Lamiaceae (Labiatae). In: *Flora of Tropical East Africa*. Edited by H. J. Beentje, S. A. Ghazanfar & R. M. Polhill. Royal Botanic Gardens, Kew.

[8] Prabhu, K.S., Lobo, R., Shirwaikar, A. A. & Shirwaikar, A. (2009). *Ocimum gratissimum*: a review of its chemical, pharmacological and ethnomedicinal properties. *The Open Complementary Medicine Journal* 1: 1–15.

[9] Raimondo, D., von Staden, L., Foden, W., Victor, J. E., Helme, N. A., Turner, R. C., Kamundi, D. A. & Manyama, P. A. (2009). *Red List of South African Plants*. *Strelitzia* 25, South African National Biodiversity Institute, Pretoria, South Africa.

[10] Royal Botanic Gardens, Kew (2017). *Seed Information Database (SID).* Version 7.1. http://data.kew.org/sid/

[11] Sulistiarini, D. (1999). *Ocimum gratissimum* L. In: *Plant Resources of South-East Asia (PROSEA) 19*: *Essential-Oil Plants*. Edited by L. P. A. Oyen & Nguyen Xuan Dung. Backhuys Publishers, Leiden, The Netherlands. pp. 140–142.

[12] WCSP (2017). *World Checklist of Selected Plant Families.* Royal Botanic Gardens, Kew. http://wcsp.science.kew.org/

Olea europaea subsp. *cuspidata* (Wall. & G.Don) Cif.

TAXONOMY AND NOMENCLATURE

FAMILY. Oleaceae

SYNONYMS. *Linociera lebrunii* Staner, *Olea africana* Mill., *O. aucheri* A.Chev. ex Ehrend., *O. chrysophylla* Lam., *O. cuspidata* Wall. ex G.Don, *O. europaea* subsp. *africana* (Mill.) P.S.Green, *O. indica* Kleinhof ex Burm.f., *O. kilimandscharica* Knobl., *O. monticola* Gand., *O. schimperi* Gand., *O. somaliensis* Baker, *O. subtrinervata* Chiov., *O. verrucosa* (Willd.) Link, and many more (see [14]).

VERNACULAR / COMMON NAMES. Brown olive, wild olive (English); mzeituni (Swahili); ejass (Boran); muthata (Embu, Kamba, Meru); molialundi (Kamba); mutamaiyu (Kikuyu); emitot (Kipsigis); kango (Luo); oloirien (Maasai); yermit (Marakwet); emdit (Nandi, Tugen); ilnyirei (Samburu); mkumbi (Taita); ethelei (Turkana).

PLANT DESCRIPTION

LIFE FORM. Tree or sometimes a shrub.

PLANT. Evergreen, 3–24 m tall [3], bole often gnarled (Figure 1), bark rough, dark brown, branches spreading (the final ones terete or, especially if sterile, subquadrangular), grey and with numerous somewhat projecting lenticels. **LEAVES.** Narrowly lanceolate to elliptic-lanceolate, acute, very acutely mucronate [12], 2–9 x 0.5–3 cm [3], subcoriaceous, glabrous on upper surface, reddish to fulvous scaly on lower surface. **FLOWERS.** Panicles axillary and terminal, flowers white to creamy yellow, scented, c. 3 mm long [3].

Figure 1 – Young plant and details of branch. (Photos: H. Migiro, KEFRI. and R. v. Blittersdorff, African Plants - A Photo Guide. www.africanplants.senckenberg.de.)

FRUITS AND SEEDS

Fruits are green, edible drupes, which turn purple or black when ripe, ellipsoid, 0.5–1 cm long [1,3]. Each fruit contains a woody stone (i.e. seed and endocarp) (Figure 2).

FLOWERING AND FRUITING. Flowering depends on rainfall and varies even within short distances. In Kenya, flowering occurs from January to October [3]. After pollination, more than one year is needed for the drupe to develop fully. Trees seed only periodically [1].

Figure 2 – External views and sections of the stones containing one seed each.
(Photos: P. Gómez-Barreiro, RBG Kew.)

DISTRIBUTION

Tropical Africa from Eritrea and East Africa to the Democratic Republic of Congo, extending south to the subtropics and South Africa. Also native to the Mascarenes, Arabian Peninsula, northern India and western China [12,14]. In Kenya, mostly in highland areas from the Taita Hills to west and northern Kenya [8].

HABITAT

Dry upland evergreen forest and forest margins, often associated with *Juniperus* and *Podocarpus*, at 950–2,500 m a.s.l. Also on rocky hillsides, lava flows and along riverbeds, occasionally forming pure stands or co-dominant [3,8].

USES

The wood is used for house construction, furniture, panelling, posts, flooring, carving, clubs and walking sticks, and makes excellent firewood and charcoal. The fruits are edible and a source of olive oil. The branches are burnt and used for flavouring soup [8]. In traditional medicine, the stem is used to treat venereal diseases and female sterility. A decoction of the bark is used to treat gonorrhoea. Bark ash is applied to wounds. A bark infusion is drunk as a vermifuge, especially for tapeworm, and is given to livestock as an anthelmintic [5]. The plant is used to lower blood pressure and infusions of the leaves are widely used as eye lotions [13]. The flowers provide nectar for bees for honey production. The tree is planted for erosion control, as an ornamental, as a windbreak and to provide shade. It is a ceremonial tree among the Maasai, used for blessings and to bring peace and good luck [8].

KNOWN HAZARDS AND SAFETY

No known hazards.

CONSERVATION STATUS

Widespread and common and not considered to be either rare or threatened. This subspecies is assessed as Least Concern (LC) in South Africa [10], but Vulnerable (VU) in Egypt [9] (at the edge of its range) according to IUCN Red List criteria. In Kenya, it is under pressure from land use changes and overexploitation. The status of this subspecies has yet to be confirmed in the IUCN Red List [4].

SEED CONSERVATION

HARVESTING. Fruits should be collected immediately after they turn purplish black to avoid predation by birds. They can be collected by climbing the tree and picking the fruits with the aid of a hook and lopping shears [1].

PROCESSING AND HANDLING. Fruits can be depulped using a mortar and pestle. The seeds can be washed in running water to remove fruit remains and then dried in the shade.

STORAGE AND VIABILITY. Seeds of this species are likely to be desiccation tolerant [11] and therefore suitable for long-term conservation in conventional seed bank conditions. Mature and properly dried seeds can be stored in airtight containers at 3°C for several years [1].

PROPAGATION

SEEDS

Dormancy and pre-treatments: Seeds of this species are reported to be physiologically dormant [2]. The endocarp should be removed before sowing [6]. Alternatively, the endocarp can be cracked with a hard object or table vice, taking care not to damage the seeds [1], or seeds can be soaked in cold water for 48 hours [8].

Figure 3 – Seedlings. (Photo: H. Migiro, KEFRI.)

Germination, sowing and planting: Further studies are needed to determine optimum temperature and light regimes for germination under laboratory controlled conditions. In the nursery, treated seeds germinate within 20–45 days (Figure 3) [1].

VEGETATIVE PROPAGATION

Can be propagated by cuttings using a rooting hormone. Leafy branch cuttings of 2–3 mm in basal diameter and trimmed to 7–10 cm in length were treated with indole-3-butyric acid (IBA). The rooting success of cuttings treated with 20 or 40 µg IBA/cutting was 75 and 90%, respectively, suggesting that vegetative propagation is an effective method [7]. Wildings can also be used.

TRADE

The timber is important on the international market and in local trade.

Authors

Desterio O. Nyamongo, Steve Davis and Efisio Mattana.

References

[1] Albrecht, J. (1993). *Tree Seed Handbook of Kenya*. GTZ Forestry Seed Centre, Muguga, Kenya.

[2] Baskin, C. C. & Baskin, J. M. (2014). *Seeds: Ecology, Biogeography, and Evolution of Dormancy and Germination*, 2nd edition. Academic Press, San Diego, USA.

[3] Beentje, H. J. (1994). *Kenya Trees, Shrubs and Lianas*. National Museums of Kenya, Nairobi, Kenya.

[4] IUCN (2017). *The IUCN Red List of Threatened Species.* Version 2017-3. http://www.iucnredlist.org/

[5] Kokwaro, J. O. (2009). *Medicinal Plants of East Africa*, 3rd edition. University of Nairobi Press, Nairobi, Kenya.

[6] Legesse Negash (1993). Investigations on the germination behaviour of wild olive seeds and the nursery establishment of the germinants. *Ethiopian Journal of Science* 16 (2): 71–81.

[7] Legesse Negash (2003). Vegetative propagation of the threatened African wild olive [*Olea europaea* L. subsp. *cuspidata* (Wall. ex DC.) Ciffieri]. *New Forests* 26 (2): 137–146.

[8] Maundu, P. & Tengnäs, B. (eds) (2005). *Useful Trees and Shrubs for Kenya*. Technical Handbook No. 35, World Agroforestry Centre, Nairobi, Kenya.

[9] National Red List (2017). *Olea europaea* ssp. *africana*. National Red List Network. http://www.nationalredlist.org/

[10] Raimondo, D., von Staden, L., Foden, W., Victor, J. E., Helme, N. A., Turner, R. C., Kamundi, D. A. & Manyama, P. A. (2009). *Red List of South African Plants. Strelitzia* 25, South African National Biodiversity Institute, Pretoria, South Africa.

[11] Royal Botanic Gardens, Kew (2017). *Seed Information Database (SID).* Version 7.1. http://data.kew.org/sid/

[12] Turrill, W. B. (1952). Oleaceae. In: *Flora of Tropical East Africa*. Edited by W. B. Turrill & E. Milne-Redhead. Crown Agents for the Colonies, London.

[13] Van Wyk, B.-E., Van Oudtshoorn, B. & Gericke, N. (2009). *Medicinal Plants of South Africa*, 2nd edition. Briza Publications, Pretoria, South Africa.

[14] WCSP (2017). *World Checklist of Selected Plant Families.* Royal Botanic Gardens, Kew. http://wcsp.science.kew.org/

Piliostigma thonningii (Schumach.) Milne-Redh.

TAXONOMY AND NOMENCLATURE

FAMILY. Leguminosae – Cercidoideae

SYNONYMS. *Bauhinia thonningii* Schumach.

VERNACULAR / COMMON NAMES. Camel's foot, monkey bread, wild bauhinia (English); mchekeche, mchikichi (Swahili, Digo); mkayamba (Giriama); mkolokolo (Kamba); mulama (Kikuyu); kumuyenjayenja (Luhya); otangalo (Luo); ol-bugoi, ol-sagararami (Masaai); mukuura (Mbeere, Tharaka); kipsakiat (Nandi); sadiandet (Sebei).

Figure 1 – Habit of *P. thonningii* (Photo: V. Otieno, KEFRI) and details of the leaves. (Photo: R. and A. Heath, Plants and People Africa.)

PLANT DESCRIPTION

LIFE FORM. Tree or occasionally a shrub.

PLANT. Deciduous, 3–10 m high (Figure 1), bark rough, dark brown to grey or black.

LEAVES. Bilobed, 5–17 x 6–19 cm, leathery, pale green with a small bristle in the notch, often folded along the midrib. **FLOWERS.** White, unisexual, panicles usually alternately leaf-opposed and axillary along branches; usually dioecious [4,5,8].

FRUITS AND SEEDS

Pods are indehiscent, 13–26 x 3–6 cm [4], flat, hairy at first, then woody, changing from green to brown when mature, and remaining on the tree for a long time. Seeds are dark brown to blackish, ovoid to ellipsoid and compressed [1,8], with a V-shaped hilum region (Figure 2), and surrounded by pulp.

FLOWERING AND FRUITING. Flowering and fruiting is irregular [1]; in Kenya, flowering can occur from January to February, from April to July and in December [4].

Figure 2 – External views of the seeds showing the V-shaped hilum (left) and internal views (right) showing the endosperm, cotyledons and embryonic axis. (Photos: P. Gómez-Barreiro, RBG Kew.)

DISTRIBUTION

Widely distributed from Senegal to Ethiopia, south to Angola, Namibia and South Africa [5]. In Kenya, it is found on the coast around Kitui, and in the Rift Valley and Nyanza provinces [1].

HABITAT

Savanna, open woodland, wooded grassland, bushland and gallery forests, from sea level to 2,150 m a.s.l. [1,4].

USES

A multipurpose tree [11]. In traditional medicine, the leaf juice is used to treat stomach pains; dry leaves are burnt and the ash used to treat snake bites; and a root infusion is used to stop prolonged menstruation, as a remedy for chest colds, and to treat venereal diseases. The bark and leaves are chewed to relieve coughs, common cold and chest pains. A root decoction is drunk to stop haemorrhage and prevent miscarriage, and a decoction of the root and bark is used for diarrhoea and snake bites [7]. The wood is used for firewood, charcoal, and in house construction. The pulp surrounding the seeds is edible. The pods and seeds yield a blue dye [4]. The tree is also used as bee forage for the production of honey, and is a valuable shade tree and used as a live fence.

KNOWN HAZARDS AND SAFETY

No known hazards.

CONSERVATION STATUS

Assessed as Least Concern (LC) in South Africa according to IUCN Red List criteria [10], this widespread species has yet to be assessed in Kenya and its status confirmed in the IUCN Red List [6].

SEED CONSERVATION

HARVESTING. Seed collection should be done immediately after the pods turn brown to prevent insect attack while still on the tree. Pods can be picked by hand from the tree [1].

PROCESSING AND HANDLING. The seeds are difficult to extract from the tough, woody pod. After drying the pods in the sun, they can be cut into small pieces. The seeds can then be extracted by pounding the pod fragments in a mortar and pestle [1].

STORAGE AND VIABILITY. Seeds of this species are reported to be desiccation tolerant [12] and after appropriate drying can therefore be stored at sub-zero temperatures for long-term conservation. Mature and properly dried seeds can also be stored in airtight containers at room temperature for several years [1].

PROPAGATION

SEEDS

Dormancy and pre-treatments: Seeds of this species are reported to be physically dormant [3]. The seeds should therefore be pre-treated to allow seed imbibition. Small amounts of seeds may be treated using a hot wire, whereas hot water treatment should be used for large quantities [9]. Alternatively, the seed coat should be nicked using a sharp tool (e.g. a scalpel, knife or nail clipper) at the distal (cotyledon) end of the seeds [1], or chemical scarification with sulphuric acid could be used [2].

Germination, sowing and planting: Under laboratory controlled conditions, scarified seeds incubated at constant 25°C in darkness germinate to percentages >80% six days after sowing, reaching their maximum germination (90–95%) after eight days [2]. Pre-treated seeds can be sown in a seedbed or in polythene tubes. Seedlings are ready for potting three weeks after germination, and can be planted out in the field after five months.

VEGETATIVE PROPAGATION

Root suckers from exposed roots can be used [8].

TRADE

No data available.

Authors

William Omondi, Victor Otieno, David Meroka and Efisio Mattana.

References

[1] Albrecht, J. (1993). *Tree Seed Handbook of Kenya*. GTZ Forestry Seed Centre, Muguga, Kenya.

[2] Ayisire, B. E., Akinro, L. A. & Amoo, S. O. (2009). Seed germination and *in vitro* propagation of *Piliostigma thonningii* — an important medicinal plant. *African Journal of Biotechnology* 8 (3): 401–404.

[3] Baskin, C. C. & Baskin, J. M. (2014). *Seeds: Ecology, Biogeography, and Evolution of Dormancy and Germination*, 2nd edition. Academic Press, San Diego, USA.

[4] Beentje, H. J. (1994). *Kenya Trees, Shrubs and Lianas*. National Museums of Kenya, Nairobi, Kenya.

[5] Brenan, J. P. M. (1967). Leguminosae Subfamily Caesalpinioideae. In: *Flora of Tropical East Africa*. Edited by E. Milne-Redhead & R. M. Polhill. Crown Agents for Oversea Governments and Administrations, London.

[6] IUCN (2017). *The IUCN Red List of Threatened Species*. Version 2017-3. http://www.iucnredlist.org/

[7] Kokwaro, J. O. (2009). *Medicinal Plants of East Africa*, 3rd edition. University of Nairobi Press, Nairobi, Kenya.

[8] Maundu, P. & Tengnäs, B. (eds) (2005). *Useful Trees and Shrubs for Kenya*. Technical Handbook No. 35, World Agroforestry Centre, Nairobi, Kenya.

[9] Msanga, H. P. (1998). *Seed Germination of Indigenous Trees in Tanzania: Including Notes on Seed Processing and Storage, and Plant Uses*. Canadian Forest Service, Northern Forestry Centre, Edmonton, Canada.

[10] Raimondo, D., von Staden, L., Foden, W., Victor, J. E., Helme, N. A., Turner, R. C., Kamundi, D. A. & Manyama, P. A. (2009). *Red List of South African Plants. Strelitzia* 25, South African National Biodiversity Institute, Pretoria, South Africa.

[11] Royal Botanic Gardens, Kew (1999–2015). *Survey of Economic Plants for Arid and Semi-Arid Lands (SEPASAL) Database*. http://apps.kew.org/sepasalweb/sepaweb

[12] Royal Botanic Gardens, Kew (2017). *Seed Information Database (SID)*. Version 7.1. http://data.kew.org/sid/

Searsia pyroides (Burch.) Moffett

TAXONOMY AND NOMENCLATURE

FAMILY. Anacardiaceae

SYNONYMS. *Rhus pyroides* Burch., *R. vulgaris* Meikle

VERNACULAR / COMMON NAMES. Mlishangwe, mrinja kondo (Swahili); mbwana nyahi, mbwanyahi (Digo); muthigiu (Embu); mutheu (Kamba); muthigio (Kikuyu); suriat (Kipsigis); omusangura (Luhya); busangura (Luhya, Bukusu dialect); awayo, sangla (Luo); olmungushi (Maasai); mubebiaiciya (Mbeere); mirimuthu (Meru); monjororioyat (Nandi); siriewo kaptamu (Pokot); sioloran (Samburu); njówaruwa (Sebei); vikunguu (Taita).

PLANT DESCRIPTION

LIFE FORM. Shrub or tree.

PLANT. Deciduous, <9 m tall [2], often thorny; young branches brownish, spreading, hairy; old branches dark greyish. **LEAVES.** Compound, petiole c. 2 cm long, terete, pubescent, somewhat longitudinally grooved above (Figure 1); leaflets 3, median leaflet 4–11 x 2.5–6 cm, elliptic or ovate to obovate or subcircular, rounded, truncate or emarginate (rarely acute) at apex, cuneate and sessile at base, or contracted into very short petiolule; lateral leaflets c. 1/3–1/2 as long as the median. **FLOWERS.** Cream or greenish yellow, in hairy, branched heads, panicles 5–20 cm long [2]; petals c. 1–1.5 mm long [4].

Figure 1 – Young plant of *S. pyroides*.
(Photo: H. Migiro, KEFRI.)

FRUITS AND SEEDS

Fruits are small subglobose drupes (Figure 2), green at first, turning reddish brown when mature, almost round, but slightly flattened when dry, <6 mm in diameter [4,6]. Seeds are small (<4 mm long), light brown (Figure 2).

FLOWERING AND FRUITING. Flowers from January to April; fruits from March to June in western parts of Kenya [6]. In the southern parts of its range (e.g. South Africa), flowering occurs from October to January.

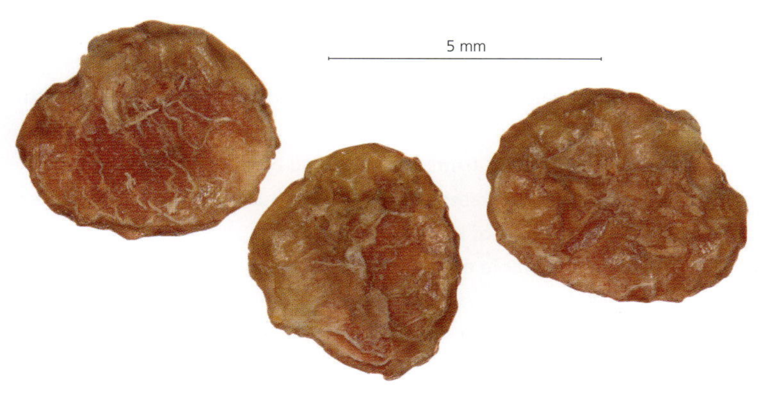

Figure 2 – Branch with unripe fruits (Photo: V. Otieno, KEFRI) and external views of the seeds. (Photo: P. Gómez-Barreiro, RBG Kew.)

DISTRIBUTION

Widely distributed from Cameroon west to Ethiopia, and south to Mozambique, Malawi, Zambia, Zimbabwe, Namibia and South Africa [4]. In Kenya, it is widespread in the Chyulu Hills, Mt Elgon, Ngong Hills, Thui Hills (Makueni), Kitui Hills, Chepararia (West Pokot) and the Lake Basin region [6].

HABITAT

Wooded grassland, thickets, or (semi-)evergreen bushland or bushed grassland, in rocky sites and dry forest margins [2], upland forest, sometimes riverine, at 800–2,700 m a.s.l. [6].

USES

The fruits are edible, and the juice extracted from fruit pulp is used to sweeten porridge. In traditional medicine, an extract from boiled fruit is drunk to stop diarrhoea. The stems are boiled and the liquid applied to wounds. The powdered root is mixed with gruel and drunk for gonorrhoea. The roots are mixed with other plants to make a drink for expectant mothers, as it is believed to make childbirth easier, as well as being used for infertility and gastrointestinal problems. The leaves are used for treating haemorrhoids; a stem decoction is used for cleaning wounds; and a root decoction is used in traditional veterinary medicine for treating diarrhoea in livestock. Young shoots and leaves are chewed and the juice swallowed as a remedy for heartburn [5]. The leaves and fruits are browsed by goats and wild game. The leaves also provide a mulch and green manure. The wood is used as firewood, and for making posts, farm implements and tool handles. *Searsia pyroides* is planted as a live fence, and its stems and branches are cut and placed around cattle sheds for protection. Small twigs are used as chewing sticks [6].

KNOWN HAZARDS AND SAFETY

The branches can be thorny and care should be taken when handling the plant. A scratch or prick from any sharp point is extremely painful and burns (hence the specific name *pyroides*, which means like fire) [9].

CONSERVATION STATUS

Assessed as Least Concern (LC) in South Africa according to IUCN Red List criteria [7], this widespread species has yet to be assessed in Kenya and its status confirmed in the IUCN Red List [3].

SEED CONSERVATION

HARVESTING. Fruits can be hand-picked directly from the crown, although it may be necessary to climb the trunk of tall trees to pick the fruits.

PROCESSING AND HANDLING. Ripe yellow greenish fruits are depulped by squeezing or rubbing gently on a wire mesh while washing in running water to remove the mucilage. Seeds are cleaned by hand-sorting, and dried to a moisture content of c. 15%.

STORAGE AND VIABILITY. No information available on seed storage behaviour, although seeds of seven other *Searsia* spp. stored at the MSB are reported to be orthodox [8], suggesting that seeds of *S. pyroides* could be desiccation tolerant. However, seeds of this species are reported to be viable for only about three months when stored at room temperatures [6], highlighting the need for further studies to characterise seed storage behaviour.

PROPAGATION

SEEDS

Dormancy and pre-treatments: Anacardiaceae, and in particular *Rhus* spp. [1], are reported to be physically dormant, so seeds of these species should be scarified before sowing. No information is available for *S. pyroides*, although seed lots of other *Searsia* spp. stored at the MSB have been scarified prior to germination [8].

Germination, sowing and planting: No information is available on optimum incubation temperature and light requirements for germination under laboratory controlled conditions. Further studies are needed to characterise seed germination. In the nursery, seeds germinate within 4–20 days with a germination rate of 50–80%. Seeds are sown in a seedbed or in polythene tubes. Seedlings are ready for potting when true leaves appear or after about four weeks. Seedlings are grown in the nursery until planted out in the field at the onset of the long rains. For woodlots, a spacing of c. 3 x 3 m is recommended. For scattered planting in cropland, trees should be planted 10–15 m apart. The species is fast growing and coppices readily.

VEGETATIVE PROPAGATION

No protocols available.

TRADE

No data available.

Authors

William Omondi, Victor Otieno, David Meroka and Efisio Mattana.

References

[1] Baskin, C. C. & Baskin, J. M. (2014). *Seeds: Ecology, Biogeography, and Evolution of Dormancy and Germination*, 2nd edition. Academic Press, San Diego, USA.

[2] Beentje, H. J. (1994). *Kenya Trees, Shrubs and Lianas*. National Museums of Kenya, Nairobi, Kenya.

[3] IUCN (2017). *The IUCN Red List of Threatened Species*. Version 2017-3. http://www.iucnredlist.org/

[4] Kokwaro, J. O. (1986). Anacardiaceae. In: *Flora of Tropical East Africa*. Edited by R. M. Polhill. A.A. Balkema, Rotterdam, The Netherlands.

[5] Kokwaro, J. O. (2009). *Medicinal Plants of East Africa*, 3rd edition. University of Nairobi Press, Nairobi, Kenya.

[6] Maundu, P. & Tengnäs, B. (eds) (2005). *Useful Trees and Shrubs for Kenya*. Technical Handbook No. 35, World Agroforestry Centre, Nairobi, Kenya.

[7] Raimondo, D., von Staden, L., Foden, W., Victor, J. E., Helme, N. A., Turner, R. C., Kamundi, D. A. & Manyama, P. A. (2009). *Red List of South African Plants. Strelitzia* 25, South African National Biodiversity Institute, Pretoria, South Africa.

[8] Royal Botanic Gardens, Kew (2017). *Seed Information Database (SID)*. Version 7.1. http://data.kew.org/sid/

[9] Watt, J. M. & Breyer-Brandwijk, M. G. (1962). *The Medicinal and Poisonous Plants of Southern and Eastern Africa*, 2nd edition. E. & S. Livingstone, Edinburgh & London.

Senegalia brevispica (Harms) Seigler & Ebinger

TAXONOMY AND NOMENCLATURE

FAMILY. Leguminosae – Caesalpinioideae

SYNONYMS. *Acacia brevispica* Harms

VERNACULAR / COMMON NAMES. Wait-a-bit thorn (English); mwarare (Swahili); aares, hammaress (Boran); mukuswi (Kamba); mwikunya (Kikuyu); ngirgirit (Kipsigis); osiri (Luo); olgirigir orok (Maasai); aiman, kiptare, korniswa (Marakwet); kaptaruu, ptar (Pokot); igirigiri (Samburu); furgorri, gorgor (Somali); munua, munwa (Tharaka); garnista (Tugen); ekurau (Turkana).

PLANT DESCRIPTION

LIFE FORM. Shrub or small tree.

PLANT. Up to 7 m tall, often semi-scandent or thicket-forming. Young branchlets densely pubescent or puberulous with many minute reddish glands. Prickles scattered, recurved or spreading. **LEAVES.** Compound, petiole 1.5 cm long, 6–18 pairs of pinnae, 20–40 leaflets in pairs, linear or oblong. **FLOWERS.** White or yellowish white in heads of 1–1.5 cm in diameter (Figure 1), arranged in racemes or aggregated into irregular terminal panicles [2].

Figure 1 – Flowers and ripe fruits. (Photos: KEFRI.)

FRUITS AND SEEDS

Fruits are purple brown, subcoriaceous, oblong, straight pods, 6–15 cm long and 1.5–3.3 cm wide (Figure 1), glabrous or pubescent with many minute reddish glands. Seeds (Figure 2) are brown, smooth, elliptic, compressed, 6–13 mm long and 6–7 mm wide, with an areole 3–9 mm long and 2–3 mm wide [2]. A rudimentary funicular aril is present [10].

FLOWERING AND FRUITING. Flowers from March to April; fruits from July to August in Kajiado (Kenya) [7].

Figure 2 – External views of the seeds showing the areole surrounded by the pleurogram and longitudinal section showing the plumule, embryonic axis and the cotyledon. (Photos: P. Gómez-Barreiro, RBG Kew.)

DISTRIBUTION

Widespread in tropical Africa, extending south to the Cape region of South Africa [2]. In Kenya, it is common in Siaya, Busia, Homa-Bay, Kajiado and Kitui counties, in the dry parts of the Rift Valley, and in humid areas of Kiambu County [1].

HABITAT

Thickets, dry bushland, woodland and forest margins [1] on a variety of soils, from rocky and stony soils to well-drained sandy loam, up to 2,100 m a.s.l.

USES

In traditional medicine, an infusion of the roots is used as a cure for intestinal worms. An extract of the roots is used as part of an aphrodisiac, rubbed into skin rashes, or used to treat female sterility and snake bites. The pounded roots, together with those of *Rauvolfia mombasiana* Stapf, are blown into the nostrils to kill throat fly maggots. In traditional veterinary medicine, an infusion of the leaves is used to treat nasal infections in calves, and an infusion of the roots is given to cattle with East Coast fever. A root decoction is given to cows to help remove the placenta [5]. The leaves, twigs and pods are browsed by goats and other livestock. The wood is used for firewood and rafters. *Senegalia brevispica* (more widely known as *Acacia brevispica*) [6] is occasionally used as a live fence [5]. Its flowers attract bees which produce a good honey.

KNOWN HAZARDS AND SAFETY

The prickles on the stems are sharp and give rise to the common name for the plant: wait-a-bit thorn [7].

CONSERVATION STATUS

Assessed as Least Concern (LC) in South Africa according to IUCN Red List criteria [9], but yet to be assessed in Kenya and to have its status confirmed in the IUCN Red List [4], although considered to be Not Threatened (nt) according to previous IUCN criteria [3].

SEED CONSERVATION

HARVESTING. Mature brown pods are collected from the crown by hand-picking. It is sometimes essential to wear protective gloves on account of the sharp prickles.

PROCESSING AND HANDLING. Pods are dried in the sun and then threshed lightly to minimise mechanical damage to the seeds. Alternatively, small amounts of seed can be extracted by hand. Seeds can be cleaned by hand-sorting, winnowing or sieving.

STORAGE AND VIABILITY. Seeds of this species are reported to be orthodox [10], and can therefore be stored in airtight containers (e.g. kilner jars and aluminium packets) in a cool dry place for short- to medium-term storage. For long-term conservation, after appropriate drying, seeds can be stored at sub-zero temperatures for many years without significant loss of viability [8].

PROPAGATION

SEEDS

Dormancy and pre-treatments: Seed scarification (chipping) and immersion in hot water, allowing the seeds to cool and soak for 24 hours, is reported to promote seed germination [7,10]. This suggests the presence of physical dormancy. However, the presence of this class of dormancy should be confirmed by further seed imbibition and germination studies.

Germination, sowing and planting: Under laboratory controlled conditions, seed lots stored at the MSB reached high germination percentages (>80%) when chipped seeds were incubated with 8 hours of light per day at constant temperatures of ≥20°C or at an alternating temperature regime of 15/30°C [10]. In the nursery, pre-treated seeds can be sown in a seedbed or in polythene tubes. Germination occurs within 7–21 days with a germination rate of 60–80%. Seedlings can be potted after three weeks, and are ready for field planting after five months. For hedgerow establishment, seedlings are planted 10–20 cm apart. *Senegalia brevispica* is relatively fast growing and coppices well, to the extent that it can become a weed in pasture as a result of its fast regeneration even after clearing and burning [7].

VEGETATIVE PROPAGATION

No protocols available.

TRADE

No data available.

Authors

William Omondi, Victor Otieno, David Meroka and Efisio Mattana.

References

[1] Beentje, H. J. (1994). *Kenya Trees, Shrubs and Lianas*. National Museums of Kenya, Nairobi, Kenya.

[2] Brenan, J. P. M. (1959). Leguminosae Subfamily Mimosoideae. In: *Flora of Tropical East Africa*. Edited by C. E. Hubbard & E. Milne-Redhead. Crown Agents for Oversea Governments and Administrations, London.

[3] ILDIS (2006–2013). *World Database of Legumes*. International Legume Database and Information Service. http://www.legumes-online.net/ildis/aweb/database.htm

[4] IUCN (2017). *The IUCN Red List of Threatened Species*. Version 2017-3. http://www.iucnredlist.org/

[5] Kokwaro, J. O. (2009). *Medicinal Plants of East Africa*, 3rd edition. University of Nairobi Press, Nairobi, Kenya.

[6] Kyalangalilwa, B., Boatwright, J. S., Daru, B. H., Maurin, O. & Van der Bank, M. (2013). Phylogenetic position and revised classification of *Acacia* s.l. (Fabaceae: Mimosoideae) in Africa, including new combinations in *Vachellia* and *Senegalia*. *Botanical Journal of the Linnean Society* 172 (4): 500–523.

[7] Maundu, P. & Tengnäs, B. (eds) (2005). *Useful Trees and Shrubs for Kenya*. Technical Handbook No. 35, World Agroforestry Centre, Nairobi, Kenya.

[8] Omondi, W., Angaine, P., Meso, M. & Otieno, V. (2011). *Seed Collection, Handling and Nursery Management: A Reference and Training Manual*. Kenya Forestry Research Institute (KEFRI), Nairobi, Kenya.

[9] Raimondo, D., von Staden, L., Foden, W., Victor, J. E., Helme, N. A., Turner, R. C., Kamundi, D. A. & Manyama, P. A. (2009). *Red List of South African Plants*. *Strelitzia* 25, South African National Biodiversity Institute, Pretoria, South Africa.

[10] Royal Botanic Gardens, Kew (2017). *Seed Information Database (SID)*. Version 7.1. http://data.kew.org/sid/

Senegalia polyacantha subsp. *campylacantha* (Hochst. ex A.Rich.) Kyal. & Boatwr.

TAXONOMY AND NOMENCLATURE

FAMILY. Leguminosae – Caesalpinioideae

SYNONYMS. *Acacia campylacantha* A.Rich., *A. polyacantha* subsp. *campylacantha* (A.Rich.) Brenan

VERNACULAR / COMMON NAMES. African catechu tree, white-stem thorn (English); mgunga, mkengewa (Swahili); kivovoa, musewa, mwelele (Kamba); kumukokwe (Luhya); ogongo, oyongo, suahowe (Luo).

PLANT DESCRIPTION

LIFE FORM. Tree.

PLANT. Deciduous, <21 m tall (Figure 1), crown flat-topped, bark fissured, whitish to yellowish or grey, scaly or papery; prickles hooked, in pairs, just below nodes. **LEAVES.** Petiole glandular, rachis pubescent, 5.5–20 cm long [6], pinnae in (6)13–40(60) pairs, leaflets in (15)26–66 pairs, linear to linear-triangular. **FLOWERS.** Cream or white, sweet-scented, sessile or nearly so, in spikes (3.5)5–12.5 cm long [3,4].

Figure 1 – General habit. (Photo: NMK.)

FRUITS AND SEEDS

Fruits are brown, glabrous or nearly so, pods dehiscent, 7–18 cm long, 1–2 cm wide. Seeds (Figure 2) are subcircular to elliptic lenticular, with a medium to small and not impressed central areole [4].

FLOWERING AND FRUITING. Time of flowering depends on the occurrence of rains. The seeds ripen c. 6 months after flowering [1].

Figure 2 – External views of the seeds showing the areole surrounded by the pleurogram and longitudinal section showing the embryonic axis and the cotyledon. (Photos: P. Gómez-Barreiro, RBG Kew.)

DISTRIBUTION

This subspecies is widespread in tropical Africa, from Gambia to Eritrea in the north to Zimbabwe in the south. It is also found in the subtropical Limpopo Province of South Africa. In Kenya, this taxon is common along the Nairobi-Thika road, in Kisumu, South Nyanza, Narok, Taita-Taveta, Machakos, Makueni and Kilifi [11].

HABITAT

Wooded grassland, deciduous woodland, bushland and riverine forest, often in poorly drained areas, at 50–1,800 m a.s.l. [3,4,5].

USES

Senegalia polyacantha yields a gum used in confectionery, but inferior to gum arabic derived mainly from *Senegalia senegal* (L.) Britton [syn. *Acacia senegal* (L.) Willd.] and *Vachellia seyal* (Delile) P.J.H.Hurter [syn. *Acacia seyal* Delile] [3,10]. The gum is also used as an adhesive. The bark is used in tanning. The roots have a strong odour and in rural areas are used to repel snakes and are placed at river crossings to ward off crocodiles [7].

KNOWN HAZARDS AND SAFETY

The prickles on the stems are sharp.

CONSERVATION STATUS

This widespread subspecies is assessed as Least Concern (LC) in South Africa according to IUCN Red List criteria [8], but its status has yet to be confirmed in the IUCN Red List [9].

Figure 3 – Cleaning seeds by winnowing. (Illustration: N. Muema, NMK.)

Figure 4 – Raising *S. polyacantha* seedlings at Nairobi Botanic Garden. (Photo: E. Mattana, RBG Kew.)

SEED CONSERVATION

HARVESTING. Timing of seed collection is very important because the pods open and release the seeds when mature. Collecting should be done immediately after the pods turn greyish. Depending on tree height, pods can be hand-picked from the crown or the tree climbed and the branches shaken to dislodge the pods [1].

PROCESSING AND HANDLING. The pods can be sun-dried and then threshed in a gunny bag (hessian sack) by using a stick. After threshing, the seeds are cleaned by winnowing (Figure 3) using a sieve or a mechanical blower [1].

STORAGE AND VIABILITY. Seeds are likely to be desiccation tolerant [12]. Mature and properly dried seeds can be stored in airtight containers at room temperature or for several years in hermetic conditions at 10°C [1].

PROPAGATION

SEEDS

Dormancy and pre-treatments: Seeds of this species are reported to have physical dormancy [2]. For good germination, the seed coat should be nicked at the distal (cotyledon) end using a sharp tool, such as a scalpel, knife or nail clipper. Large quantities of seeds should be treated by pouring boiling water on them in a vessel, then soaking the seeds for 24 hours in the water as it cools [1].

Germination, sowing and planting: Seed germination experiments carried out on seed lots stored at the MSB under laboratory controlled conditions revealed high germination percentages (>80%) when chipped seeds where incubated at temperatures of 20–25°C or at an alternating temperature regime of 15/30°C [12]. In the nursery, seeds germinate within 10–20 days in ideal conditions. Expected germination rate of mature healthy and properly treated seed lots is between 60–90% [1]. Germinated seedlings (Figure 4) should be transplanted into bags when several leaves have formed [6].

VEGETATIVE PROPAGATION

No protocols available.

TRADE

Mostly local trade in rural areas in the Makueni, Kitui and Machakos Districts.

Authors

Veronicah Ngumbau, Patrick Muthoka, Steve Davis and Efisio Mattana.

References

[1] Albrecht, J. (1993). *Tree Seed Handbook of Kenya*. GTZ Forestry Seed Centre, Muguga, Kenya.

[2] Baskin, C. C. & Baskin, J. M. (2014). *Seeds: Ecology, Biogeography, and Evolution of Dormancy and Germination*, 2nd edition. Academic Press, San Diego, USA.

[3] Beentje, H. J. (1994). *Kenya Trees, Shrubs and Lianas.* National Museums of Kenya, Nairobi, Kenya.

[4] Brenan, J. P. M. (1959). Leguminosae Subfamily Mimosoideae. In: *Flora of Tropical East Africa*. Edited by C. E. Hubbard & E. Milne-Redhead. Crown Agents for Oversea Governments and Administrations, London.

[5] Brenan, J. P. M. (ed.) (1970). Leguminosae (Mimosoideae). In: *Flora Zambesiaca*, Vol. 3, part 1. Flora Zambesiaca Managing Committee, Royal Botanic Gardens, Kew.

[6] Coates Palgrave, K. & Coates Palgrave, M. (2002). *Trees of Southern Africa*, 3rd edition. Struik Publishers, Cape Town, South Africa.

[7] Dlamini, M. D. (2005). *Senegalia polyacantha* (Willd.) Seigler & Ebinger subsp. *campylacantha* (Hochtst. [*sic*] ex A.Rich.) Kyal. & Boatwr. South African National Biodiversity Institute (SANBI), South Africa. http://pza.sanbi.org/

[8] Foden, W. & Potter, L. (2005). *Senegalia polyacantha* (Willd.) Seigler & Ebinger subsp. *campylacantha* (Hochst. ex A.Rich.) Kyal. & Boatwr. In: *National Assessment: Red List of South African Plants.* Version 2017.1. http://redlist.sanbi.org/

[9] IUCN (2017). *The IUCN Red List of Threatened Species.* Version 2017-3. http://www.iucnredlist.org/

[10] Kyalangalilwa, B., Boatwright, J. S., Daru, B. H., Maurin, O. & Van der Bank, M. (2013). Phylogenetic position and revised classification of *Acacia* s.l. (Fabaceae: Mimosoideae) in Africa, including new combinations in *Vachellia* and *Senegalia*. *Botanical Journal of the Linnean Society* 172 (4): 500–523.

[11] Maundu, P. & Tengnäs, B. (eds) (2005). *Useful Trees and Shrubs for Kenya*. Technical Handbook No. 35, World Agroforestry Centre, Nairobi, Kenya.

[12] Royal Botanic Gardens, Kew (2017). *Seed Information Database (SID).* Version 7.1. http://data.kew.org/sid/

Senegalia senegal (L.) Britton

TAXONOMY AND NOMENCLATURE

FAMILY. Leguminosae – Caesalpinioideae

SYNONYMS. *Acacia circummarginata* Chiov., *A. cufodontii* Chiov., *A. oxyosprion* Chiov., *A. rupestris* Boiss., *A. senegal* (L.) Willd., *A. spinosa* Marloth & Engl., *A. volkii* Suess., *Mimosa senegal* L.

VERNACULAR / COMMON NAMES. Gum arabic (English); kikwata, mgunga (Swahili); dimitu (Boran); chikwata (Digo); king'ole (Kamba); kiluor (Luo); olderekesi (Maasai); mung'othi (Mbeere); chemangayan (Pokot); eldekeci (Samburu); adadgeti (Somali); ekunoit (Turkana).

PLANT DESCRIPTION

LIFE FORM. Shrub or tree.

PLANT. Deciduous, <12(–15) m high, low-branching (Figure 1); bark yellow to brown in young trees, becoming grey, gnarled and cracked on older trees; young branches densely pubescent, soon glabrescent; prickles 3, just below nodes, with central prickle hooked downwards and laterals curved upwards, or else in pairs or solitary. **LEAVES.** Compound, rachis ± pubescent, pinnae 1.5–2.5 cm long, with 5–25 pairs of leaflets per pinna, leaflets oval, 1–9 x 0.5–3 mm. **FLOWERS.** White or cream, fragrant, sessile, in spikes 2–10 cm long [6].

Figure 1 – Tree with ripe fruits. (Photo: S. Sanogo, IER.)

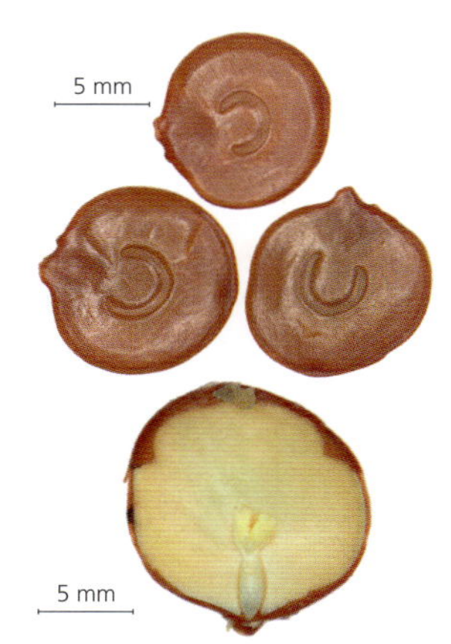

Figure 2 – External views of the seeds showing the areole surrounded by the pleurogram and longitudinal section showing the embryonic axis and the cotyledon. (Photos: P. Gómez-Barreiro, RBG Kew.)

FRUITS AND SEEDS

Pods are usually grey brown, sometimes pale or dark brown, dehiscent, densely to sparsely pubescent [6], <9 cm long and 1–3.5 cm wide, 5–6(15)-seeded [10]. Seeds ± subcircular-lenticular, with a small to medium central areole (Figure 2) [6].

FLOWERING AND FRUITING. In its native area of occurrence, *Senegalia senegal* flowers throughout the year, with a peak in February and March, followed by a seeding peak in June and July [1].

DISTRIBUTION

Widespread in tropical Africa from Mauritania and Senegal in the west to Ethiopia and Somalia in the north-east, through Central Africa and tropical East Africa to South Africa (Northern Provinces and KwaZulu-Natal) in the subtropics [17]. Also occurs in Oman, Pakistan and India [12]. Introduced to Egypt, Australia and elsewhere.

HABITAT

Widely distributed in the semi-arid and arid zones [1], mainly on sandy soils (or slightly loamy sands), in areas receiving 300–700 mm annual rainfall (tolerating as little as 100 mm rainfall and up to 11 months of drought), at altitudes ranging from 100–1,950 m a.s.l.

USES

Senegalia senegal (more widely known as *Acacia senegal*) [13] is the main source of gum arabic which has been used for at least 4,000 years in the preparation of food, in human and veterinary medicine, and in crafts and cosmetics. Gum exudation occurs naturally in response to wounds [2]. A section of bark is wounded or stripped off to induce gum formation for collection [14]. All gum arabic harvested in Kenya comes from wild sources and is collected mainly by women and children [18]. Between 60% and 75% of world production of gum arabic is used in the food industry (Figure 3) and in human and animal medicine [5]. *S. senegal* is also a valuable source of fuelwood, charcoal, and timber for house building and fencing, and is a source of fibre and fodder for goats, sheep and camels [14]. In the pharmaceutical sector, extracts from the plant are used in the manufacture of emulsions, as a binder and coating for tablets, as an ingredient in cough medicines, and in the manufacture of cosmetics. In traditional medicine, gum arabic is extensively used to treat coughs, diarrhoea, dysentery, haemorrhage and inflammation [15]. The seeds are edible and widely consumed throughout arid regions of East Africa. The flowers attract bees which produce an amber honey with a mild aroma that granulates rapidly. *S. senegal* is also used for reforestation, in soil reclamation, as a windbreak and for sand dune fixation [10]. It improves soil fertility by nitrogen-fixation and nutrient cycling [16].

Figure 3 – Gum exuding from tree (Photo: S. Sanogo, IER) and edible arabic gum pieces available in the market (Photo: E. Mattana, RBG Kew.)

KNOWN HAZARDS AND SAFETY

Gum arabic has been evaluated toxicologically as a safe food additive. The prickles on the stems are sharp.

CONSERVATION STATUS

This widespread species is assessed as Least Concern (LC) in Namibia [7] and South Africa [17], but its status has yet to be confirmed in the IUCN Red List [11].

SEED CONSERVATION

HARVESTING. Pods should be harvested before they open, by shaking the branches over a tarpaulin spread on the ground. Pods are often collected when they are still green to minimise insect damage [12].

PROCESSING AND HANDLING. The pods can be sun-dried and then threshed in a gunny bag (hessian sack) by using a stick. After threshing, seeds can be cleaned by winnowing using a sieve or a mechanical blower [1].

STORAGE AND VIABILITY. Seeds of this species are reported to be orthodox [19] and after appropriate drying can therefore be stored at sub-zero temperatures for long-term conservation. In addition, mature and properly dried seeds can be stored in airtight containers for at least one year at room temperature and for several years at 10°C [1].

PROPAGATION

SEEDS

Dormancy and pre-treatments: Unlike others in the genus (see [4]) the seed coat of *S. senegal* is permeable to water and, therefore, seeds of this species do not have physical dormancy [1,4,9,12]. Nevertheless, 14 minutes soaking in 95% sulphuric acid, or a pre-germination soaking in water for 12–24 hours (Figure 4), accelerates and synchronises germination [9].

Germination, sowing and planting: High germination percentage (100%) is reported under laboratory controlled conditions for scarified seeds that are incubated in agar at constant 20–25°C in the light [19]. In nursery conditions, this species is easy to propagate from seed (Figure 5). Seeds should be sown in polypots or in 30 cm long tubes, 2–4 seeds per tube, and thinned to one seedling after 4–6 weeks. Frequent root pruning is necessary during the first two years. For intercropping, a spacing of 10 x 10 m is recommended. For gum production, plants can be raised in the nursery and transplanted, or direct seeded in plantations at a spacing of c. 4 x 4 m [12].

VEGETATIVE PROPAGATION

Can be propagated by stem cuttings taken during the rainy season and treated with hormone rooting powder [3,8].

Figure 4 – Soaking in hot water as a pre-treatment to break seed dormancy before sowing. (Illustration: N. Muema, NMK.)

Figure 5 – Seedlings propagated in the nursery. (Photo: T. Ulian, RBG Kew.)

TRADE

Commercial gum is collected from the wild (in Kenya from Wajir, Mandera, Isiolo, Marakwet, Garissa and Samburu) mainly by children and women. It is usually picked for export to the Far East and Europe. The gum trade in Kenya is less lucrative than that in Sudan and Somalia, mainly due to the poor quality of the gum resulting from the various grades and types of gum being mixed [14]. The flowers are a good source of honey, which is an important source of nutrition and income generation to support rural livelihoods in marginal lands.

Authors

Veronicah Ngumbau, Patrick Muthoka, Tiziana Ulian, Steve Davis and Efisio Mattana.

References

[1] Albrecht, J. (1993). *Tree Seed Handbook of Kenya*. GTZ Forestry Seed Centre, Muguga, Kenya.

[2] Awouda, E. H. M. (1974). *Production and Supply of Gum Arabic*. Forest Department, Khartoum, Sudan.

[3] Badji, S., Ndiaye, I., Danthu, P. & Colonna, J. P. (1991). Vegetative propagation studies of gum arabic trees. 1. Propagation of *Acacia senegal* (L.) Willd. using lignified cuttings of small diameter with eight nodes. *Agroforestry Systems* 14 (3): 183–191.

[4] Baskin, C. C. & Baskin, J. M. (2014). *Seeds: Ecology, Biogeography, and Evolution of Dormancy and Germination*, 2nd edition. Academic Press, San Diego, USA.

[5] Boer, E. (2002). *Acacia senegal* (L.) Willd. In: *PROTA (Plant Resources of Tropical Africa/Ressources Végétales de l'Afrique Tropicale)*. Edited by L. P. A. Oyen & R. H. M. J. Lemmens. Wageningen, The Netherlands. http://www.prota4u.org

[6] Brenan, J. P. M. (1959). Leguminosae Subfamily Mimosoideae. In: *Flora of Tropical East Africa*. Edited by C. E. Hubbard & E. Milne-Redhead. Crown Agents for Oversea Governments and Administrations, London.

[7] Craven, P. & Loots, S. (2009). Namibia. In: *Southern African Plant Red Data Lists*. Edited by J. S. Golding. Southern African Botanical Diversity Network Report No. 14, SABONET, Pretoria, South Africa. pp. 61–92.

[8] Danthu, P., Leblanc, J. M., Badji, S. & Colonna, J. P. (1992). Vegetative propagation studies of gum arabic trees. 2. The vegetative propagation of adult *Acacia senegal*. *Agroforestry Systems* 19 (1): 15–25.

[9] Danthu, P., Roussel, J., Dia, M. & Sarr, A. (1992). Effect of different pre-treatments on the germination of *Acacia senegal* seeds. *Seed Science and Technology* 20: 111–117.

[10] Duke, J. A. (1981). *Handbook of Legumes of World Economic Importance*. Plenum Press, New York, USA.

[11] IUCN (2017). *The IUCN Red List of Threatened Species*. Version 2017-3. http://www.iucnredlist.org/

[12] Jøker, D. (2000). *Acacia senegal* (L.) Willd. *Seed Leaflet* No. 5. Danida Forest Seed Centre, Humlebaek, Denmark.

[13] Kyalangalilwa, B., Boatwright, J. S., Daru, B. H., Maurin, O. & Van der Bank, M. (2013). Phylogenetic position and revised classification of *Acacia* s.l. (Fabaceae: Mimosoideae) in Africa, including new combinations in *Vachellia* and *Senegalia*. *Botanical Journal of the Linnean Society* 172 (4): 500–523.

[14] Maundu, P. M., Ngugi, G. W. & Kabuye, C. H. S. (1999). *Traditional Food Plants of Kenya*. Kenya Resource Centre for Indigenous Knowledge (KENRIK), National Museums of Kenya, Nairobi, Kenya.

[15] Morton, J. F. (1977). *Major Medicinal Plants: Botany, Culture, and Uses*. C.C. Thomas, Springfield, Illinois, USA.

[16] Orwa, C., Mutua, A., Kindt, R., Jamnadass, R. & Simons, A. (2009). *Agroforestree Database: A Tree Reference and Selection Guide*. Version 4.0. World Agroforestry Centre, Kenya. http://www.worldagroforestry.org/output/agroforestree-database

[17] Raimondo, D., von Staden, L., Foden, W., Victor, J. E., Helme, N. A., Turner, R. C., Kamundi, D. A. & Manyama, P. A. (2009). *Red List of South African Plants*. *Strelitzia* 25, South African National Biodiversity Institute, Pretoria, South Africa.

[18] Royal Botanic Gardens, Kew (1999–2015). *Survey of Economic Plants for Arid and Semi-Arid Lands (SEPASAL) Database*. http://apps.kew.org/sepasalweb/sepaweb

[19] Royal Botanic Gardens, Kew (2017). *Seed Information Database (SID)*. Version 7.1. http://data.kew.org/sid/

Senna didymobotrya (Fresen.) H.S.Irwin & Barneby

TAXONOMY AND NOMENCLATURE

FAMILY. Leguminosae – Caesalpinioideae

SYNONYMS. *Cassia didymobotrya* Fresen., *C. nairobensis* Hort. ex L.H.Bailey, *C. nairobiensis* L.H.Bailey, *C. verdickii* De Wild.

VERNACULAR / COMMON NAMES. Popcorn bush (English); mwinu (Swahili); inyumganai (Kamba); mwino (Kikuyu); senetwet (Kipsigis, Marakwet, Nandi); mobeno (Kisii); luvinu (Luhya); ovino (Luo); osenetoi (Maasai); kirao (Meru); mbinu, mshua (Taita); senetiet (Tugen).

Figure 1 – General habit and flower of *S. didymobotrya*. (Photo: V. Otieno, KEFRI.)

PLANT DESCRIPTION

LIFE FORM. Shrub or tree.

PLANT. Rather soft-wooded, 2–4 m tall, foetid when fresh, branches dark brown, sparsely pubescent, rarely subglabrous. **LEAVES.** Simply paripinnate, narrowly oblong elliptic, <25 cm long, with 6–18 pairs of leaflets, each 2–6.5 x 0.6–2.5 cm. **FLOWERS.** In erect axillary racemes <50 cm long, bright yellow, petals 5, slightly unequal, at first incurved, later more spreading, ovate to obovate, each c. 2–3 x 1–1.5 cm, with a slender claw c. 1 mm long (Figure 1) [2,4,11].

FRUITS AND SEEDS

Fruits are flat, yellow brown legumes, containing c. 20 seeds [8], corrugated, very tardily dehiscent along sutures [11]. Seeds (Figure 2) are brown, c. 7 x 4 mm, oblong, flattened, apiculate at the proximal end, with a narrowly obovate to oblong pleurogram in the centre of each face [11].

FLOWERING AND FRUITING. Flowers in Kenya from January to February and from June to December [2].

Figure 2 – External views of the seeds showing the pleurogram and the pointed hilar end; longitudinal and cross-sections of the seeds showing the cotyledons, the embryonic axis and the endosperm. (Photos: P. Gómez-Barreiro, RBG Kew.)

DISTRIBUTION

Widespread in tropical Africa [4,11]. In Kenya, mainly recorded from the west, south and central regions [2]. Introduced and naturalised in many parts of Asia, Australasia, Central America, and on some Pacific Ocean islands [5].

HABITAT

Riverine, by lakeshores, damp forest edges [2], swamps, roadsides and ruderal sites [11], at 700–2,250 m a.s.l [2].

USES

Senna didymobotrya is widely used in traditional medicine. A decoction of the leaves, stems and roots is strongly purgative and emetic. The leaves are used to treat gonorrhoea, back pain and stomach problems. The roots are used as an antidote for poisonings. A decoction of the roots and leaves is used to induce vomiting in treating malarial fever and headaches. In traditional veterinary medicine, the leaves, stems and roots are used for treating various cattle diseases [7]. The stems and branches are used as firewood [8]. The plant is used as a green manure and cover crop, and provides shade in tea plantations. The species is widely planted as an ornamental [9].

KNOWN HAZARDS AND SAFETY

Poisonous to humans and livestock [12]. Decoctions of the leaves, stems and roots can cause severe vomiting and diarrhoea, and can be fatal if doses are exceeded [7]. The leaves are poisonous to cattle and sheep [12]. This species can be invasive when introduced to areas outside its natural range, forming extensive impenetrable thickets and out-competing native vegetation [3].

CONSERVATION STATUS

This widespread species is not considered to be either rare or threatened. It has been assessed as Lower Risk-Least Concern (LR-lc) according to previous IUCN Red List criteria [11], but its current status has yet to be confirmed in the IUCN Red List [6].

SEED CONSERVATION

HARVESTING. Mature seeds are harvested by hand-picking the pods immediately before natural dispersal.

PROCESSING AND HANDLING. Mature dry pods shatter easily, thereby dispersing the seeds. Dry pods can be threshed in a gunny bag (hessian sack) with a stick, or by means of a flailing thresher [8].

STORAGE AND VIABILITY. Seeds are reported to be desiccation tolerant and after appropriate drying can therefore be stored for long-term conservation at sub-zero temperatures [10]. If seeds are kept dry and free of insects, they can be stored at room temperature for three years without significant loss of viability [8].

PROPAGATION

SEEDS

Dormancy and pre-treatments: Seeds of this species are reported to be physically dormant [1]. They should be scarified (e.g. by manually chipping with a scalpel) before sowing [10].

Germination, sowing and planting: Germination tests carried out under laboratory controlled conditions on seed lots stored at the MSB resulted in high germination rates (>80%) when chipped seeds were incubated with 8 hours of light per day, either at constant temperatures of 20°C or 25°C or under an alternating temperature regime of 15/30°C [10].

VEGETATIVE PROPAGATION

No protocols available.

TRADE

No data available.

Authors

Desterio O. Nyamongo, Steve Davis and Efisio Mattana.

References

[1] Baskin, C. C. & Baskin, J. M. (2014). *Seeds: Ecology, Biogeography, and Evolution of Dormancy and Germination*, 2nd edition. Academic Press, San Diego, USA.

[2] Beentje, H. J. (1994). *Kenya Trees, Shrubs and Lianas.* National Museums of Kenya, Nairobi, Kenya.

[3] BioNET-EAFRINET (2017). *Senna didymobotrya* (African senna). In: *BioNET-EAFRINET Keys and Fact Sheets: Invasive Plants.* http://keys.lucidcentral.org/keys/v3/eafrinet/plants.htm

[4] Brenan, J. P. M. (1967). Leguminosae Subfamily Caesalpinioideae. In: *Flora of Tropical East Africa.* Edited by E. Milne-Redhead & R. M. Polhill. Crown Agents for Oversea Governments and Administrations, London.

[5] ILDIS (2006–2013). *World Database of Legumes.* International Legume Database and Information Service. http://www.legumes-online.net/ildis/aweb/database.htm

[6] IUCN (2017). *The IUCN Red List of Threatened Species.* Version 2017-3. http://www.iucnredlist.org/

[7] Kokwaro, J. O. (2009). *Medicinal Plants of East Africa*, 3rd edition. University of Nairobi Press, Nairobi, Kenya.

[8] Msanga, H. P. (1998). *Seed Germination of Indigenous Trees in Tanzania: Including Notes on Seed Processing and Storage, and Plant Uses.* Canadian Forest Service, Northern Forestry Centre, Edmonton, Canada.

[9] Orwa, C., Mutua, A., Kindt, R., Jamnadass, R. & Simons, A. (2009). *Agroforestree Database: A Tree Reference and Selection Guide.* Version 4.0. World Agroforestry Centre, Kenya. http://www.worldagroforestry.org/output/agroforestree-database

[10] Royal Botanic Gardens, Kew (2017). *Seed Information Database (SID).* Version 7.1. http://data.kew.org/sid/

[11] Timberlake, J. R., Pope, G. V., Polhill, R. M. & Martins, E. S. (eds) (2007). Leguminosae (Caesalpinoideae). In: *Flora Zambesiaca*, Vol. 3, part 2. Flora Zambesiaca Managing Committee, Royal Botanic Gardens, Kew.

[12] Watt, J. M. & Breyer-Brandwijk, M. G. (1962). *The Medicinal and Poisonous Plants of Southern and Eastern Africa*, 2nd edition. E. & S. Livingstone, Edinburgh & London.

Sesbania sesban (L.) Merr.

TAXONOMY AND NOMENCLATURE

FAMILY. Leguminosae – Papilionoideae

SYNONYMS. *Aeschynomene sesban* L., *Sesbania aegyptiaca* Poir., *S. confaloniana* (Chiov.) Chiov., *S. pubescens* sensu auct.

VERNACULAR / COMMON NAMES. Common sesban, Egyptian rattle pod (English); daisa (Boran); kinukamuhondo (Giriama); munyongo (Kamba); natiatia (Keiyo); mwethia (Kikuyu); omoisabisabi (Kisii); kumusubasubi, lukhule (Luhya); chisubasubi (Luhya, Bukusu dialect); oyieko, sawosawo (Luo); walbaiyondet (Nandi); loiyangalani (Samburu).

PLANT DESCRIPTION

LIFE FORM. Tree or shrub.

PLANT. Short-lived, soft-wooded, 1–9 m tall [3], stems usually pubescent at first, becoming glabrous. **LEAVES.** Compound, 2–18.5 cm long, rachis <12 cm long, 6–27 pairs of leaflets, 0.5–2.6 cm x 2–5.5 mm, oblong, obtuse to emarginated at apex, glabrous or somewhat pubescent beneath, stipules 7 mm (Figure 1). **FLOWERS.** Pale yellow spotted with violet or uniformly yellow or suffused with purple [6,13]; racemes 15 cm long with 2–20 flowers, petals <1.8 cm long [3].

Figure 1 – Branchlets with pods and leaves. (Photo: D. O. Nyamongo, KALRO.)

FRUITS AND SEEDS

Pods (Figure 1) are straight or slightly curved, 20–30 cm long, but usually shorter, and 0.4 cm wide [6,13]. Seeds (Figure 2) olive green or brown, usually mottled purplish or black, subcylindric, c. 4 x 2–2.25 x 2 mm [13], with an axial-bent embryo and endosperm present [12].

FLOWERING AND FRUITING. Flowers almost continuously throughout the year, with a peak in Loitokitok during February and March; seeds mature in June and July. In Siaya, Kitale and Kisumu, flowering occurs in October. In Kakamega, flowering occurs in November; seeds mature in March [1].

Figure 2 – External views and cross-section of the seeds showing the cotyledons and the endosperm. (Photos: P. Gómez-Barreiro, RBG Kew.)

DISTRIBUTION

Widely distributed from Senegal to Somalia and south to South Africa; also cultivated throughout tropical Africa and Asia [5,9]. In Kenya, it is widely distributed in Kakamega, Kisumu, Siaya, Kitale, Kisii, South Nyanza, Nyeri and Loitokitok [1].

HABITAT

River and stream margins, lakeshores, wet and swampy ground; only away from water in areas of high and frequent rainfall, at 30–2,200 m a.s.l. [6,13].

USES

In traditional medicine, the leaves are used as a poultice to treat swellings and inflammation. Ground leaves are used for stomach problems. A leaf infusion is used to treat pain after childbirth and to promote placental expulsion. In traditional veterinary medicine, the leaves are used to produce a paste to increase milk production in cows and to treat eye inflammation in livestock [8]. The wood is used as firewood and for making charcoal. The pods and leaves are used for fodder. Soap can be produced from the leaves, and the bark is a source of fibre (e.g. for making rope and fishing nets) [1]. *Sesbania sesban* is used as a shade tree (e.g. in coffee and tea plantations) [3], green manure and soil stabiliser, and for intercropping [11].

KNOWN HAZARDS AND SAFETY

The seeds contain canavanine which is toxic [4] and the leaf meal is poisonous to poultry [11].

CONSERVATION STATUS

This widespread species is not considered to be either rare or threatened. Assessed as Least Concern (LC) in South Africa according to IUCN Red List criteria [14], but its status has yet to be confirmed in the IUCN Red List [7].

SEED CONSERVATION

HARVESTING. Mature brown pods can be hand-picked from the crown (Figure 3).

PROCESSING AND HANDLING. Pods can be sun-dried and seeds extracted by light threshing. Pods that do not split can be opened by hand. After extraction, seeds can be cleaned by hand-sorting, winnowing or sieving [1].

STORAGE AND VIABILITY. Seeds are reported to be desiccation tolerant [12] and after appropriate drying can therefore be stored at sub-zero temperatures for long-term conservation. Seeds can be stored in airtight containers in a cool dry place for short- to medium-term storage without significant loss of viability [1].

Figure 3 – Seed collecting. (Photo: A. McRobb, © RBG Kew.)

PROPAGATION

SEEDS

Dormancy and pre-treatments: Seeds of this species are reported to be physically dormant [2]. Before sowing, they should be soaked in hot water at 80°C for 8 minutes [15] or scarified by chipping with a scalpel [12].

Germination, sowing and planting: Germination tests carried out under laboratory controlled conditions on seed lots stored at the MSB achieved high germination rates (>85%) when chipped. Seeds were incubated in the light at constant temperatures between 10°C and 25°C [12]. In the nursery, seeds can germinate within 2–10 days. The expected germination rate for mature, healthy and properly treated seed lots is 60–80% [1].

VEGETATIVE PROPAGATION

Can be propagated using wildings [10]. Propagation from stem cuttings is possible, but not widely practiced.

TRADE

No data available.

Authors

Desterio O. Nyamongo, Steve Davis and Efisio Mattana.

References

[1] Albrecht, J. (1993). *Tree Seed Handbook of Kenya*. GTZ Forestry Seed Centre, Muguga, Kenya.

[2] Baskin, C. C. & Baskin, J. M. (2014). *Seeds: Ecology, Biogeography, and Evolution of Dormancy and Germination*, 2nd edition. Academic Press, San Diego, USA.

[3] Beentje, H. J. (1994). *Kenya Trees, Shrubs and Lianas*. National Museums of Kenya, Nairobi, Kenya.

[4] Burkill, H. M. (1995). *The Useful Plants of West Tropical Africa, Vol. 3: Families J–L*. 2nd edition. Royal Botanic Gardens, Kew.

[5] Conservatoire et Jardin botaniques & South African National Biodiversity Institute (SANBI) (2016). *African Plant Database*. Version 3.4.0. http://www.ville-ge.ch/musinfo/bd/cjb/africa/

[6] Gillett, J. B., Polhill, R. M. & Verdcourt, B. (1971). Leguminosae Subfamily Papilionoideae. In: *Flora of Tropical East Africa*. Edited by E. Milne-Redhead & R. M. Polhill. Crown Agents for Oversea Governments and Administrations, London.

[7] IUCN (2017). *The IUCN Red List of Threatened Species*. Version 2017-3. http://www.iucnredlist.org/

[8] Kokwaro, J. O. (2009). *Medicinal Plants of East Africa*, 3rd edition. University of Nairobi Press, Nairobi, Kenya.

[9] Maundu, P. & Tengnäs, B. (eds) (2005). *Useful Trees and Shrubs for Kenya*. Technical Handbook No. 35, World Agroforestry Centre, Nairobi, Kenya.

[10] Mbuya, L. P., Msanga, H. P., Ruffo, C. K., Birnie, A. & Tengnäs, B. (1994). *Useful Trees and Shrubs for Tanzania: Identification, Propagation and Management for Agricultural and Pastoral Communities*. Regional Soil Conservation Unit (RCSU) and Swedish International Development Authority (SIDA), Nairobi, Kenya.

[11] Orwa, C., Mutua, A., Kindt, R., Jamnadass, R. & Simons, A. (2009). *Agroforestree Database: A Tree Reference and Selection Guide*. Version 4.0. World Agroforestry Centre, Kenya. http://www.worldagroforestry.org/output/agroforestree-database

[12] Royal Botanic Gardens, Kew (2017). *Seed Information Database (SID)*. Version 7.1. http://data.kew.org/sid/

[13] Timberlake, J. R., Pope, G. V., Polhill, R. M. & Martins, E. S. (eds) (2007). Leguminosae (Papilionoideae). In: *Flora Zambesiaca*, Vol. 3, part 3. Flora Zambesiaca Managing Committee, Royal Botanic Gardens, Kew.

[14] van der Colff, D. (2015). *Sesbania sesban* (L.) Merr. subsp. *sesban*. In: *National Assessment: Red List of South African Plants*. Version 2017.1. http://redlist.sanbi.org/

[15] Wang, Y. R. & Hanson, J. (2008). An improved method for breaking dormancy in seeds of *Sesbania sesban*. *Experimental Agriculture* 44 (2): 185–195.

Stereospermum kunthianum Cham.

TAXONOMY AND NOMENCLATURE

FAMILY. Bignoniaceae

SYNONYMS. *Bignonia lanata* R.Br. ex Fresen., *Dolichandrone smithii* Baker, *Stereospermum arguezona* A.Rich., *S. arnoldianum* De Wild., *S. cinereoviride* K.Schum., *S. dentatum* A.Rich., *S. integrifolium* A.Rich.

VERNACULAR / COMMON NAMES. Pink jacaranda (English); mti-sumu (Swahili); mwagaivu, ndondu (Digo); mahorlu (Luhya); nyariango, pololok (Luo); nyakabur (Turkana).

PLANT DESCRIPTION

LIFE FORM. Tree or shrub.

PLANT. Deciduous, <20 m tall; bark grey to whitish, rarely dark brown, usually flaking in plaques (Figure 1). **LEAVES.** Compound, 4–13 x 2–5 cm long, opposite, leaflets elliptic or narrowly elliptic, glabrous to tomentose. **FLOWERS.** Fragrant, showy, mauve to whitish, usually pinkish with red streaks on lower corolla lobes, in large, drooping panicles on long peduncles (Figure 2); corolla tube <3 cm long, corolla lobes spreading, 3–4 cm in diameter; calyx bell-shaped, 2–5-lobed; stamens 4, enclosed within corolla tube; ovary linear-oblong, 2-chambered [2,6].

Figure 1 – General habit of *S. kunthianum*. (Photo: V. Otieno, KEFRI.)

Figure 2 – Flowers and fruits. (Photos: V. Otieno, KEFRI.)

FRUITS AND SEEDS

Fruits are flat capsules or paired pods, <60 x 1 cm, cylindric, greenish purple, reddish brown to dark brown, pendulous, spirally twisted, smooth (Figure 2), splitting into 2 valves to release many flat, long, narrow, winged seeds (Figure 3) [4].

FLOWERING AND FRUITING.

Flowering is reported to occur in Kenya from February to March, in July, and from October to December [1]. In the Sifuyo area of Kenya (Siaya County), flowers and mature fruits can be found almost simultaneously in April.

Figure 3 – External views of the winged seeds. (Photo: V. Otieno, KEFRI.)

DISTRIBUTION

Widely distributed throughout tropical Africa [2,9]. In Kenya, it is common along the southern coast, in the west, and Turkana [6].

HABITAT

Deciduous woodland, sandy savanna woodland, riverine forest [5], rocky bushland, wooded grassland and secondary bush [1], at 20–2,050 m a.s.l. [2].

USES

In traditional medicine, the bark and fruits are used to treat coughs. An infusion of the leaves is used for washing wounds and for treating malaria and ulcers. A root decoction is used for treating venereal diseases [4]. The wood is used for making poles, furniture and tool handles, and for firewood and low-quality charcoal. The flowers provide bee forage for the production of honey [6]. *Stereospermum kunthianum* has potential for planting as an ornamental [7].

KNOWN HAZARDS AND SAFETY

No known hazards.

CONSERVATION STATUS

This widespread species has yet to be assessed according to IUCN Red List criteria in Kenya and its status confirmed in the IUCN Red List [3]. Trees are being overexploited for fuelwood; protection, regeneration and planting should therefore be encouraged [6].

SEED CONSERVATION

HARVESTING. Fruits should be collected before they split open, otherwise the seeds will be dispersed by wind [6]. Greyish brown capsules can be collected from the crown by spreading a net or canvas on the ground and then climbing the tree to hand-pick the capsules. Alternatively, branches can be shaken to dislodge the fruits.

PROCESSING AND HANDLING. Mature capsules can be sun-dried to allow splitting and release of the seeds. Capsules that do not split can be either threshed or opened by hand. Seeds can be cleaned by hand-sorting.

STORAGE AND VIABILITY. Seeds are reported to be orthodox and after appropriate drying can therefore be stored at sub-zero temperatures for long-term conservation [8]. Seeds are short-lived in open storage at room temperature, remaining viable for only three months [6].

PROPAGATION

SEEDS

Dormancy and pre-treatments: No information available on seed dormancy. However, seed lots stored at the MSB did not require pre-treatments before germination [8]. No pre-treatments are required before sowing in the nursery [6].

Germination, sowing and planting: Experiments under laboratory controlled conditions on seed lots stored at the MSB highlighted high germination percentages (100%) for untreated seeds, incubated in the light at constant 20°C [8]. In the nursery, seeds germinate within 4–10 days from sowing, with a germination rate of 70–80%. Seeds can be sown in a seedbed or in polythene tubes. Seedlings are ready for potting when true leaves appear or after about four weeks. Seedlings are grown in the nursery until planted out in the field at the onset of the long rains.

Weeding is essential before planting. For woodlots, a spacing of c. 3 x 3 m is recommended. For scattered planting in cropland, trees should be 10–15 m apart. The species is fairly fast growing [6].

VEGETATIVE PROPAGATION

Can be propagated from root suckers [6].

TRADE

Products are sold in local markets. No data are available on international trade.

Authors

William Omondi, Victor Otieno, David Meroka and Efisio Mattana.

References

[1] Beentje, H. J. (1994). *Kenya Trees, Shrubs and Lianas*. National Museums of Kenya, Nairobi, Kenya.

[2] Bidgood, S., Verdcourt, B. & Vollesen, K. (2006). Bignoniaceae. In: *Flora of Tropical East Africa*. Edited by H. J. Beentje & S. A. Ghazanfar. Royal Botanic Gardens, Kew.

[3] IUCN (2017). *The IUCN Red List of Threatened Species*. Version 2017-3. http://www.iucnredlist.org/

[4] Kokwaro, J. O. (2009). *Medicinal Plants of East Africa*, 3rd edition. University of Nairobi Press, Nairobi, Kenya.

[5] Launert, E. (ed.) (1988). Bignoniaceae. In: *Flora Zambesiaca*, Vol. 8, part 3. Flora Zambesiaca Managing Committee, Royal Botanic Gardens, Kew.

[6] Maundu, P. & Tengnäs, B. (eds) (2005). *Useful Trees and Shrubs for Kenya*. Technical Handbook No. 35, World Agroforestry Centre, Nairobi, Kenya.

[7] Orwa, C., Mutua, A., Kindt, R., Jamnadass, R. & Simons, A. (2009). *Agroforestree Database: A Tree Reference and Selection Guide*. Version 4.0. World Agroforestry Centre, Kenya. http://www.worldagroforestry.org/output/agroforestree-database

[8] Royal Botanic Gardens, Kew (2017). *Seed Information Database (SID)*. Version 7.1. http://data.kew.org/sid/

[9] WCSP (2017). *World Checklist of Selected Plant Families*. Royal Botanic Gardens, Kew. http://wcsp.science.kew.org/

Tetradenia riparia (Hochst.) Codd

TAXONOMY AND NOMENCLATURE

FAMILY. Lamiaceae

SYNONYMS. *Basilicum myriostachyum* (Benth.) Kuntze, *B. riparium* (Hochst.) Kuntze, *Gumira ferruginea* (A.Rich.) Kuntze, *Iboza riparia* (Hochst.) N.E.Br., *Moschosma myriostachyum* Benth., *M. riparium* Hochst., *Plectranthus riparius* Hochst., *Premna ferruginea* A.Rich.

VERNACULAR / COMMON NAMES. Ginger bush, misty plume bush, river ginger-bush (English); thivea (Kamba); omorako (Kisii); oring-lagaldes (Maasai); lonwa (Marakwet); lonuo (Pokot); ngeliot (Samburu).

Figure 1 – Leaves of *T. riparia*. (Photo: D. O. Nyamongo, KALRO.)

PLANT DESCRIPTION

LIFE FORM. Shrub or tree.

PLANT. Deciduous, dioecious, 1–3(5) m tall, freely branched, stems semi-succulent, 4-angled at first and glandular pubescent. **LEAVES.** Petiolate, ovate oblong to rotund, sparsely or glandular-pubescent on both surfaces; veins often prominent below, apex rounded, base rounded, margins coarsely crenate (Figure 1); strongly aromatic (hence the vernacular name, ginger bush). **FLOWERS.** Terminal panicles of white to mauve flowers, diffusely branched, <8 cm long (in male spikes), 2.5 cm long (in female spikes), appearing usually after the leaves are shed; corolla 3–3.5 mm long (in male flowers), 2–2.5 mm (in female flowers), tube funnel-shaped [3,8].

FRUITS AND SEEDS

Nutlets oblong, pale brown (Figure 2), 0.6 mm long [3].

FLOWERING AND FRUITING. Flowers recorded in Kenya from February to April, and from June to August [2].

Figure 2 – External views of the nutlets. (Photo: P. Gómez-Barreiro, RBG Kew.)

DISTRIBUTION

Tropical East Africa (including Kenya) to Southern Africa [3,12]. In Kenya, recorded mostly from the west, south and central regions [2].

HABITAT

Brachystegia woodland [8], bushland on rocky slopes [2], forest margins and along streams [7,11], at 30–2,250 m a.s.l. [2,4].

USES

Used for flavouring and seasoning and to marinate meat [6]. In traditional medicine, leaf infusions are mainly used for respiratory ailments and mouth ulcers [10]. The roots are used to treat schistosomiasis, indigestion and rheumatism. The leaves are used as an emetic. A leaf concoction is drunk for oedema and as an anthelmintic. Crushed leaves are used by the Pokot to treat cataracts [2]. In Southern Africa, the Zulu have many uses for the plant including the relief of chest complaints, stomach ache and malaria [1]. A root decoction is used to treat cattle with East Coast fever [6]. *Tetradenia riparia* is also used for soil conservation and as a live fence [7].

KNOWN HAZARDS AND SAFETY

No known hazards.

CONSERVATION STATUS

This widespread and common species is assessed according to IUCN Red List criteria as Least Concern (LC) in tropical East Africa [8] and South Africa [11], but its status has yet to be confirmed in the IUCN Red List [5].

SEED CONSERVATION

HARVESTING. Seeds can be harvested by stripping the ripe fructescences by hand.

PROCESSING AND HANDLING. Seeds can be separated by rubbing the infructescences between the hands or against a rough surface, then winnowing or seed blowing.

STORAGE AND VIABILITY. Seeds of this species are reported to be desiccation tolerant [9] and can therefore be stored at sub-zero temperatures for long-term conservation.

PROPAGATION

SEEDS

Dormancy and pre-treatments: No information is available on seed dormancy. Seed propagation is not common practice as plants can be propagated vegetatively.

Germination, sowing and planting: Seed germination experiments carried out on seed lots stored at the MSB revealed high germination percentages (>90%) for seeds incubated at constant temperatures of 15–35°C [9].

VEGETATIVE PROPAGATION

Can be propagated by division or cuttings [7]. Cuttings root easily in river sand [1].

TRADE

No data available.

Authors

Desterio O. Nyamongo, Steve Davis and Efisio Mattana.

References

[1] Aubrey, A. (2001). *Tetradenia riparia* (Hochst.) Codd. South African National Biodiversity Institute (SANBI), South Africa. http://pza.sanbi.org/

[2] Beentje, H. J. (1994). *Kenya Trees, Shrubs and Lianas*. National Museums of Kenya, Nairobi, Kenya.

[3] Codd, L. E. (1983). The genus *Tetradenia* Benth. (Lamiaceae). I. African species. *Bothalia* 14 (2): 177–183.

[4] Conservatoire et Jardin botaniques & South African National Biodiversity Institute (SANBI) (2016). *African Plant Database*. Version 3.4.0. http://www.ville-ge.ch/musinfo/bd/cjb/africa/

[5] IUCN (2017). *The IUCN Red List of Threatened Species*. Version 2017-3. http://www.iucnredlist.org/

[6] Kokwaro, J. O. (2009). *Medicinal Plants of East Africa*, 3rd edition. University of Nairobi Press, Nairobi, Kenya.

[7] Maundu, P. & Tengnäs, B. (eds) (2005). *Useful Trees and Shrubs for Kenya*. Technical Handbook No. 35, World Agroforestry Centre, Nairobi, Kenya.

[8] Paton, A. J., Bramley, G., Ryding, O., Polhill, R. M., Harvey, Y. B., Iwarson, M., Willis, F., Phillipson, P. B., Balkwill, K., Lukhoba, C. W., Otiend, D. F. & Harley, R. M. (2009). Lamiaceae (Labiatae). In: *Flora of Tropical East Africa*. Edited by H. J. Beentje, S. A. Ghazanfar & R. M. Polhill. Royal Botanic Gardens, Kew.

[9] Royal Botanic Gardens, Kew (2017). *Seed Information Database (SID)*. Version 7.1. http://data.kew.org/sid/

[10] Van Wyk, B.-E., Van Oudtshoorn, B. & Gericke, N. (2009). *Medicinal Plants of South Africa*, 2nd edition. Briza Publications, Pretoria, South Africa.

[11] von Staden, L. (2016). *Tetradenia riparia* (Hochst.) Codd. In: *National Assessment*: *Red List of South African Plants*. Version 2017.1. http://redlist.sanbi.org/

[12] WCSP (2017). *World Checklist of Selected Plant Families*. Royal Botanic Gardens, Kew. http://wcsp.science.kew.org/

Vachellia tortilis (Forssk.) Galasso & Banfi

TAXONOMY AND NOMENCLATURE

FAMILY. Leguminosae–Caesalpinioideae

SYNONYMS. *Acacia tortilis* (Forssk.) Hayne, *Mimosa tortilis* Forssk.

VERNACULAR / COMMON NAMES. Umbrella thorn (English); mgunga (Swahili); dadacha (Boran); mugaa (Embu, Mbeere, Meru); mulaa (Kamba); chebitet (Kipsigis); ol-gorete (Maasai); ses (Marakwet, Pokot); sesya (Pokot); ltepes (Samburu); abak (Somali); sietsiet (Tugen); ewoi (Turkana).

PLANT DESCRIPTION

LIFE FORM. Tree.

PLANT. Thorny, <4 m tall when growing in drylands, up to 20 m in riparian vegetation; crown narrow in young trees, later spreading, flat-topped or umbrella-like (Figure 1); bark longitudinally fissured, dark grey; spines in pairs at nodes, of two kinds: straight, 3–8 cm long, and hooked, <7 mm long. **LEAVES.** Biparipinnate; rachis short, 2–10 pairs of pinnae, with 6–19 pairs of leaflets. **FLOWERS.** Whitish or cream, fragrant, in axillary heads [3,4].

Figure 1 – Habit of *V. tortilis*. (Photo: V. Otieno, KEFRI.)

FRUITS AND SEEDS

Pods are indehiscent, pale brown, spirally twisted or contorted, glabrous or pubescent [4]. Seeds are olive to red brown, smooth, elliptic, compressed, 7 x 4.5–6 mm with an areole of 5–6 x 3–4 mm (Figure 2) [4].

FLOWERING AND FRUITING. Flowering occurs in two distinct periods prior to the start of the short and long rains. In the Rift Valley, the first flowering season occurs in September before the short rains; the second flowering season occurs in April–May before the long rains. Seeds mature in February and in August–September, for the two flowering seasons, respectively. In the Eastern Province of Kenya, flowering occurs in October before the short rains and in February–March before the long rains. In this region, seeds mature in January–February and in August–September [1].

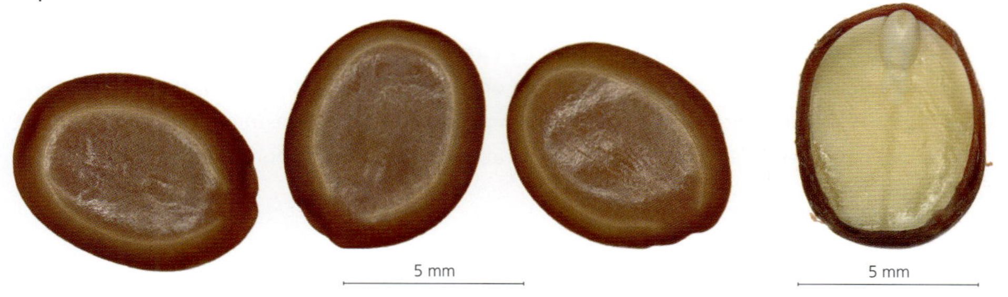

Figure 2 – External views of the seeds showing the areole surrounded by the pleurogram, and longitudinal section showing the embryonic axis and the cotyledon. (Photos: P. Gómez-Barreiro, RBG Kew.)

DISTRIBUTION

Widespread throughout the savanna and dry zones of Africa, from Algeria and Egypt in the north to South Africa; also native in Israel and the Arabian Peninsula. *Vachellia tortilis* has been introduced to India, Iran and Pakistan [6].

HABITAT

Savanna and dry bushland, sometimes occurring as pure stands or mixed with other species, especially *Commiphora* and *Terminalia* spp., semi-desert scrub, and along streams [3]. Favours alkaline soils and sand dunes [11] in areas with 200–900 mm annual rainfall and temperatures up to 50°C, but can grow in areas receiving as little as 40 mm annual rainfall [8]. The long taproot and numerous lateral roots enable the tree to tolerate drought [11].

USES

The Turkana remove the seeds and use the pods to make porridge, while the Maasai eat the immature seeds [11]. The gum is also edible [3]. *Vachellia tortilis* is an important source of fodder for cattle, and the leaves and fruits are browsed by goats and other livestock. Pokot and Maasai use a bark decoction to treat stomach ache and diarrhoea [3]. The wood is used for house construction, and for making boxes, moisture-proof plywood, gun and rifle parts, furniture and farm implements. It is suitable for firewood and charcoal. The bark provides tannins and dyes; powdered bark is used as a disinfectant for healing wounds [11]. *Vachellia tortilis* is also used for soil erosion control and dune stabilisation [13].

KNOWN HAZARDS AND SAFETY

The spines on the stems are sharp.

CONSERVATION STATUS

This widespread species is not considered to be either rare or threatened. Assessed as Least Concern (LC) in South Africa [12], but its status has yet to be confirmed in the IUCN Red List [7].

SEED CONSERVATION

HARVESTING. Pods can be collected when they change colour from green to yellow brown or light brown. Pods should be harvested directly from the tree by shaking the branches over a tarpaulin on the ground. Pods which have been lying on the ground for some time should not be harvested as the seeds may be infested [8].

PROCESSING AND HANDLING. Pods can be sun-dried under ambient conditions, and seeds extracted by pounding the pods in a mortar. Pods can also be placed inside a sisal sack and pounded with a flailing thresher (Figure 3). After threshing, the material should be sieved and cleaned by winnowing (Figure 4) or by using an air screen cleaner [8].

STORAGE AND VIABILITY. Seeds are reported to be orthodox [14] and after appropriate drying can therefore be stored at sub-zero temperatures for long-term conservation. In addition, properly dried seeds can be stored in airtight containers for at least one year at room temperature and for several years at 10°C [1].

Figure 3 – Seed processing by threshing. (Illustration: N. Muema, NMK.)

Figure 4 – Seed cleaning by winnowing. (Illustration: N. Muema, NMK.)

PROPAGATION

SEEDS

Dormancy and pre-treatments: Seeds of this species are reported to have physical dormancy [2]. Seed dormancy can be broken by soaking the seeds in warm water, preferably at 35–40°C for 24 hours prior to germination. Alternatively, seeds can be scarified by soaking in 50% concentrated sulphuric acid solution for 40–50 minutes, followed by washing in cold water and then drying overnight prior to sowing. Seeds of various sizes may differ in their scarification requirements [8].

Germination, sowing and planting: High germination percentages (>85%) have been achieved in the laboratory under controlled conditions for scarified seeds incubated with 8 or 12 hours of light per day at temperatures of 20–25°C [14]. In the nursery, seeds can be sown in pots or sown directly in the ground [11]. Without pre-treatment, seeds germinate poorly, attaining a germination rate of only 2–5% after 60 days [1].

VEGETATIVE PROPAGATION

Preliminary experiments indicate that propagation by stem cuttings is possible using hormone rooting powder [5].

TRADE

Vachellia tortilis (more widely known as *Acacia tortilis*) [9] is the most important acacia among rural communities in Kenya. Pods are sold for fodder (in Lodwar in the Turkana region and in Mandera in the North Eastern Province) and for human food. Firewood and charcoal derived from *V. tortilis* are widely sold in small market centres in Kenya [10].

Authors

Veronicah Ngumbau, Patrick Muthoka, Tiziana Ulian, Steve Davis and Efisio Mattana.

References

[1] Albrecht, J. (1993). *Tree Seed Handbook of Kenya*. GTZ Forestry Seed Centre, Muguga, Kenya.

[2] Baskin, C. C. & Baskin, J. M. (2014). *Seeds: Ecology, Biogeography, and Evolution of Dormancy and Germination*, 2nd edition. Academic Press, San Diego, USA.

[3] Beentje, H. J. (1994). *Kenya Trees, Shrubs and Lianas*. National Museums of Kenya, Nairobi, Kenya.

[4] Brenan, J. P. M. (1959). Leguminosae Subfamily Mimosoideae. In: *Flora of Tropical East Africa*. Edited by C. E. Hubbard & E. Milne-Redhead. Crown Agents for Oversea Governments and Administrations, London.

[5] Hassan Elnour Adam & Mohamed Elnour Taha (2008). *Vegetative Propagation Study Using IAA on Some of Sudanese Forest Trees in Kordofan, Sudan*. Conference Poster, Technical University of Dresden, Germany. www.tropentag.de/2008/abstracts/posters/552.pdf

[6] ILDIS (2006–2013). *World Database of Legumes*. International Legume Database and Information Service. http://www.legumes-online.net/ildis/aweb/database.htm

[7] IUCN (2017). *The IUCN Red List of Threatened Species.* Version 2017-3. http://www.iucnredlist.org/

[8] Jøker, D. (2000). *Acacia tortilis* (Forssk.) Hayne. *Seed Leaflet* No. 19. Danida Forest Seed Centre, Humlebaek, Denmark.

[9] Kyalangalilwa, B., Boatwright, J. S., Daru, B. H., Maurin, O. & Van der Bank, M. (2013). Phylogenetic position and revised classification of *Acacia* s.l. (Fabaceae: Mimosoideae) in Africa, including new combinations in *Vachellia* and *Senegalia*. *Botanical Journal of the Linnean Society* 172 (4): 500–523.

[10] Maundu, P. M., Ngugi, G. W. & Kabuye, C. H. S. (1999). *Traditional Food Plants of Kenya.* Kenya Resource Centre for Indigenous Knowledge (KENRIK), National Museums of Kenya, Nairobi, Kenya.

[11] Orwa, C., Mutua, A., Kindt, R., Jamnadass, R. & Simons, A. (2009). *Agroforestree Database: A Tree Reference and Selection Guide*. Version 4.0. World Agroforestry Centre, Kenya. http://www.worldagroforestry.org/output/agroforestree-database

[12] Raimondo, D., von Staden, L., Foden, W., Victor, J. E., Helme, N. A., Turner, R. C., Kamundi, D. A. & Manyama, P. A. (2009). *Red List of South African Plants. Strelitzia* 25, South African National Biodiversity Institute, Pretoria, South Africa.

[13] Royal Botanic Gardens, Kew (1999–2015). *Survey of Economic Plants for Arid and Semi-Arid Lands (SEPASAL) Database*. http://apps.kew.org/sepasalweb/sepaweb

[14] Royal Botanic Gardens, Kew (2017). *Seed Information Database (SID).* Version 7.1. http://data.kew.org/sid/

Vangueria madagascariensis J.F.Gmel.

TAXONOMY AND NOMENCLATURE

FAMILY. Rubiaceae

SYNONYMS. *Canthium edule* (Vahl) Baill., *C. maleolens* Chiov., *Dondisia foetida* Hassk., *Vangueria acutiloba* Robyns, *V. commersonii* Jacq., *V. cymosa* C.F.Gaertn., *V. edulis* Vahl, *V. floribunda* Robyns, *V. robynsii* Tennant

VERNACULAR / COMMON NAMES. Smooth wild-medlar, Spanish tamarind, tamarind-of-the-Indies (English); mviru (Swahili); buriri (Boran); kikomoa (Kamba); mubiru (Kikuyu, Mbeere); msada (Kinyamwezi); komolik (Kipsigis); omuogi (Kisii); anyuka olgumi (Luo); olgumei (Maasai); komol (Marakwet); mubiru (Meru); komolwo (Pokot); ngoromusui (Samburu); kamolwet (Sebei); komolik (Tugen).

PLANT DESCRIPTION

LIFE FORM. Shrub or small tree.

PLANT. Deciduous, often multi-stemmed, crown sometimes spreading, 1.5–15 m tall (Figure 1); stems glabrous, longitudinally ridged, bark pale to dark brown or grey. **LEAVES.** Narrowly to broadly elliptic or elliptic lanceolate, 8–28 x 3.2–15 cm, apex acute to shortly acuminate, base cuneate, rounded, or less often subcordate, glabrous or sometimes young leaves pilose beneath, adult leaves sparsely pubescent. **FLOWERS.** Inflorescence 7–10-flowered, pubescent; corolla greenish, yellow or cream, tube c. 3–5 mm, lobes 3.5–4.5 mm long [10].

Figure 1 – General habit of *V. madagascariensis*. (Photo: D. O. Nyamongo, KALRO.)

FRUITS AND SEEDS

Fruits are green or greenish yellow to brownish or light reddish (Figure 2), subglobose, 2.5–5 cm in diameter, with 4–5 pyrenes (Figure 3) and woody endocarp [10].

FLOWERING AND FRUITING. Flowers throughout the year [1]; fruits available from April to May in Kitui, Kiambu and Narok, and from August to September in Baringo, Kiambu and West Pokot [5].

Figure 2 – Unripe fruits. (Photo: D. O. Nyamongo, KALRO.)

Figure 3 – External views of the pyrenes and cross-section showing the woody endocarp, endosperm and cotyledons. (Photos: P. Gómez-Barreiro, RBG Kew.)

DISTRIBUTION

Widespread in subhumid to semi-arid parts of Africa, from tropical West Africa to East Africa, and south to South Africa. Also found in Madagascar and cultivated in parts of tropical Asia [10,11].

HABITAT

Evergreen forest, riverine forest, woodland, bushland, savanna and sometimes on rock outcrops [10], from sea level to 2,400 m a.s.l. [5].

USES

The fruit is edible and used for flavouring beer [1,6]. The foliage is browsed by cattle [2]. The wood is used for making tool handles and poles for huts [6]. In traditional medicine, the roots and bark are used as an anthelminthic. The flowers attract bees, making the tree useful for honey production [7].

KNOWN HAZARDS AND SAFETY

No known hazards.

CONSERVATION STATUS

This widespread species is not considered to be either rare or threatened. Assessed as Least Concern (LC) in South Africa [8], but its status has yet to be confirmed in the IUCN Red List [3].

SEED CONSERVATION

HARVESTING. Ripe fruits can be collected by hand-picking from the crown, or by spreading a net or canvas under the tree and shaking the branches to dislodge the fruits.

PROCESSING AND HANDLING. Dry fruits can be depulped by soaking in cold water, then squeezing to release the pyrenes. Alternatively, unripe fruits can be left to ripen and then rubbed on a wire mesh.

STORAGE AND VIABILITY. Seeds of this species are probably orthodox [4,9].

PROPAGATION

SEEDS

Dormancy and pre-treatments: Pyrenes should be cracked or scarified [4,9].

Germination, sowing and planting: No information is available on optimum incubation temperature and light requirements for germination. However, experiments carried out on seed lots stored at the MSB showed high germination percentages for scarified seeds incubated at constant 20°C in the light [9].

VEGETATIVE PROPAGATION

Can be coppiced [7].

TRADE

Fruits are sold in local markets. No significant trade has been reported.

Authors

Desterio O. Nyamongo and Efisio Mattana.

References

[1] Beentje, H. J. (1994). *Kenya Trees, Shrubs and Lianas.* National Museums of Kenya, Nairobi, Kenya.

[2] Burkill, H. M. (1997). *The Useful Plants of West Tropical Africa, Vol. 4: Families M–R.* 2nd edition. Royal Botanic Gardens, Kew.

[3] IUCN (2017). *The IUCN Red List of Threatened Species.* Version 2017-3. http://www.iucnredlist.org/

[4] Maara, N. T., Karachi, M. & Ahenda, J. O. (2006). Effects of pre-germination treatments, desiccation and storage temperature on germination of *Carissa edulis, Vangueria madagascariensis* and *Ximenia americana* seeds. *Journal of Tropical Forest Science* 18 (2): 124–129.

[5] Maundu, P. & Tengnäs, B. (eds) (2005). *Useful Trees and Shrubs for Kenya.* Technical Handbook No. 35, World Agroforestry Centre, Nairobi, Kenya.

[6] Maundu, P. M., Ngugi, G. W. & Kabuye, C. H. S. (1999). *Traditional Food Plants of Kenya.* Kenya Resource Centre for Indigenous Knowledge (KENRIK), National Museums of Kenya, Nairobi, Kenya.

[7] Orwa, C., Mutua, A., Kindt, R., Jamnadass, R. & Simons, A. (2009). *Agroforestree Database: A Tree Reference and Selection Guide.* Version 4.0. World Agroforestry Centre, Kenya. http://www.worldagroforestry.org/output/agroforestree-database

[8] Raimondo, D., von Staden, L., Foden, W., Victor, J. E., Helme, N. A., Turner, R. C., Kamundi, D. A. & Manyama, P. A. (2009). *Red List of South African Plants. Strelitzia* 25, South African National Biodiversity Institute, Pretoria, South Africa.

[9] Royal Botanic Gardens, Kew (2017). *Seed Information Database (SID).* Version 7.1. http://data.kew.org/sid/

[10] Verdcourt, B. & Bridson, D. (1991). Rubiaceae (Part 3). In: *Flora of Tropical East Africa.* Edited by R. M. Polhill. A.A. Balkema, Rotterdam, The Netherlands.

[11] WCSP (2017). *World Checklist of Selected Plant Families.* Royal Botanic Gardens, Kew. http://wcsp.science.kew.org/

Vernonia amygdalina Delile

TAXONOMY AND NOMENCLATURE

FAMILY. Compositae

SYNONYMS. *Bracheilema paniculatum* R.Br., *Cacalia amygdalina* Kuntze, *Cheliusia abyssinica* Sch. Bip. ex A.Rich., *Decaneurum amygdalinum* DC., *Gymnanthemum amygdalinum* (Delile) Sch.Bip. ex Walp., *Vernonia adenosticta* Fenzl ex Walp., *V. eritreana* Klatt, *V. giorgii* De Wild., *V. randii* S.Moore, *V. vogeliana* Benth., *V. weisseana* Muschl.

VERNACULAR / COMMON NAMES. Bitter leaf, bitter-tea vernonia (English); omororia (Kisii); musuritsa (Luhya); olusia (Luo); cheburiandat (Nandi).

Figure 1 – General habit of *V. amygdalina*. (Photo: KALRO.)

PLANT DESCRIPTION

LIFE FORM. Shrub or small tree.

PLANT. Much-branched, spreading, coppice-growing, 0.5–10 m tall, trunk <40 cm in diameter (Figure 1); bark grey or brown, smooth, becoming longitudinally fissured, slash green becoming black on exposure; stems pubescent, hairs asymmetrical, T-shaped. **LEAVES.** Petiolate, elliptic, lanceolate or (broadly) ovate, 4–15(28) x 1.2–4(15) cm, base cuneate or rounded, margins minutely denticulate to coarsely serrate, apex shortly acuminate or apiculate, thinly pubescent and glabrescent except for midrib above, finely pubescent (especially in the veins) and often glabrescent beneath. **FLOWERS.** Several capitula in terminal compound corymbiform cymes; florets 10–24 per head, sweet-scented, corolla white, occasionally tinged pale lilac, mauve or pink [3] (Figure 2).

Figure 2 – Inflorescence of *V. amygdalina* (Photo: V. Otieno, KEFRI.)

FRUITS AND SEEDS

Achenes are 1.7–3.5 mm long, 10-ribbed, thinly pubescent, usually glandular; outer pappus of linear scales, often scant, caducous, 0.2–1.3 mm long, inner of cream or brownish bristles, 4–7 mm long, distally flattened [3].

FLOWERING AND FRUITING. Flowers have been recorded in April, from August to September, and from November to December [2]. However, in Bungoma and Kakamega, flowering occurs mostly from December to March, and fruiting occurs from March to April [6].

DISTRIBUTION

Widely distributed in tropical Africa, including tropical East Africa, and extending south to South Africa. Also native in Yemen [3]. In Kenya, it is common in western counties [2,6]. Widely cultivated in West Africa [10].

HABITAT

Margins of forests, rivers and lakes, woodland, wooded grassland, grassland, secondary bushland, and abandoned cultivated areas where it may form thickets, at 300–2,300 m a.s.l. [3,6].

USES

In traditional medicine, an infusion of the roots or leaves is used as an anti-trematode. The leaf juice is used to treat fever, and pounded roots or leaves are used to treat fresh wounds. The leaves are used as an antidote for snake bites. In veterinary medicine, leaf juice is used as ear drops for deworming livestock [5]. The wood is used as firewood. *Vernonia amygdalina* is planted as a live fence, for soil conservation, and as an ornamental [6]. Widely cultivated in West Africa for its edible leaves [10].

KNOWN HAZARDS AND SAFETY

No known hazards.

CONSERVATION STATUS

This widespread species is not considered to be either rare or threatened. Assessed as Least Concern (LC) in South Africa [7], but its status has yet to be confirmed in the IUCN Red List [4].

SEED CONSERVATION

HARVESTING. The mature achenes can be hand-picked.
PROCESSING AND HANDLING. Achenes are processed by hand-sorting. The pappus is removed along with any other plant material and damaged seeds.
STORAGE AND VIABILITY. Seeds of this species are reported to be desiccation tolerant [8] and after appropriate drying can therefore be stored in the seed bank at sub-zero temperatures for long-term conservation.

PROPAGATION

SEEDS

Dormancy and pre-treatments: Seeds of this species are reported to be non-dormant [1]. No pre-treatments are required [6].
Germination, sowing and planting: Under laboratory controlled conditions, seed incubated at 15°C with a photoperiod of 8 hours of light per day achieved 80% final germination [8]. When incubated under a range of temperatures (10–30°C), germination ranged from c. 45 to c. 55% with an optimum between 15–25°C. Fluctuating temperatures did not improve the final germination rate compared to that achieved with constant temperatures. Furthermore, seeds seem to be indifferent to light conditions, although exposure to light filtered through green leaves inhibited germination [9]. In the nursery, seeds collected from dry flower heads are broadcast on nursery beds prepared with humus-rich soil and shaded from excessive heat and

sunlight. Watering is essential during dry periods. Seeds germinate after 2–3 weeks, and 4–6 weeks after emergence, the seedlings are transplanted (among other crops in home gardens, as a hedge or live fence, or planted in rows in commercial fields) [10].

VEGETATIVE PROPAGATION

Stem cuttings are the preferred method of propagation by farmers. In West Africa, home gardens may contain plants with varying degrees of bitterness and other selected characteristics. Cuttings may be planted erect or at angle of 45° to obtain more sideshoots [10].

TRADE

In West and Central Africa, bitter leaf is an important vegetable usually grown for home consumption and less often for sale at the market. However, processed (dried or frozen) leaves are exported from West Africa to Europe. No trade statistics are available [10]. The timber is sold in the international market and also at a local level.

Authors

Desterio O. Nyamongo, Steve Davis and Efisio Mattana.

References

[1] Baskin, C. C. & Baskin, J. M. (2014). *Seeds: Ecology, Biogeography, and Evolution of Dormancy and Germination*, 2nd edition. Academic Press, San Diego, USA.

[2] Beentje, H. J. (1994). *Kenya Trees, Shrubs and Lianas*. National Museums of Kenya, Nairobi, Kenya.

[3] Beentje, H. J. & Smith, S. A. L. (eds) (2000). Compositae (Part 1). In: *Flora of Tropical East Africa*. A.A. Balkema, Rotterdam, The Netherlands.

[4] IUCN (2017). *The IUCN Red List of Threatened Species*. Version 2017-3. http://www.iucnredlist.org/

[5] Kokwaro, J. O. (2009). *Medicinal Plants of East Africa*, 3rd edition. University of Nairobi Press, Nairobi, Kenya.

[6] Maundu, P. & Tengnäs, B. (eds) (2005). *Useful Trees and Shrubs for Kenya*. Technical Handbook No. 35, World Agroforestry Centre, Nairobi, Kenya.

[7] Raimondo, D., von Staden, L., Foden, W., Victor, J. E., Helme, N. A., Turner, R. C., Kamundi, D. A. & Manyama, P. A. (2009). *Red List of South African Plants. Strelitzia* 25, South African National Biodiversity Institute, Pretoria, South Africa.

[8] Royal Botanic Gardens, Kew (2017). *Seed Information Database (SID)*. Version 7.1. http://data.kew.org/sid/

[9] Teketay, D. & Granström, A. (1997). Germination ecology of forest species from the highlands of Ethiopia. *Journal of Tropical Ecology* 13 (6): 805–831.

[10] Ucheck Fomum, F. (2004). *Vernonia amygdalina* Delile. In: *PROTA (Plant Resources of Tropical Africa/ Ressources Végétales de l'Afrique Tropicale)*. Edited by G. J. H. Grubben & O. A. Denton. Wageningen, The Netherlands. http://www.prota4u.org

Vitex doniana Sweet

TAXONOMY AND NOMENCLATURE

FAMILY. Lamiaceae

SYNONYMS. *Vitex chariensis* A.Chev., *V. cienkowskii* Kotschy & Peyr., *V. cuneata* Schumach. & Thonn., *V. dewevrei* De Wild. & T.Durand, *V. homblei* De Wild., *V. hornei* Hemsl., *V. pachyphylla* Baker, *V. paludosa* Vatke, *V. umbrosa* G.Don ex Sabine

VERNACULAR / COMMON NAMES. Black plum (English); mfudu (Swahili); muhuru (Kamba, Kikuyu); muekelwet (Kipsigis); mutahuru (Kisii); muholu, omuhutu (Luhya); jwelu, kalemba (Luo); muburu (Mbeere); tirkirwa (Pokot).

PLANT DESCRIPTION

LIFE FORM. Tree.

PLANT. Deciduous, much-branched, 4–8 m tall (Figure 1), bark rough, pale brown or greyish white, with narrow vertical cracks. **LEAVES.** Compound, opposite, petiole 5–17 cm long, 3–6 foliolate, leaflets 7–23 x 3–9 cm, median largest. **FLOWERS.** Inflorescences axillary (Figure 2), densely flowered, 2.5–10 cm long, corolla 8–12 mm long, densely tomentose outside, tube white to violet, lobes mauve [6,8].

Figure 1 – General habit of *V. doniana*. (Photo: H. Migiro, KEFRI.)

Figure 2 – Inflorescence of *V. doniana*. (Photo: H. Migiro, KEFRI.)

FRUITS AND SEEDS

Mature fruits are c. 1.5–3 cm long, purplish black, with a starchy black pulp. Each fruit consists of a hard, conical pyrene (stone) [6] with 1–4 seeds [7].

FLOWERING AND FRUITING. Flowers in January to February, April to May and December [3]. Fruits in July and August, and from December to February [6].

DISTRIBUTION

Widespread throughout tropical Africa from Senegal to Cameroon, and in East Africa and Angola [8]; also native to Madagascar and Comoros [10]. In Kenya, found mainly in the west; rare in Central Province and on the south coast [3,6].

HABITAT

Evergreen forest margins, swamp forest margins and riverine forest in high rainfall areas, and in coastal forest often on termite mounds, from near sea level to 1,750 m a.s.l. [8].

USES

The fruits are edible and eaten raw or made into jam. The leaves, fruits and seeds are browsed by cattle [6]. In traditional medicine, a decoction of the cooked roots is used to treat back pain in women. Pounded leaves are added to grain beer; the leaf juice is also used as a remedy for eye diseases. A bark infusion is used to clean fresh wounds; dried bark powder is mixed with hot water and drunk as a tea for epilepsy. A root decoction is used as an anthelmintic [5]. The wood is used for house building, poles, carvings, firewood and charcoal. *Vitex doniana* is planted for shade and as a boundary marker. Its flowers attract bees, and its leaves provide a good mulch [6].

KNOWN HAZARDS AND SAFETY

No known hazards.

CONSERVATION STATUS

This widespread species is assessed as Least Concern (LC) according to IUCN Red List criteria [4].

SEED CONSERVATION

HARVESTING. Mature purplish black fruits can be collected from the crown by spreading a net or canvas under the tree and climbing the tree to hand-pick the fruits; alternatively, the branches are shaken to dislodge the fruits. Lopping shears are used to cut small fruiting branches.

PROCESSING AND HANDLING. For extracting small quantities of stones, the skin and pulp can be removed with a knife and the nuts washed in water. For large quantities, the fruits may be soaked in water for 24 hours then mixed in a concrete mixer with gravel in a ratio of 1:2 by weight and large amounts of water. After some time in the mixer, the water, fruit pulp and skin fragments are poured off and the gravel is picked out by hand. Alternatively, fruits can be soaked in water for 24 hours and the stones extracted using a flailing thresher [7].

STORAGE AND VIABILITY. No information is available on seed storage behaviour [9]. Congeneric species are reported to have either orthodox or recalcitrant seeds [9], highlighting the need for further studies on the physiological responses of seeds of this species to desiccation. At room temperature, dried seeds can be stored for two years without significant loss of viability [7].

PROPAGATION

SEEDS

Dormancy and pre-treatments: Seed of this species are reported to be physiologically dormant [2]. Stones should be soaked in water for 24 hours, changing the water after 12 hours before sowing [7]. Further studies are needed to confirm the class of dormancy.

Germination, sowing and planting: No information is available on optimum conditions for germination under laboratory controlled conditions. In the nursery, germination is good but sporadic, reaching 30% after five weeks and 60% after eight weeks from sowing. Each stone may produce up to four seedlings [7]. Seedlings that are 4–6 weeks old can be pricked out, planted in pots and grown in the nursery for up to four months. Weeding is essential before planting and until the seedlings are established. Planting should be in prepared pits. For woodlots, a spacing of c. 5 x 5 m is recommended. For scattered planting in cropland, seedlings should be planted 15–20 m apart. Trees should be pruned regularly and coppice well.

VEGETATIVE PROPAGATION

Can be propagated from root suckers [6] and wildings. Experiments on stem cuttings have been carried out; further research is needed [1].

TRADE

Fruits are sold in local markets. No data are available on international trade.

Authors

Desterio O. Nyamongo and Efisio Mattana.

References

[1] Achigan Dako, E. G., N'Danikou, S., Tchokponhoue, D. A., Assogba Komlan, F., Larwanou, M., Sognon Vodouhe, R. & Ahanchede, A. (2014). Sustainable use and conservation of *Vitex doniana* Sweet: unlocking the propagation ability using stem cuttings. *Journal of Agriculture and Environment for International Development* 108 (1): 43–62.

[2] Baskin, C. C. & Baskin, J. M. (2014). *Seeds: Ecology, Biogeography, and Evolution of Dormancy and Germination*, 2nd edition. Academic Press, San Diego, USA.

[3] Beentje, H. J. (1994). *Kenya Trees, Shrubs and Lianas*. National Museums of Kenya, Nairobi, Kenya.

[4] IUCN (2017). *The IUCN Red List of Threatened Species*. Version 2017-3. http://www.iucnredlist.org/

[5] Kokwaro, J. O. (2009). *Medicinal Plants of East Africa*, 3rd edition. University of Nairobi Press, Nairobi, Kenya.

[6] Maundu, P. & Tengnäs, B. (eds) (2005). *Useful Trees and Shrubs for Kenya*. Technical Handbook No. 35, World Agroforestry Centre, Nairobi, Kenya.

[7] Msanga, H. P. (1998). *Seed Germination of Indigenous Trees in Tanzania: Including Notes on Seed Processing and Storage, and Plant Uses*. Canadian Forest Service, Northern Forestry Centre, Edmonton, Canada.

[8] Pope, G. V. & Martins, E. S. (eds) (2005). Verbenaceae. In: *Flora Zambesiaca*, Vol. 8, part 7. Flora Zambesiaca Managing Committee, Royal Botanic Gardens, Kew.

[9] Royal Botanic Gardens, Kew (2017). *Seed Information Database (SID)*. Version 7.1. http://data.kew.org/sid/

[10] WCSP (2017). *World Checklist of Selected Plant Families*. Royal Botanic Gardens, Kew. http://wcsp.science.kew.org/

Vitex keniensis Turrill

TAXONOMY AND NOMENCLATURE

FAMILY. Lamiaceae

SYNONYMS. *Vitex balbi* Chiov.

VERNACULAR / COMMON NAMES. Meru oak (English); mfuu (Swahili); muhuru (Kikuyu); muuru (Meru).

PLANT DESCRIPTION

LIFE FORM. Tree.

PLANT. Deciduous, <35 m tall (Figure 1), crown rounded, bole straight, bark grey to pale brown, with narrow vertical grooves. **LEAVES.** Compound, leaflets 5, <25 cm long, elliptic, base often unequal, light green and sandpapery above, pale green and densely hairy beneath. **FLOWERS.** 7–8 mm long, white or purplish, largest lobe dark mauve, flowers in axillary dichasia 12–18 cm long [3,6].

Figure 1 – General habit of *V. keniensis*. (Photo: D. O. Nyamongo, KALRO.)

FRUITS AND SEEDS

The fruits are drupes, roughly parallel-sided, c. 1.5 cm long, with a ± fleshy mesocarp and a hard-celled endocarp. Drupes are shiny and black when ripe. Endocarps (Figure 2) are very hard, slightly elongated, with a rounded apex at the side opposite the base, c. 1.3 x 1 cm. Each fruit consists of a hard pyrene (stone) with 1–4 seeds [7].

FLOWERING AND FRUITING. Flowers usually between January and April (and also recorded in May and November) [3]. Seeds take five months to ripen and are mostly available between June and October [1].

Figure 2 – Pyrenes after removing the fleshy mesocarp. (Photo: D. O. Nyamongo, KALRO.)

DISTRIBUTION

Tropical East Africa (including Kenya) and Mozambique [9].

HABITAT

Moist evergreen forest at 1,300–2,000 m [3], such as the montane forests east of Mt Kenya at 1,500–1,850 m a.s.l. Also planted outside of its natural range at 1,500–2,200 m a.s.l. on the south-eastern and southern slopes of Mt Kenya, on the northern slopes of Mt Kilimanjaro and Mt Elgon, in the Western Highlands, in Londiani and the Nandi Hills, and in Kakamega Forest [1].

USES

The fruits are edible [1], but in most areas they are only eaten in times of food shortage [5]. The light wood is of high quality but not durable and only moderately resistant to termites. It is mainly used for furniture, panelling, cabinet work and as fuelwood [1]. *Vitex keniensis* is a popular ornamental tree and is sometimes planted as a windbreak. The leaves provide a useful mulch [5].

KNOWN HAZARDS AND SAFETY

No known hazards.

CONSERVATION STATUS

Assessed as Vulnerable (VU) according to IUCN Red List criteria; it is threatened by overexploitation [4].

SEED CONSERVATION

HARVESTING. Although fruits should be collected only when they turn brown (i.e. when the seeds are mature), they are often predated by monkeys and birds, so green fruits that have grown to final size are also collected (by climbing the trees when c. 20% of the fruits have turned brown) [5].

PROCESSING AND HANDLING. Fruits can be packed in gunny bags (hessian sacks) and should be processed within a few days of collection. The pulp can be removed with a flailing thresher without previous soaking or by rubbing on a wire mesh [5]. Alternatively, a concrete mixer can be used to mix the fruits with ballast, tumbling for about six hours. After the pulp has been removed, the pyrenes are dried in the shade. Seeds cannot be extracted from the hard endocarp; the pyrene is the storage and germination unit. This results in up to four seedlings germinating per stone [5,6].

STORAGE AND VIABILITY. Seeds of this species are likely to be orthodox [8]. Mature and properly dried seeds can be stored in airtight containers at 3°C in a coldstore for at least one year [1].

Figure 3 – Seedlings propagated in nursery. (Photo: H. Migiro, KEFRI.)

PROPAGATION

SEEDS

Dormancy and pre-treatments: It is likely that seeds of this species have some level of physiological dormancy [2]. However, this should be confirmed experimentally under laboratory controlled conditions. No pre-treatments are required before sowing in nursery conditions, but soaking in cold water for 24 hours has been reported to improve germination [5].

Germination, sowing and planting: Under optimal conditions, seeds germinate within 18–50 days depending on the medium. In nursery conditions (Figure 3), the expected germination rate of mature and healthy seed lots is 40–60%.

VEGETATIVE PROPAGATION

No protocols available.

TRADE

The timber has little importance on the international market; most of the trade is for local use.

Authors

Desterio O. Nyamongo and Efisio Mattana.

References

[1] Albrecht, J. (1993). *Tree Seed Handbook of Kenya.* GTZ Forestry Seed Centre, Muguga, Kenya.

[2] Baskin, C. C. & Baskin, J. M. (2014). *Seeds: Ecology, Biogeography, and Evolution of Dormancy and Germination*, 2nd edition. Academic Press, San Diego, USA.

[3] Beentje, H. J. (1994). *Kenya Trees, Shrubs and Lianas.* National Museums of Kenya, Nairobi, Kenya.

[4] IUCN (2017). *The IUCN Red List of Threatened Species.* Version 2017-3. http://www.iucnredlist.org/

[5] Jøker, D. & Mngulwi, F. (2000). *Vitex keniensis* Turrill. *Seed Leaflet* No. 38. Danida Forest Seed Centre, Humlebaek, Denmark.

[6] Maundu, P. & Tengnäs, B. (eds) (2005). *Useful Trees and Shrubs for Kenya.* Technical Handbook No. 35, World Agroforestry Centre, Nairobi, Kenya.

[7] Msanga, H. P. (1998). *Seed Germination of Indigenous Trees in Tanzania: Including Notes on Seed Processing and Storage, and Plant Uses.* Canadian Forest Service, Northern Forestry Centre, Edmonton, Canada.

[8] Royal Botanic Gardens, Kew (2017). *Seed Information Database (SID).* Version 7.1. http://data.kew.org/sid/

[9] WCSP (2017). *World Checklist of Selected Plant Families.* Royal Botanic Gardens, Kew. http://wcsp.science.kew.org/

Zanthoxylum gilletii (De Wild.) P.G.Waterman

TAXONOMY AND NOMENCLATURE

FAMILY. Rutaceae

SYNONYMS. *Fagara amaniensis* Engl., *F. discolor* Engl., *F. gilletii* De Wild., *F. inaequalis* Engl., *F. iturensis* Engl., *F. kivuensis* Lebrun ex Gilbert, *F. macrophylla* (Oliv.) Engl., *F. melanorhachis* Hoyle, *F. obliquefoliolata* Engl., *F. rigidifolia* Engl., *F. tessmannii* Engl.

VERNACULAR / COMMON NAMES. African satinwood, white African mahogany (English); muchagasa, muchagatha (Kikuyu); sogowait (Kipsigis); shikhoma (Luhya); sogomaitha (Luo); sagawoita, sagowat (Nandi).

Figure 1 – General habit and close-up of the prickle-bearing protuberances and flowers. (Photos: H. Gaya, KEFRI and T. Ulian, RBG Kew.)

PLANT DESCRIPTION

LIFE FORM. Tree.

PLANT. Deciduous, <35 m tall, bole clear up to 15(–25) m (Figure 1), with many woody, prickle-bearing protuberances <3 cm long (Figure 1), but old trees often lack these; wood scented.

LEAVES. Alternate, clustered at ends of branches, imparipinnately compound, <120(–150) cm long, leaflets 13–27(–51), 8–30 x 4–10 cm; stipules absent. **FLOWERS.** Inflorescence a terminal or axillary pyramidal panicle, 20–40 cm long; flowers cream or yellow (Figure 1), unisexual (monoecious), in clusters, pentamerous, almost sessile, petals 1–2.5 mm long [2,4].

FRUITS AND SEEDS

Fruits are globose follicles 3.5–6 mm in diameter, reddish, glandular pitted, dehiscent, 1(–2)-seeded. Seeds globose, 2.5–3.5 mm in diameter, black and shiny [4,7].

FLOWERING AND FRUITING. Flowers sporadically in January and February, and from July to November. Fruits mature in November and December. Trees start flowering when 10 years old and may flower and fruit annually or biennially [7,9]. Seeds ripen about three months after flowering [1].

DISTRIBUTION

Widely distributed from Guinea and Sierra Leone east to Kenya and south to northern Angola, Zimbabwe and Mozambique [5,7]. In Kenya, it is mostly found in the west [2].

HABITAT

Moist forest, evergreen rainforest, riverine and gallery forest, wooded savanna, as well as secondary forest on old farmland, up to 2,400 m a.s.l. [5].

USES

The wood is used for timber production as well as for carpentry and all heavy construction works. The bark decoction is drunk to treat constipation, complicated gastrointestinal conditions, common cold and fever [8]. *Zanthoxylum gilletii* is occasionally planted as an ornamental shade tree [5].

KNOWN HAZARDS AND SAFETY

The prickles on the trunk and branches are sharp.

CONSERVATION STATUS

This widespread species is assessed as Vulnerable (VU) in Zimbabwe according to IUCN Red List criteria [6], but its status has yet to be confirmed in the IUCN Red List [3]. Although widespread, there is some concern over genetic erosion. Loggers target large trees with straight trunks, and these have become scarce in many regions [7].

SEED CONSERVATION

HARVESTING. The mature, red brownish fruits are collected by climbing the tree before the capsules open and the seeds disperse [1].

PROCESSING AND HANDLING. After drying the fruits in the sun for one or two days, the seeds are extracted by shaking the dried inflorescences. The extracted seeds should not be exposed to direct sun [1].

STORAGE AND VIABILITY. Seed storage behaviour is not yet known, and there are unresolved seed dormancy problems [1]. Seeds are reported to be recalcitrant [8] or probably orthodox [10]. In West Africa, germination is rapid and viability short. Seeds that are kept in the shade can be stored for up to two months [8].

PROPAGATION

SEEDS

Dormancy and pre-treatments: No successful treatment has been developed to date. It is assumed that the oily seed coat plays a role in dormancy [1], but the class of dormancy is currently unknown. The oily and hard seed coat contributes to the often poor germination. Washing seeds

Figure 2 – Seedling propagated in nursery. (Photo: H. Gaya, KEFRI.)

thoroughly with a soap solution improves germination percentage and rate [8].

Germination, sowing and planting: Without treatment, seeds germinate poorly (2–5% after 60 days [1]). Seedlings (Figure 2) are light-demanding and natural regeneration may be abundant in large gaps in the forest and in regrowth on old farmland. Germination starts three weeks after sowing. Germination rates of 20–50% have been reported, but in western Kenya, final percentages up to 80% have been reached in 75–120 days [7].

VEGETATIVE PROPAGATION

In West Africa, wildings are commonly used for planting [7].

TRADE

The timber is of little importance on the international market; most of the trade is for local use [7].

Authors

Efisio Mattana and Tiziana Ulian.

References

[1] Albrecht, J. (1993). *Tree Seed Handbook of Kenya*. GTZ Forestry Seed Centre, Muguga, Kenya.

[2] Beentje, H. J. (1994). *Kenya Trees, Shrubs and Lianas*. National Museums of Kenya, Nairobi, Kenya.

[3] IUCN (2017). *The IUCN Red List of Threatened Species*. Version 2017-3. http://www.iucnredlist.org/

[4] Kokwaro, J. O. (1982). Rutaceae. In: *Flora of Tropical East Africa*. Edited by R. M. Polhill. A.A. Balkema, Rotterdam, The Netherlands.

[5] Lebrun, J.-P. & Stork, A. L. (2010). *Tropical African Flowering Plants, Ecology and Distribution*. Vol. 5 – Buxaceae – Simaroubaceae. Conservatoire et Jardin Botaniques de la Ville de Genève, Switzerland.

[6] Mapaura, A. & Timberlake, J. R. (2002). Zimbabwe. In: *Southern African Plant Red Data Lists*. Edited by J. S. Golding. Southern African Botanical Diversity Network Report No. 14, SABONET, Pretoria, South Africa. pp. 157–182.

[7] Okeyo, M. M. (2008). *Zanthoxylum gilletii* (De Wild.) P.G.Waterman. In: *PROTA 7 (1)*: *Timbers/Bois d'œuvre 1*. [CD-Rom]. Edited by D. Louppe, A. A. Oteng-Amoako & M. Brink. PROTA, Wageningen, The Netherlands.

[8] Orwa, C., Mutua, A., Kindt, R., Jamnadass, R. & Simons, A. (2009). *Agroforestree Database: A Tree Reference and Selection Guide*. Version 4.0. World Agroforestry Centre, Kenya. http://www.worldagroforestry.org/output/agroforestree-database

[9] Royal Botanic Gardens, Kew (1999–2015). *Survey of Economic Plants for Arid and Semi-Arid Lands (SEPASAL) Database*. http://apps.kew.org/sepasalweb/sepaweb

[10] Royal Botanic Gardens, Kew (2017). *Seed Information Database (SID)*. Version 7.1. http://data.kew.org/sid/

Seed processing at IER in Sikasso.
(Photo: T. Ulian, RBG Kew.)

MALI

Alchornea cordifolia (Schumach. & Thonn.) Müll.Arg.

TAXONOMY AND NOMENCLATURE

FAMILY. Euphorbiaceae

SYNONYMS. *Alchornea cordata* Benth., *Schousboea cordifolia* Schumach. & Thonn.

VERNACULAR / COMMON NAMES. Christmas bush, dovewood (English); arbre de djeman (French); konossasa, kounaninkala, moridaba (Bambara); gimii, ngimii (Fula); diangba (Malinke).

PLANT DESCRIPTION

LIFE FORM. Tree or shrub.

PLANT. Evergreen, dioecious, laxly branched, scandent or climbing (Figure 1), 3–5(8) m high, much-branched. **LEAVES.** Simple, alternate, ovate, 5–25 x 3–15 cm, apex acute, margins dentate, base cordate. **FLOWERS.** Male and female inflorescences <45 cm long; male inflorescence an axillary panicle, female inflorescence an axillary spike or lax panicle; flowers small, c. 2 mm, greenish white, petals absent [2,7].

Figure 1 – General habit of *A. cordifolia*. (Photo: S. Sanogo, IER.)

FRUITS AND SEEDS

Fruits are bi- or tri-lobed capsules (Figure 2), somewhat flattened, c. 1 cm in diameter, reddish when ripe, with two persistent styles [2]. Seeds (Figure 3) are ovoid-ellipsoid, c. 6 mm long, smooth, bright red [7].

FLOWERING AND FRUITING. In West Africa, flowers during the first half of the dry season [2].

Figure 2 – Unripe fruits. (Photo: S. Sanogo, IER.)

Figure 3 – External views and longitudinal and cross-sections of the seeds showing the endosperm and the spatulate embryo. (Photos: P. Gómez-Barreiro, RBG Kew.)

DISTRIBUTION

Tropical Africa, from Senegal to Kenya and Tanzania, southwards throughout Central Africa to Angola [14].

HABITAT

Savanna, riverine forest, riverbanks and swamps, often forming thickets in disturbed unburned places, at 650–1,500 m a.s.l. [5].

USES

This species has many medicinal uses throughout its range [2,4,7]. Extracts from the leaves are used to treat diarrhoea [1,10] and may be beneficial in treating inflammatory diseases [9]. The leaves are also used for coughs. The roots and leaves are purgative and used for treating malaria and skin diseases [10]. Plant extracts have significant hypoglycaemic and erythropoietin effects in diabetic animals [8]. The leaves and stem bark are used to treat urinary, respiratory and gastrointestinal disorders [7,10]. In West Africa, the leaves are a source of fodder for small ruminants and poultry [7]. The fruits and leaves provide a black dye used for mats, fabrics and leather [12]. *Alchornea cordifolia* is used as an alley crop for *in situ* mulch production in banana and maize plantations in West and East Africa, and is also used as a windbreak around crops [7].

KNOWN HAZARDS AND SAFETY

No known hazards.

CONSERVATION STATUS

This widespread species is assessed as Least Concern (LC) in Burkina Faso on the basis of its distribution and habitat [12], but its status has yet to be confirmed in the IUCN Red List [6].

SEED CONSERVATION

HARVESTING. Ripe fruits can be collected directly from the plants.
PROCESSING AND HANDLING. Ripe fruits should be kept in paper bags and stored in a shaded and dry place until delivered to the seed bank where seed extraction can take place.
STORAGE AND VIABILITY. Seeds are reported to be orthodox [11]. Therefore, after appropriate drying, they can be stored at sub-zero temperatures for long-term conservation.

PROPAGATION

SEEDS

Dormancy and pre-treatments: No information on seed dormancy is available. However, physiological dormancy has been reported for three congeneric species [3]. Further research is needed.
Germination, sowing and planting: No information is available on seed germination under laboratory controlled conditions; the optimum incubation conditions are unknown. However, in the nursery, germination takes 3–12 weeks when seeds are planted directly in moist soil [7].

VEGETATIVE PROPAGATION

Can be propagated very easily from stem cuttings, which root in nine weeks [7].

PHYTOCHEMICAL ANALYSES

Phytochemical analyses of leaf decoctions, carried out by DMT/INRSP in Mali as part of the MGU–Useful Plants Project, helped to highlight the presence of very strong antibacterial and low antifungal and antiplasmodial properties. The lack of toxicity of decoctions of the leaves and bark has been confirmed by the analyses; coumarins, flavonoids, saponins, tannins, mucilage and leucoanthocyanins were found to be present in high amounts [13].

TRADE

The leaves, root bark and fruits are sold in local markets in Ghana and Burkina Faso from November to January [7].

Authors

Sidi Sanogo, Abdoul K. Sanogo, Bokary A. Kelly, Rokia Sanogo, Stéphane Rivière, Efisio Mattana and Tiziana Ulian.

References

[1] Agbor, G. A., Léopold, T. & Jeanne, N. Y. (2004). The antidiarrhoeal activity of *Alchornea cordifolia* leaf extract. *Phytotherapy Research* 18 (11): 873–876.

[2] Arbonnier, M. (2004). *Trees, Shrubs and Lianas of West African Dry Zones*. CIRAD, France; Margraf Publishers, Weikersheim, Germany; and Muséum national d'histoire naturelle (MNHN), Paris, France.

[3] Baskin, C. C. & Baskin, J. M. (2014). *Seeds: Ecology, Biogeography, and Evolution of Dormancy and Germination*, 2nd edition. Academic Press, San Diego, USA.

[4] Burkill, H. M. (1994). *The Useful Plants of West Tropical Africa, Vol. 2: Families E–I*. 2nd edition. Royal Botanic Gardens, Kew.

[5] Conservatoire et Jardin botaniques & South African National Biodiversity Institute (SANBI) (2016). *African Plant Database*. Version 3.4.0. http://www.ville-ge.ch/musinfo/bd/cjb/africa/

[6] IUCN (2017). *The IUCN Red List of Threatened Species*. Version 2017-3. http://www.iucnredlist.org/

[7] Mavar-Manga, H., Lejoly, J., Quetin-Leclercq, J. & Schmelzer, G. H. (2007). *Alchornea cordifolia* (Schumach. & Thonn.) Müll.Arg. In: *PROTA (Plant Resources of Tropical Africa/Ressources Végétales de l'Afrique Tropicale)*. Edited by G. H. Schmelzer & A. Gurib-Fakim. Wageningen, The Netherlands. http://www.prota4u.org

[8] Mohammed, R. K., Ibrahim, S., Atawodi, S. E., Eze, E. D., Suleiman, J. B., Ugwu, M. N. & Malgwi, I. S. (2013). Anti-diabetic and haematological effects of n-butanol fraction of *alchornea* [*sic*] *cordifolia* leaf extract in streptozotocin-induced diabetic wistar rats. *Scientific Journal of Biological Sciences* 2 (3): 45–53.

[9] Osadebe, P. O. & Okoye, F. B. C. (2003). Anti-inflammatory effects of crude methanolic extract and fractions of *Alchornea cordifolia* leaves. *Journal of Ethnopharmacology* 89 (1): 19–24.

[10] Quattrocchi, U. (2012). *CRC World Dictionary of Medicinal and Poisonous Plants: Common Names, Scientific Names, Eponyms, Synonyms, and Etymology, Vol. I: A–B*. CRC Press, Boca Raton, Florida, USA.

[11] Royal Botanic Gardens, Kew (2017). *Seed Information Database (SID)*. Version 7.1. http://data.kew.org/sid/

[12] Sacande, M., Sanou, L. & Beentje, H. (2012). *Guide d'Identification des Arbres du Burkina Faso*. Royal Botanic Gardens, Kew.

[13] Sanogo, A. K., Sanogo, S. & Sanogo, R. (2010). *Rapport Projet UPP. Atelier de Fin de Phase du Projet MGU–Useful Plants Project (UPP)*. MGU–Useful Plants Project (UPP), Mali.

[14] WCSP (2017). *World Checklist of Selected Plant Families*. Royal Botanic Gardens, Kew. http://wcsp.science.kew.org/

Anthocleista djalonensis A.Chev.

TAXONOMY AND NOMENCLATURE

FAMILY. Gentianaceae

SYNONYMS. *Anthocleista kerstingii* Gilg ex Volkens

VERNACULAR / COMMON NAMES. Cabbage tree (English); fèrètadèbè, samatoulo (Bambara); fèrèta lafira, kogan (Malinke); dugu sudo, sonvige, sunyige (Senufo).

PLANT DESCRIPTION

LIFE FORM. Tree.

PLANT. Up to 15 m tall, trunk <40 cm in diameter (Figure 1); twigs sometimes with two erect spines or small cushions above leaf axils. **LEAVES.** Opposite, simple, entire, oblong-elliptical to obovate-elliptical 9–35(115) x 5–17(50) cm, base cordate, rounded or cuneate, apex rounded, petiole 1–9 cm. **FLOWERS.** Bisexual, regular with 4 free sepals, tube cylindric, 20–32 mm long, corolla lobes 11–14, oblong, lanceolate, white or creamy [2].

Figure 1 – General habit of *A. djalonensis*. (Photo: S. Sanogo, IER.)

FRUITS AND SEEDS

Ellipsoid berries (Figure 2), 3.5–5 x 2–3.5 cm, rounded at the apex, thick-walled, dark green, many seeded. Seeds (Figure 3) are obliquely ovoid, 2.5 x 1.5–2 x 1 mm [2].

FLOWERING AND FRUITING. Flowers from August to December [4].

Figure 2 – Fruiting branches. (Photo: S. Sanogo, IER.)

Figure 3 – External views, and longitudinal and cross-sections, of the seeds showing the endosperm and the linear embryo. (Photos: P. Gómez-Barreiro, RBG Kew.)

DISTRIBUTION

Tropical West Africa, from Mali and Senegal to Cameroon [2,7].

HABITAT

Dry places, savanna or thickets, from sea level to 500 m a.s.l. [2].

USES

In West Africa, the root is widely used as a diuretic and strong purgative [8]. The roots and leaves have antipyretic properties [9]. The leaves are used to treat jaundice, diarrhoea and dysentery [5]. The roots are used to treat impotence, liver and heart disorders, and hernias [4]. The bark has antimicrobial activity and is strongly purgative [5]. The root bark is used to treat haemorrhoids [11].

KNOWN HAZARDS AND SAFETY

No known hazards.

CONSERVATION STATUS

Assessed as Least Concern (LC) in Burkina Faso on the basis of its distribution and habitat [8], but its status has yet to be confirmed in the IUCN Red List [3]. The species has disappeared from many areas of Mali as a result of over-harvesting for medicinal uses [2,11].

SEED CONSERVATION

HARVESTING. Ripe fruits can be harvested from March to August, depending on the area. However, the fruits mature at different times, even on individual trees. Fruits do not change colour when ripe, but mature fruits are eaten by birds, so bird predation is a maturity indicator [9].

PROCESSING AND HANDLING. Before extracting the seeds, harvested fruits should be allowed to after-ripen for a few days to enable complete and homogeneous maturity [9]. The seeds can be extracted from the pulp by manual pressing or depulping with a knife, and then cleaned by extensive washing under running water.

STORAGE AND VIABILITY. Seeds of this species are reported to be orthodox [6]. Therefore, after appropriate drying, they can be stored at sub-zero temperatures for long-term conservation. However, low viability percentages (<10%) have been recorded following drying at 15% RH and freezing at –20°C for approximately 1.5 years at the MSB [6], suggesting that seeds of this species could be short-lived in seed bank conditions. This needs to be confirmed by further studies.

PROPAGATION

SEEDS

Dormancy and pre-treatments: No information is available on seed dormancy and pre-treatments. Congeneric species are reported to be either physiologically dormant or non-dormant, highlighting the need for further research to determine the class of dormancy present in the seeds of *Anthocleista djalonensis* [1].

Germination, sowing and planting: No information is available on seed germination under laboratory controlled conditions. In the nursery, a good germination rate (80–90%) is achieved

Figure 4 – Reintroduction plot of *A. djalonensis* in Zégoua. These plants are fruiting at 6 years old. (Photo: S. Sanogo, IER.)

when freshly extracted seeds are sown in a propagator using a sandy soil substrate [9]. Seedlings can be transplanted two weeks after germination. A very fine spray should be used for watering. Transplanted seedlings should be watered twice daily until the planting season. Seedlings must be at least 40 cm high before transplanting to the field [9]. Preferable planting sites are damp areas (banks, shallows) with >1,000 mm annual rainfall. For on-farm planting, seedlings should be planted in holes measuring 50 x 50 x 50 cm. During very dry months (December to April), 15 litres of water should be applied daily for 15 days [9]. In Zégoua, south Mali (rainfall 1,000–1,200 mm) plants grown in a plantation reached 2.5 m high in four years and fruited in six years (Figure 4) [10]. *Anthocleista djalonensis* is fire sensitive, so weeding is recommended in the dry season and a firewall should be cleared around planting areas if possible [10].

VEGETATIVE PROPAGATION

No protocols available.

PHYTOCHEMICAL ANALYSES

Phytochemical analyses of leaf and bark extracts were carried out by DMT/INRSP. Important chemical compounds found included coumarins, flavonoids, saponins, tannins and leucoanthocyanins. Strong antipyretic and antioxidant activities were detected.

TRADE

No data are available, despite all parts of the tree being used by therapists and herbalists in Mali.

Authors

Sidi Sanogo, Abdoul K. Sanogo, Bokary A. Kelly, Rokia Sanogo, Stéphane Rivière, Efisio Mattana and Tiziana Ulian.

References

[1] Baskin, C. C. & Baskin, J. M. (2014). *Seeds: Ecology, Biogeography, and Evolution of Dormancy and Germination*, 2nd edition. Academic Press, San Diego, USA.

[2] de Ruijter A. (2007). *Anthocleista djalonensis* A.Chev. In: *PROTA (Plant Resources of Tropical Africa/ Ressources Végétales de l'Afrique Tropicale)*. Edited by G. H. Schmelzer & A. Gurib-Fakim. Wageningen, The Netherlands. http://www.prota4u.org

[3] IUCN (2017). *The IUCN Red List of Threatened Species.* Version 2017-3. http://www.iucnredlist.org/

[4] Malgras, D. (1992). *Arbres et Arbustes Guérisseurs des Savanes Maliennes*. Agence de Cooperation Culturelle et Technique, Paris, France.

[5] Quattrocchi, U. (2012). *CRC World Dictionary of Medicinal and Poisonous Plants: Common Names, Scientific Names, Eponyms, Synonyms, and Etymology, Vol. I: A–B.* CRC Press, Boca Raton, Florida, USA.

[6] Royal Botanic Gardens, Kew (2017). *Seed Information Database (SID).* Version 7.1. http://data.kew.org/sid/

[7] Sacande, M., Sanogo, S. & Beentje, H. (2016). *Guide d'Identification des Arbres du Mali*. Royal Botanic Gardens, Kew.

[8] Sacande, M., Sanou, L. & Beentje, H. (2012). *Guide d'Identification des Arbres du Burkina Faso*. Royal Botanic Gardens, Kew.

[9] Sanogo, A. K., Sanogo, S. & Sidibé, S. I. (2011). *Conservation et Gestion Durable des Plantes Utiles aux Communautés Rurales*. Rapport de campagne 17, Commission scientifique, IER, Mali.

[10] Sanogo, A. K., Sanogo, S. & Sidibé, S. I. (2013). *Conservation et Gestion Durable des Plantes Utiles aux Communautés Rurales*. Rapport de campagne 19, Comité de Programme, IER, Mali.

[11] Sanogo, A. K., Sanogo, S. & Sidibé, S. I., Senou, O. & Dembélé, M. (2007). *Identification d'Espèces de Plantes Utiles aux Communautés Rurales*. Rapport d'enquête, IER, DMT and Royal Botanic Gardens, Kew.

Bobgunnia madagascariensis
(Desv.) J.H.Kirkbr. & Wiersema

TAXONOMY AND NOMENCLATURE

FAMILY. Leguminosae – Papilionoideae

SYNONYMS. *Swartzia madagascariensis* Desv., *S. marginata* Benth., *S. sapinii* De Wild., *Tounatea madagascariensis* (Desv.) Baill.

VERNACULAR / COMMON NAMES. Snake bean, white ironwood (English); petit dim (French); samagwara, samakata (Bambara); sinsènè (Bobo); samagara (Malinke); gugeriki, ohti (Peul); nakonon (Senufo).

PLANT DESCRIPTION

LIFE FORM. Tree or shrub.

PLANT. Deciduous, <18 m (Figure 1), bark deeply fissured or reticulate, grey or brown.

LEAVES. Petiole and rachis together are (3)6–15(18) cm long, leaflets (3)5–9(13), alternate, occasionally opposite, usually elliptic or broadly elliptic. **FLOWERS.** Racemes <5(8) cm long, sometimes branched, with up to 10 flowers, calyx globose, splitting for about half its length into 2–5 lobes, reflexing at anthesis, petals white with a basal yellow patch inside, flowers sweet-scented [18].

FRUITS AND SEEDS

Pods (Figure 2) cylindric, usually somewhat bumpy and twisted, hard, dark brown to black, pendent, indehiscent [18], 6–30 x 1.3–1.8 cm [6], containing 10–15 seeds in a sticky yellow tissue [11]. Seeds (Figure 3) are oblong-reniform, compressed, with a small hilum below the radical lobe, pale brown or greyish, shiny [18].

FLOWERING AND FRUITING. In West Africa, flowers at the end of the dry season or after the first rains, soon after coming into leaf [2].

DISTRIBUTION

Tropical Africa, from Senegal, Mali and Chad to Tanzania, southwards to Central and Southern Africa. Despite the specific epithet, this species does not occur in Madagascar [7].

HABITAT

Deciduous woodland or wooded grassland, mainly on sandy soils [18], at 50–2,000 m a.s.l. [6].

USES

Bobgunnia madagascariensis has many uses in traditional medicine [10,15]. For example, a fruit decoction is used as an emetic and to treat schistosomiasis, leprosy and earache. The roots and seed pods are used as an abortifacient; the roots are also used as an anthelminthic, and to treat snake bites and venereal diseases. Powdered fruits, seeds and root bark are used as a fish poison. The seed pods are ground to a powder and used as a repellent to protect stored seeds from predation by insects, termites and weevils [13]. The wood makes excellent firewood and charcoal [10].

KNOWN HAZARDS AND SAFETY

The leaves and fruits are poisonous to cattle and sheep [4,15]. The powder from crushed pods has been associated with fatal human poisoning [5]. Aqueous extracts of the fruit have been shown to

Figure 1 – General habit of *B. madagascariensis*. (Photo: S. Sanogo, IER.)

Figure 2 – Ripe (black) and unripe (green) fruits. (Photo: S. Sanogo, IER.)

have deleterious effects on mice and could have deleterious health implications for humans and other animals [14]. Phytochemical analyses carried out at DMT/INRSP have also shown that the root bark powder has some toxicity (see 'Phytochemical analyses', below).

CONSERVATION STATUS

This widespread species is assessed as Least Concern (LC) in West Africa [17] and Namibia [12] on the basis of its distribution and habitat, but its status has yet to be confirmed in the IUCN Red List [8].

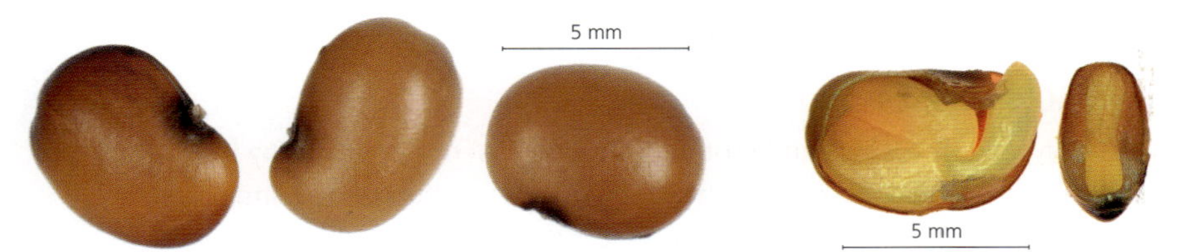

Figure 3 – External views, and longitudinal and cross-sections, of the seeds showing the embryo and endosperm. (Photos: P. Gómez-Barreiro, RBG Kew.)

SEED CONSERVATION

HARVESTING. Fruits are indehiscent and should be collected directly from the plants.

PROCESSING AND HANDLING. Pods can be sun-dried and pounded in a mortar to extract the seeds which can be cleaned by shaking and winnowing [11].

STORAGE AND VIABILITY. Seeds are reported to be orthodox [16]. Therefore, after appropriate drying, they can be stored at sub-zero temperatures for long-term conservation. At room temperature, seeds can be stored for three years without significant loss of viability if kept dry and free of insects [11].

PROPAGATION

SEEDS

Dormancy and pre-treatments: Seeds of this species are reported to be physically dormant [3] and should therefore be scarified to allow water imbibition. Soaking the seeds in hot water (10 minutes at 100°C) was identified as the best pre-treatment, although dry heat treatment at 80°C for 30 minutes and sulphuric acid (98%) treatment for 10 minutes were also successful in breaking seed dormancy [1]. Alternatively, seeds can be scarified by manually chipping with a scalpel [16].

Germination, sowing and planting: Germination experiments carried out under laboratory controlled conditions on seed lots stored at the MSB showed 100% germination for chipped seeds incubated at constant 25°C, with a photoperiod of 8 hours of light per day [16]. In the nursery, pre-treated seeds germinated uniformly (Figures 4&5), reaching 80% germination after 25 days from sowing [11].

VEGETATIVE PROPAGATION

Experiments indicate that stem cuttings can be rooted when treated with indole-3-butyric acid (IBA); a concentration of 300 ppm IBA gave the best results. *Bobgunnia madagascariensis* can also be propagated from root suckers [10]. A study in Mali found that 33% of tree stumps re-sprouted and 100% of tree stumps had basal shoots [9].

Figure 4 – Seedlings, two weeks after sowing in sand. (Photo: S. Sanogo, IER.)

Figure 5 – Propagation at the nursery of Sikasso. (Photo: R. Dackouo, IER.)

PHYTOCHEMICAL ANALYSES

Phytochemical analyses of the leaves and root bark (Figure 6) were carried out by DMT/INRSP. The results indicate that *B. madagascariensis* has antibacterial, antifungal and antiplasmodial properties. The species is rich in polyphenolic substances (tannins and flavonoids), which are partly responsible for the antioxidant activity. However, analysis of the root bark decoction showed some toxicity, so root bark powder should be used with great caution.

Figure 6 – Collecting roots for phytochemical analyses. (Photo: S. Sanogo, IER.)

TRADE

The wood is mainly traded locally but occasionally enters the international market (e.g. timber exports from Tanzania) [10]. No data are available for Mali.

Authors

Sidi Sanogo, Abdoul K. Sanogo, Bokary A. Kelly, Rokia Sanogo, Stéphane Rivière, Efisio Mattana, Steve Davis and Tiziana Ulian.

References

[1] Amri, E. (2010). The effect of pre-sowing seed treatment on germination of snake bean (*Swartzia madagascariensis*); a reported medicinal plant. *Research Journal of Agriculture and Biological Sciences* 6 (4): 557–561.

[2] Arbonnier, M. (2004). *Trees, Shrubs and Lianas of West African Dry Zones*. CIRAD, France; Margraf Publishers, Weikersheim, Germany; and Muséum national d'histoire naturelle (MNHN), Paris, France.

[3] Baskin, C. C. & Baskin, J. M. (2014). *Seeds: Ecology, Biogeography, and Evolution of Dormancy and Germination*, 2nd edition. Academic Press, San Diego, USA.

[4] Burkill, H. M. (1995). *The Useful Plants of West Tropical Africa, Vol. 3: Families J–L*. 2nd edition. Royal Botanic Gardens, Kew.

[5] Coates Palgrave, K. & Coates Palgrave, M. (2002). *Trees of Southern Africa*, 3rd edition. Struik Publishers, Cape Town, South Africa.

[6] Conservatoire et Jardin botaniques & South African National Biodiversity Institute (SANBI) (2016). *African Plant Database*. Version 3.4.0. http://www.ville-ge.ch/musinfo/bd/cjb/africa/

[7] ILDIS (2006–2013). *World Database of Legumes*. International Legume Database and Information Service. http://www.legumes-online.net/ildis/aweb/database.htm

[8] IUCN (2017). *The IUCN Red List of Threatened Species*. Version 2017-3. http://www.iucnredlist.org/

[9] Kelly, B. A. (1995). *Régime Taillis sous Futaie dans la Forêt Classée de Farako*. Note technique, ARFS 19. MDRE, IER, CRRAS and ARFS, Mali.

[10] Mojeremane, W. (2012). *Bobgunnia madagascariensis* (Desv.) J.H.Kirkbr. & Wiersema. In: *PROTA (Plant Resources of Tropical Africa/Ressources Végétales de l'Afrique Tropicale)*. Edited by R. H. M. J. Lemmens, D. Louppe & A. A. Oteng-Amoako. Wageningen, The Netherlands. http://www.prota4u.org

[11] Msanga, H. P. (1998). *Seed Germination of Indigenous Trees in Tanzania: Including Notes on Seed Processing and Storage, and Plant Uses*. Canadian Forest Service, Northern Forestry Centre, Edmonton, Canada.

[12] National Red List (2017). *Bobgunnia madagascariensis*. National Red List Network. http://www.nationalredlist.org/

[13] Natural Resources Institute (2017). *Bobgunnia madagascariensis.* Optimising Pesticidal Plants: Technology Information, Outreach and Networks (OPTIONs) Project. Natural Resources Institute (NRI), University of Greenwich, UK. http://projects.nri.org/options/background/plants-database/bobgunnia-madagascariensis

[14] Nyahangare, E. T., Hove, T., Mvumi, B. M., Hamudikuwanda, H., Belmain, S. R., Madzimure, J. & Stevenson, P. C. (2012). Acute mammalian toxicity of four pesticidal plants. *Journal of Medicinal Plants Research* 6 (13): 2674–2680.

[15] Royal Botanic Gardens, Kew (1999–2015). *Survey of Economic Plants for Arid and Semi-Arid Lands (SEPASAL) Database*. http://apps.kew.org/sepasalweb/sepaweb

[16] Royal Botanic Gardens, Kew (2017). *Seed Information Database (SID).* Version 7.1. http://data.kew.org/sid/

[17] Sacande, M., Sanou, L. & Beentje, H. (2012). *Guide d'Identification des Arbres du Burkina Faso.* Royal Botanic Gardens, Kew.

[18] Timberlake, J. R., Pope, G. V., Polhill, R. M. & Martins, E. S. (eds) (2007). Leguminosae (Papilionoideae). In: *Flora Zambesiaca*, Vol. 3, part 3. Flora Zambesiaca Managing Committee, Royal Botanic Gardens, Kew.

Carapa procera DC.

TAXONOMY AND NOMENCLATURE

FAMILY. Meliaceae

SYNONYMS. *Carapa guineensis* Sweet ex A.Juss., *C. gummiflua* C.DC., *C. touloucouna* Guill. & Perr.

VERNACULAR / COMMON NAMES. Crabwood (English); kobi (Bambara); gobi (Pular); dougo-bi (Senufo).

PLANT DESCRIPTION

LIFE FORM. Tree.

PLANT. Up to 30 m tall (Figure 1) [8], pachycaulous, slash with sticky red sap [10]. **LEAVES.** Variable in shape, compound, imparipinnate, clustered at end of branches, 6–18 pairs of opposite leaflets, oblong to elliptic, <40 x 16 cm, petiole and rachis 25–150 cm long [14]. **FLOWERS.** Inflorescences are loose panicles <75 cm long, flowers fragrant, white and purple, petals 5–6, 5–6 mm long [14].

Figure 1 – General habit of *C. procera*. (Photo: R. Dackouo, IER.)

FRUITS AND SEEDS

Fruits (Figure 2) are rhomboid or globular capsules of 10 cm in diameter, mucronate at the top, with 5 dehiscent valves [8]. Seeds (Figure 2) are dark brown, shiny, c. 3 cm long [14].

FLOWERING AND FRUITING. In the Sikasso region of Mali, plants flower from January to March during the dry season; fruits ripen in May and June [15].

Figure 2 – Unripe fruits (left) and ripe fruits split open (right) showing the seeds. (Photos: R. Dackouo, IER.)

DISTRIBUTION

Tropical West Africa, from Senegal to Angola; also occurs in the Amazon basin in South America [4].

HABITAT

Riverine forest, lowland rainforest and montane forest, at 1–2,450 m a.s.l. [4].

USES

The seed oil is used in human and veterinary traditional medicine [5], for example against ticks and for treating wounds in cattle [1]. It is also used for phytosanitary treatment of cotton as part of organic and fair trade cotton production [16], and has potential for use in manufacturing cosmetic products [7]. The seed oil is edible but bitter ('oil crab') [14], the seeds themselves being emetic [3]. The oil is used for making soap (Figure 3) and candles. The tree is planted for shade in villages of Senegal [4].

Figure 3 – Workshops on seed extraction and the preparation of soap from oil. (Photos: S. Sanogo, IER.)

KNOWN HAZARDS AND SAFETY

The seeds are mildly poisonous and can cause vomiting [3].

CONSERVATION STATUS

Assessed as Least Concern (LC) on the basis of its distribution and habitat [12,13], but its status has yet to be confirmed in the IUCN Red List [6]. In Mali, it is protected by forestry legislation [12].

SEED CONSERVATION

HARVESTING. Fruits can be collected from the trees when the first capsules begin to open [14].

PROCESSING AND HANDLING. Capsules open when drying; seeds can then be manually removed from the capsules [14].

STORAGE AND VIABILITY. Seeds are recalcitrant and do not tolerate drying. They cannot be stored at sub-zero temperatures for long-term conservation [11]. For short-term conservation, they can be stored in sawdust and moist vermiculite to retain humidity [15].

PROPAGATION

SEEDS

Dormancy and pre-treatments: Seeds are non-dormant [2] and do not therefore require specific pre-treatments prior to sowing [14].

Germination, sowing and planting: Seeds should be incubated at 26–31°C and typically germinate within four weeks [15]. Freshly harvested seeds sown in a clay loam soil produce good-sized plants in three months (height >50 cm) [15] (Figure 4).

VEGETATIVE PROPAGATION

Rooting success from cuttings is reported to be low. A success rate of c. 20% has been achieved using 10 cm cuttings with one pair of leaves, hormone rooting powder, and a substrate of 60% peat and 40% polystyrene [9].

Figure 4 – Plantation of 3-year-old plants at Ifola in the Sikasso region. (Photo: R. Dackouo, IER.)

TRADE

The sale price of *Carapa procera* oil varies between 1,000 CFA francs and 3,000 CFA francs per litre in Sikasso (Mali).

Authors

Sidi Sanogo, Abdoul K. Sanogo, Bokary A. Kelly, Stéphane Rivière, Efisio Mattana and Tiziana Ulian.

References

[1] Bah, M. S. (1993). The importance of traditional veterinary medicine (TVM) in animal health programmes. In: *Research for Development of Animal Traction in West Africa. Proceedings of the 4th Workshop of West Africa Animal Traction Network, 9–13 July 1990, Kano, Nigeria*. Edited by P. R. Lawrence, K. Lawrence, J. T. Dijkman & P. H. Starkey. International Livestock Centre for Africa (ILCA), Addis Ababa, Ethiopia. pp. 33–36.

[2] Baskin, C. C. & Baskin, J. M. (2014). *Seeds: Ecology, Biogeography, and Evolution of Dormancy and Germination*, 2nd edition. Academic Press, San Diego, USA.

[3] Burkill, H. M. (1997). *The Useful Plants of West Tropical Africa, Vol. 4: Families M–R*. 2nd edition. Royal Botanic Gardens, Kew.

[4] Conservatoire et Jardin botaniques & South African National Biodiversity Institute (SANBI) (2016). *African Plant Database*. Version 3.4.0. http://www.ville-ge.ch/musinfo/bd/cjb/africa/

[5] Guèye, M., Kenfack, D. & Forget, P.-M. (2010). Importance socio-culturelle, potentialités économiques et thérapeutiques du *Carapa* (Meliaceae) au Sénégal. In: *Systématique et Conservation des Plantes Africaines*. Edited by X. van der Burgt, L. van der Maesen & J. M. Onana. Royal Botanic Gardens, Kew. pp. 359–367.

[6] IUCN (2017). *The IUCN Red List of Threatened Species.* Version 2017-3. http://www.iucnredlist.org/

[7] Minzangi, K., Kaaya, A. N., Kansiime, F., Tabuti, J. R. S., Samvura, B. & Grahl-Nielsen, O. (2011). Fatty acid composition of seed oils from selected wild plants of Kahuzi-Biega National Park and surroundings, Democratic Republic of Congo. *African Journal of Food Science* 5 (4): 219–226.

[8] Ngwamashi, E. (2009). *Inventaire des Espèces Ligneuses Locales Pour le Reboisement à des Fins Énergétiques*. University of Kinshasa, Democratic Republic of Congo.

[9] Pangou, S. V., De Zoysa, N. & Lechon, G. (2011). Comparison between field performance of cuttings and seedlings of *Carapa procera* DC. (Meliaceae). *International Research Journal of Plant Science* 2 (9): 281–287.

[10] Quattrocchi, U. (2012). *CRC World Dictionary of Medicinal and Poisonous Plants: Common Names, Scientific Names, Eponyms, Synonyms, and Etymology, Vol. II: C–D*. CRC Press, Boca Raton, Florida, USA.

[11] Royal Botanic Gardens, Kew (2017). *Seed Information Database (SID).* Version 7.1. http://data.kew.org/sid/

[12] Sacande, M., Sanogo, S. & Beentje, H. (2016). *Guide d'Identification des Arbres du Mali*. Royal Botanic Gardens, Kew.

[13] Sacande, M., Sanou, L. & Beentje, H. (2012). *Guide d'Identification des Arbres du Burkina Faso*. Royal Botanic Gardens, Kew.

[14] Sanogo, S. & Sacandé, M. (2007). *Carapa procera* DC. *Seed Leaflet* No. 136. Edited by L. Schmidt. Forest & Landscape Denmark, Hørsholm, Denmark.

[15] Sanogo, S., Sacandé, M., Van Damme, P. & Ndiaye, I. (2013). Caractérisation, germination et conservation des graines de *Carapa procera* DC. (Meliaceae), une espèce utile en santé humaine et animale. *Biotechnology, Agronomy, Society* and *Environment* 17 (2): 321–331.

[16] Sanogo, Y. & Favreau, L. (2007). *Commerce Équitable et Développement Durable: la Filière Coton au Mali.* Série Recherches No. 4. Université du Québec en Outaouais and Centre Canadien d'Étude et de Coopération Internationale (CECI), Canada.

Coix lacryma-jobi L.

TAXONOMY AND NOMENCLATURE

FAMILY. Poaceae

SYNONYMS. *Coix lacryma* L., *C. ovata* Stokes, *C. pendula* Salisb., *Lithagrostis lacryma-jobi* (L.) Gaertn.

VERNACULAR / COMMON NAMES. Job's tear (English); l'herbe à chapelet, larme de Job, larme de la vierge (French); baya-kissè (Bambara); dougo-konon (Senufo).

Figure 1 – Culms of *C. lacryma-jobi*. (Photo: S. Sanogo, IER.)

PLANT DESCRIPTION

LIFE FORM. Annual herb.

PLANT. Culms <3 m long. **LEAVES.** Linear-lanceolate, cauline, ligulate, membrane without cilia, base cordate, 10–45 x 2–7 cm (Figure 1) [2]. **FLOWERS.** Inflorescences monoecious, subtended by the same spatheole; inflorescence comprising racemes bearing a triad of spikelets [3].

FRUITS AND SEEDS

The pseudofruits comprise the female inflorescence enclosed in (when ripe) a hardened, black or cream white and very shiny spatheole (Figure 2). Caryopsis with an adherent pericarp, orbicular, the embryo as long as the caryopsis [3].

FLOWERING AND FRUITING. Flowering and fruiting are strictly linked to rainfall patterns.

Figure 2 – External views of the hardened spatheole (above) and of the naked caryopses.
(Photos: P. Gómez-Barreiro, RBG Kew.)

DISTRIBUTION

Native of Asia but now introduced and naturalised throughout the tropics, including in both tropical West Africa and East Africa [8].

HABITAT

Stream banks and moist places [4]. In West Africa, often found around villages and on old cultivation sites [1].

USES

The hard-shelled pseudofruits are mainly used as beads for necklaces (Figure 3), rosaries, for decoration, and as rattles in dances. Cultivation for food (grain) in West Africa is limited. The grass is valued in West Africa for providing good-quality grazing for cattle [1].

Figure 3 – Pseudofruits (spatheoles) used to make necklaces at the Sikasso seed fair in Mali. (Photo: S. Sanogo, IER.)

KNOWN HAZARDS AND SAFETY

No known hazards.

CONSERVATION STATUS

This widespread species is not considered to be either rare or threatened, but has yet to be assessed throughout its range. The population in Sri Lanka has been assessed as Vulnerable (VU) according to IUCN Red List criteria [6]. Its status has yet to be confirmed in the IUCN Red List [5].

Figure 4 – Seed harvesting in the nursery at Sikasso. (Photo: S. Sanogo, IER.)

SEED CONSERVATION

HARVESTING. Pseudofruits can be hand-picked directly from the plant (Figure 4).

PROCESSING AND HANDLING. Pseudofruits can be separated and cleaned by hand by winnowing or sieving. The pseudofruit is the conservation unit as the seeds cannot be extracted easily from the spatheole (Figure 2).

STORAGE AND VIABILITY. Seeds are likely to be orthodox [7]. Therefore, after appropriate drying, they can be stored at sub-zero temperatures for long-term conservation.

PROPAGATION

SEEDS

Dormancy and pre-treatments: No information is available on seed dormancy.

Germination, sowing and planting: Under laboratory controlled conditions, seed germination experiments carried out on a seed lot stored at the MSB found c. 80% germination when seeds were sown in cycles of 35/20°C with 8 hours of light per day [7].

VEGETATIVE PROPAGATION

No protocols available.

TRADE

No data available.

Authors

Sidi Sanogo, Abdoul K. Sanogo, Bokary A. Kelly, Stéphane Rivière, Efisio Mattana, Steve Davis and Tiziana Ulian.

References

[1] Burkill, H. M. (1994). *The Useful Plants of West Tropical Africa, Vol. 2: Families E–I.* 2nd edition. Royal Botanic Gardens, Kew.

[2] Clayton, W. D., Phillips, S. M. & Renvoize, S. A. (1974). Gramineae (Part 2). In: *Flora of Tropical East Africa*. Edited by R. M. Polhill. Crown Agents for Oversea Governments and Administrations, London.

[3] Clayton, W. D., Vorontsova, M. S., Harman, K. T. & Williamson, H. (2002 onwards). *GrassBase — The Online World Grass Flora*. https://www.kew.org/data/grassbase/index.html

[4] Hutchinson, J. & Dalziel, J. M. (1972). *Flora of West Tropical Africa*, Vol. 3, part 1. Crown Agents for Oversea Governments and Administrations, London.

[5] IUCN (2017). *The IUCN Red List of Threatened Species*. Version 2017-3. http://www.iucnredlist.org/

[6] Ministry of Environment (MOE) (2012). *The National Red List 2012 of Sri Lanka; Conservation Status of the Fauna and Flora*. Ministry of Environment, Colombo, Sri Lanka.

[7] Royal Botanic Gardens, Kew (2017). *Seed Information Database (SID).* Version 7.1. http://data.kew.org/sid/

[8] WCSP (2017). *World Checklist of Selected Plant Families.* Royal Botanic Gardens, Kew. http://wcsp.science.kew.org/

Cola cordifolia (Cav.) R.Br.

TAXONOMY AND NOMENCLATURE

FAMILY. Malvaceae

SYNONYMS. *Southwellia cordifolia* (Cav.) Spach, *Sterculia cordifolia* Cav.

VERNACULAR / COMMON NAMES. Mandinka kola (English); tabayer (French); ntaba, ntabanôgô (Bambara, Malinke); wongo (Minyanka); tabai (Peul); woma, wongo (Senufo).

PLANT DESCRIPTION

LIFE FORM. Tree.

PLANT. Up to 10–25 m tall (Figure 1), crown dense, trunk with buttresses, fluted at base, stout, <1 m in diameter, bark grey brown, peeling into papery flakes, slash white striped russet red [1]. **LEAVES.** Bunched at end of branches, 5–15 x 4–13 cm, very broadly ovate-cordate, almost entire or 3-5-lobed, apex acuminate. **FLOWERS.** Yellow, shortly pedicellate and apetalous, clustered in terminal panicles <9 cm long, appearing before the leaves [1].

Figure 1 – *C. cordifolia* growing on a sacred site in the land of Woroni, Sikasso region.
(Photo S. Sanogo, IER.)

FRUITS AND SEEDS

Follicles 3–5, c. 10 cm long, spreading, subsessile, oblong-ovoid, with a horn-like often curled, sharp-pointed apex, golden tomentose and finely longitudinally ridged outside, opening widely after dehiscence (Figure 2). Seeds numerous, c. 2 x 1 cm, oblong-ellipsoid, with a whitish aril at one end and a smooth, dull blackish testa (Figures 2&3).

FLOWERING AND FRUITING. Flowers from February to May in the Sikasso region of Mali; fruits from March to June. Maximum seed dispersal occurs from July to August [6].

Figure 2 – Fruits and fresh seeds. (Photos: S. Sanogo, IER.)

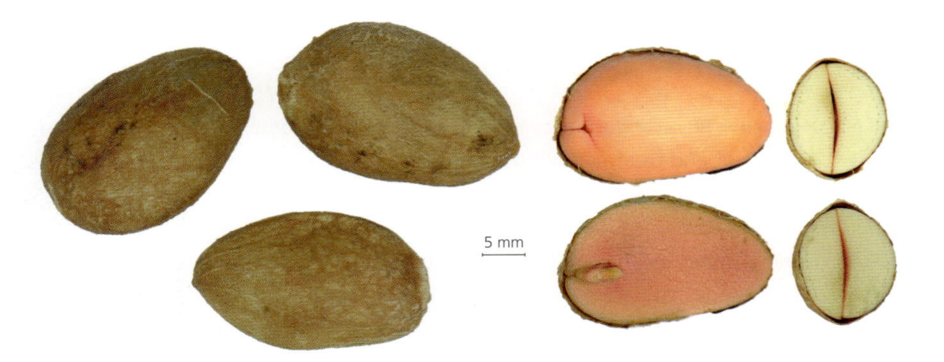

Figure 3 – External views, and longitudinal and cross-sections, of dry seeds showing the embryo and cotyledons. (Photos: P. Gómez-Barreiro, RBG Kew.)

DISTRIBUTION

Tropical West Africa, from Senegal to Burkina Faso [1].

HABITAT

Savanna, gallery forest, rainforest edges and dry forests [1,5].

USES

The fruits are edible and are subject to significant trade during the rainy season. The leaves are used for the packaging of shea butter (*Parkia biglobosa* (Jacq.) G.Don) and baobab (*Adansonia digitata* L.) [6,10]. In traditional medicine, the bark is used as a laxative, and for treating lung disorders and venereal diseases. The leaves are used to treat leprosy [10]. A detailed study has revealed the presence of pectin in the leaves and bark, the components of which have important immune-modulating activity [2]. Aqueous extracts of the bark and leaves are traditionally used in Mali for treating wounds, abdominal pain, gastritis and gastric ulcers [3]. Crushed bark and branch bases are used locally for abscesses and gangrene [13]. The wood is used in the construction of buildings, for making kitchenware and children's toys, and as firewood. *Cola cordifolia* is an ornamental species and used for soil conservation in agriculture and forestry [10]. It is also planted for shade, and is one of the most common palaver trees in villages, where communities socialise and discuss matters [9].

KNOWN HAZARDS AND SAFETY

No known hazards.

CONSERVATION STATUS

Assessed as Least Concern (LC) in Burkina Faso on the basis of its distribution and habitat [11], but its status has yet to be confirmed in the IUCN Red List [8].

SEED CONSERVATION

HARVESTING. Ripe fruits can be harvested from the tree using sticks [7].

PROCESSING AND HANDLING. Seeds can be extracted manually from the follicles and depulped by extensive washing with water [7].

STORAGE AND VIABILITY. Seeds of this species are reported as likely to be recalcitrant [11], and cannot therefore be stored at sub-zero temperatures for long-term conservation.

PROPAGATION

SEEDS

Dormancy and pre-treatments: Physical dormancy has been reported for some *Cola* spp. [4]. Soaking seeds in water for 48 hours [6] and treatment with sulphuric acid [10] have been reported to give high germination percentages, suggesting that physical dormancy could occur in *C. cordifolia*. This should be confirmed by further studies.

Germination, sowing and planting: No information is available on seed germination under laboratory controlled conditions. In Côte d'Ivoire, good results were obtained by planting in the open, with a spacing of 3 x 3 m [10]. Trees planted in groves and village forests (Figure 4) showed 60% survival rate after four years and an annual increment in height of 50 cm [12].

VEGETATIVE PROPAGATION

No information is available. As part of the domestication of local fruit trees, *C. cordifolia* has been used as a rootstock for *C. nitida* Vahl ex H.West using cleft and side grafting techniques. Grafted *C. nitida* produce fruit five years after planting (Figure 5).

TRADE

The fruits of *Cola cordifolia* are sold locally in southern Mali.

Figure 4 – 6-month-old plant in a therapist's plot in Zégoua. (Photo: S. Sanogo, IER.)

Figure 5 – *C. nitida* in fruit with *C. cordifolia* as rootstock. (Photo: S. Sanogo, IER.)

Authors

Sidi Sanogo, Abdoul K. Sanogo, Bokary A. Kelly, Stéphane Rivière, Rokia Sanogo and Efisio Mattana.

References

[1] Arbonnier, M. (2004). *Trees, Shrubs and Lianas of West African Dry Zones*. CIRAD, France; Margraf Publishers, Weikersheim, Germany; and Muséum national d'histoire naturelle (MNHN), Paris, France.

[2] Austarheim, I. (2011). *Cola cordifolia* (Cav.) R.Br. [Malvaceae]. School of Pharmacy, Universitas Oloensis, Norway. http://www.mn.uio.no/farmasi/english/research/projects/maliplants/medicinal-plants/studied-in-norway/cola-cordifolia.html

[3] Austarheim, I., Mahamane, H., Sanogo, R., Togola, A., Khaledabadi, M., Vestrheim, A. C., Inngierdingen, K. T., Michaelsen, T. E., Diallo, D. & Paulsen, B. S. (2012). Anti-ulcer polysaccharides from *Cola cordifolia* bark and leaves. *Journal of Ethnopharmacology* 143 (1): 221–227.

[4] Baskin, C. C. & Baskin, J. M. (2014). *Seeds: Ecology, Biogeography, and Evolution of Dormancy and Germination*, 2nd edition. Academic Press, San Diego, USA.

[5] Conservatoire et Jardin botaniques & South African National Biodiversity Institute (SANBI) (2016). *African Plant Database*. Version 3.4.0. http://www.ville-ge.ch/musinfo/bd/cjb/africa/

[6] Cuny, P., Sanogo, S. & Sommer, N. (1998). *Arbres du Domaine Soudanien: Leurs Usages et Leur Multiplication*. IER, Mali and Intercoopération, Switzerland.

[7] Food and Agriculture Organization of the United Nations (FAO) (2001). *Situation des Ressources Génétique Forestières du Mali*. FAO, Rome, Italy.

[8] IUCN (2017). *The IUCN Red List of Threatened Species*. Version 2017-3. http://www.iucnredlist.org/

[9] Mugnier, J. (2008). *Nouvelle Flore Illustrée du Sénégal et Régions Voisines. Les Dicotylédones*, Tome 1. Agroservices, France.

[10] PROTA4U (2017). *Cola cordifolia* (Cav.) R.Br. In: *PROTA (Plant Resources of Tropical Africa/Ressources Végétales de l'Afrique Tropicale)*. Wageningen, The Netherlands. http://www.prota4u.org

[11] Sacande, M., Sanou, L. & Beentje, H. (2012). *Guide d'Identification des Arbres du Burkina Faso*. Royal Botanic Gardens, Kew.

[12] Sanogo, A. K., Sanogo, S. & Sidibé, S. I. (2012). *Conservation et Gestion Durable des Plantes Utiles aux Communautés*. Rapport annuel de recherche, Comité de Programme, IER, Mali.

[13] Société Française d'Ethnopharmacologie (2010). Banque de données Prélude. *Cola cordifolia* R.Br. http://www.ethnopharmacologia.org/recherche-dans-prelude/?plant_id=3500

Combretum micranthum G.Don

TAXONOMY AND NOMENCLATURE

FAMILY. Combretaceae

SYNONYMS. *Combretum altum* Guill. & Perr. ex DC., *C. floribundum* Engl. & Diels, *C. parviflorum* Rchb. ex DC., *C. raimbaultii* Heckel

VERNACULAR / COMMON NAMES. Bush tea (English); kinkeliba (French); kôlôbè (Bambara); tallika (Fula); kinkelliba, kou lomkalan (Mandinka).

PLANT DESCRIPTION

LIFE FORM. Shrub, tree or liana.

PLANT. 2–10 m tall (Figure 1), up to 20 m when a liana twining around nearby trees.

LEAVES. Opposite or 3-verticillate, variable in shape, elliptic or ovate, 5–10 x 2.5–5 cm wide (Figure 2); apex acuminate, base cuneate, glabrous with hair tufts in veins. **FLOWERS.** Whitish, 2 mm in diameter, calyx with rust-coloured scales (Figure 2), in spike-like axillary racemes, 3–5 cm long, stalk scaly [1,11].

Figure 1 – General habit of *C. micranthum*. (Photo: S. Sanogo, IER.)

Figure 2 – Flowering (left) and fruiting (right) branches. (Photos: S. Sanogo, IER.)

FRUITS AND SEEDS

Four-winged, indehiscent, one-seeded fruits (Figure 3), 12–15 mm long, ± glabrous, brown when ripe [1].

FLOWERING AND FRUITING. Flowering and fruiting occur during the second half of the dry season and into the wet season, usually before the first leaves appear or as the leaves open [11].

Figure 3 – External views of the fruits and seeds, and cross-section of the fruit, showing the folded embryo. (Photos: P. Gómez-Barreiro, RBG Kew.)

DISTRIBUTION

Tropical West Africa, widely distributed throughout the Sahel from Senegal and Mauritania to Nigeria and Niger; common, gregarious and locally abundant [1].

HABITAT

Savanna, in dry sites on sandstone, clay, laterite, crystalline rocks and skeletal soils [11]; frequently found on termitaria (even though the roots are very susceptible to termite attack [3]). Found in areas with 300–1,500 mm annual rainfall, from sea level to 1,000 m a.s.l. [11].

USES

Used widely in traditional medicine [9]. A tea made from the leaves is used as a tonic [13]. The leaves also have antipyretic, diuretic, antidiarrhoeal and cholagogic properties. The roots are a vermifuge and are used to treat wounds, fever, syphilis, enuresis and female infertility [4,5]. Hepatic protective properties have been confirmed by studies undertaken by DMT/INRSP in Mali [11]. The flowers attract bees which produce a good honey [7,11]. The wood is used for roofs, furniture and frames, and in a similar way to rattan for making baskets [1,7]; it also makes excellent firewood. Leafy branches are browsed by cattle, sheep and goats [1,4].

KNOWN HAZARDS AND SAFETY

No known hazards.

CONSERVATION STATUS

Assessed as Least Concern (LC) in Burkina Faso on the basis of its distribution and habitat [12], but its status has yet to be confirmed in the IUCN Red List [6].

SEED CONSERVATION

HARVESTING. Fruits can be harvested directly from the plants. The optimum harvesting season in the Sikasso region of Mali is from December to January. It is advisable to collect the fruits before they are fully ripe as the ripe fruits of many *Combretum* spp. are attacked by insects [11].

PROCESSING AND HANDLING. Seeds are usually not extracted from the fruits before sowing because extracted seeds are more fragile and prone to damage [11]. The fruit is therefore the conservation unit.

STORAGE AND VIABILITY. Seeds of this species are reported to be orthodox [10]. Therefore, after appropriate drying, they can be stored at sub-zero temperatures for long-term conservation. When stored at 4°C, seeds can survive for four years with germination >96% [11].

PROPAGATION

SEEDS

Dormancy and pre-treatments: Seeds germinate easily without any pre-treatments, suggesting that seeds of this species may be non-dormant. Nevertheless, little information is available on seed germination under laboratory controlled conditions. Further research is needed to determine the class of dormancy (if any).

Germination, sowing and planting: Germination experiments carried out on seed lots stored at the MSB revealed high germination percentages for seeds incubated at 25°C (≥86%) [10].

VEGETATIVE PROPAGATION

Naturally regenerates from coppice shoots [3] and by layering [2].

TRADE

The leaves, and to a lesser extent the stem bark and roots, are commonly sold in local markets. According to studies by DMT/INRSP in Mali, the leaves are used in the preparation of an improved traditional medicine (MTA) called Hépatisane [8].

Authors

Sidi Sanogo, Abdoul K. Sanogo, Bokary A. Kelly, Stéphane Rivière, Efisio Mattana, Steve Davis and Tiziana Ulian.

References

[1] Arbonnier, M. (2004). *Trees, Shrubs and Lianas of West African Dry Zones*. CIRAD, France; Margraf Publishers, Weikersheim, Germany; and Muséum national d'histoire naturelle (MNHN), Paris, France.

[2] Bationo, B-A., Karim, S., Bellefontaine, R., Saadou, M., Guinko, S., Ichaou, A. & Bouhari, A. (2005). Le marcottage terrestre: une technique économique pour la régénération de certains ligneux tropicaux. *Sécheresse* 16 (4): 309–311.

[3] Bognounou, F., Tigabu, M., Savadogo, P., Thiombiano, A., Boussim, I. J., Oden, P. C. & Guinko, S. (2010). Regeneration of five Combretaceae species along a latitudinal gradient in Sahelo-Sudanian zone of Burkina Faso. *Annals of Forest Science* 67 (3): 306.

[4] Burkill, H. M. (1985). *The Useful Plants of West Tropical Africa, Vol. 1: Families A–D*. 2nd edition. Royal Botanic Gardens, Kew.

[5] Fortin, D., Lö, M. & Maynart, G. (1997). *Plantes Médicinales du Sahel*. Enda Editions, Dakar, Senegal.

[6] IUCN (2017). *The IUCN Red List of Threatened Species.* Version 2017-3. http://www.iucnredlist.org/

[7] Mugnier, J. (2008). *Nouvelle Flore Illustrée du Sénégal et Régions Voisines. Les Dicotylédones*, Tome 1. Agroservices, France.

[8] Organisation Ouest Africaine de la Santé (OOAS) (2013). *Pharmacopée Afrique de l'Ouest*. OOAS/ West African Health Organization (WAHO), Bobo Dioulasso, Burkina Faso.

[9] Royal Botanic Gardens, Kew (1999–2015). *Survey of Economic Plants for Arid and Semi-Arid Lands (SEPASAL) Database*. http://apps.kew.org/sepasalweb/sepaweb

[10] Royal Botanic Gardens, Kew (2017). *Seed Information Database (SID).* Version 7.1. http://data.kew.org/sid/

[11] Sacandé, M. & Sanon, M. (2007). *Combretum micranthum* G.Don. *Seed Leaflet* No. 129. Edited by L. Schmidt. Millennium Seed Bank Project (MSBP), Wakehurst Place, Royal Botanic Gardens, Kew.

[12] Sacande, M., Sanou, L. & Beentje, H. (2012). *Guide d'Identification des Arbres du Burkina Faso*. Royal Botanic Gardens, Kew.

[13] von Maydell, H.-J. (1990). *Trees and Shrubs of the Sahel: Their Characteristics and Uses*. Josef Margraf, Weikersheim, Germany.

Entada africana Guill. & Perr.

TAXONOMY AND NOMENCLATURE

FAMILY. Leguminosae – Caesalpinioideae

SYNONYMS. *Entada sudanica* Schweinf., *Entadopsis sudanica* (Schweinf.) G.C.C.Gilbert & Bou

VERNACULAR / COMMON NAMES. Entada d'Afrique (French); didi diaba (Bambara); samanéré (Bambara, Jula); bitiki (Bobo); balayoro (Dogon); diali kamba (Malinke); zanenge (Minyanka, Senufo); mbatari (Peul).

PLANT DESCRIPTION

LIFE FORM. Tree or shrub.

PLANT. Up to 12 m (Figure 1), bole low-branching, <30 cm in diameter, crown spreading, open, bark very rough [2,3]. **LEAVES.** Alternate, bipinnate, drooping, glabrous, 15–45 cm long, pinnae (1)4–6(9) pairs, leaflets (8)24 pairs per pinna. **FLOWERS.** White or cream turning yellowish, in spike-like racemes 5–10 cm long [2].

Figure 1 – General habit of *E. africana*. (Photo: S. Sanogo, IER.)

FRUITS AND SEEDS

Fruits are flat brownish pods with undulating margins (Figure 2), 15–40 x 5–8 cm. Seeds are dark brown-black with the areole surrounded by a pleurogram (Figure 3). They are arranged individually inside the pod segments [2].

FLOWERING AND FRUITING. In southern Mali, flowering occurs from April to May and fruiting from April to June [4].

Figure 2 – Fruits at seed dispersal. (Photo: S. Sanogo, IER.)

Figure 3 – External views of the seeds showing the areole surrounded by the pleurogram, and longitudinal section. (Photos: P. Gómez-Barreiro, RBG Kew.)

DISTRIBUTION

Widespread, from Senegal and Mali east to Ethiopia and Uganda, and south to the Democratic Republic of Congo [5].

HABITAT

Savanna, degraded secondary vegetation, wooded scrub in flooded hollows and along stream banks, on deep sandy or rocky soils, at 450–1,100 m a.s.l. [3]. Sensitive to fires [10].

USES

In traditional medicine, an infusion of the leaves or roots is taken as a tonic [10]. The plant is also used for treating hepatitis and stomach problems [4]. The leaves, bark and roots are used to treat fever (including malarial fever). The leaves are also used to prevent wound infections [10]. Extracts of the stem bark have antioxidant and hepato-protective activities [7]. The plant has potential as an antidote against various pathogens because of its emetic properties. In Central Africa, the plant is used as a fish poison [11]. The wood is used to make tool handles, ladders, mortars and domestic utensils [2]. The leaves have good forage quality, and the flowers provide a good source of nectar for honey bees [10].

KNOWN HAZARDS AND SAFETY

No known hazards.

CONSERVATION STATUS

Assessed as Least Concern (LC) in Burkina Faso on the basis of its distribution and habitat [9], but its status has yet to be confirmed in the IUCN Red List [6].

SEED CONSERVATION

HARVESTING. Fruits can be collected directly from the plants (Figure 4).

PROCESSING AND HANDLING. Seeds can be extracted from the fruits by hand.

STORAGE AND VIABILITY. Seeds are reported to be orthodox [8]. Therefore, after appropriate drying, they can be stored at sub-zero-temperatures for long-term conservation.

PROPAGATION

SEEDS

Dormancy and pre-treatments: Seeds of this species are reported to germinate easily without any pre-treatments and can therefore be considered to be non-dormant. However, scarification enhances germination, suggesting that further studies should be carried out to confirm whether dormancy is present [12].

Germination, sowing and planting: Seeds of this species germinate well in both light and dark conditions [12]. Under laboratory controlled conditions, chipped seeds that are incubated at constant temperatures of 20°C and 25°C reached 85% and 90% of final germination, respectively [8]. Under nursery conditions (Figure 5), the observed germination rate of untreated seeds conserved by farmers is 75% [1] (Figure 6).

VEGETATIVE PROPAGATION

No information available.

Figure 4 – Fruit harvesting. (Photo: S. Sanogo, IER.)

Figure 5 – Weeding of seedlings in the village nursery of Yanfolila. (Photo: S. Sanogo, IER.)

Figure 6 – 18-month-old plant in the arboretum of Zégoua. (Photo: S. Sanogo, IER.)

TRADE

Significant amounts of stem bark and roots are collected by therapists and herbalists in Mali. However, no trade statistics are available.

Authors

Sidi Sanogo, Abdoul K. Sanogo, Bokary A. Kelly, Rokia Sanogo, Stéphane Rivière, Efisio Mattana, Steve Davis and Tiziana Ulian.

References

[1] Alexandre, D.-Y. (1991–1992). *Régénération de la Forêt du Nazinon*. Notes au projet BKF 89/011. ORSTOM, Burkina Faso.

[2] Arbonnier, M. (2004). *Trees, Shrubs and Lianas of West African Dry Zones*. CIRAD, France; Margraf Publishers, Weikersheim, Germany; and Muséum national d'histoire naturelle (MNHN), Paris, France.

[3] Conservatoire et Jardin botaniques & South African National Biodiversity Institute (SANBI) (2016). *African Plant Database*. Version 3.4.0. http://www.ville-ge.ch/musinfo/bd/cjb/africa/

[4] DNEF (2009). *Répertoire des Espèces Forestières Ligneuses des Régions de Mopti, Tombouctou et Gao*. Rapport final. Direction Nationale des Eaux et Forêts (DNEF), Mali.

[5] ILDIS (2006–2013). *World Database of Legumes*. International Legume Database and Information Service. http://www.legumes-online.net/ildis/aweb/database.htm

[6] IUCN (2017). *The IUCN Red List of Threatened Species*. Version 2017-3. http://www.iucnredlist.org/

[7] Njayou, F. N., Aboudi, E. C. E., Tandjang, M. K., Tchana, A. K., Ngadjui, B. T. & Moundipa, P. F. (2013). Hepatoprotective and antioxidant activities of stem bark extract of *Khaya grandifoliola* (Welw) CDC [*sic*] and *Entada africana* Guill. et Perr. *Journal of Natural Products* 6: 73–80.

[8] Royal Botanic Gardens, Kew (2017). *Seed Information Database (SID)*. Version 7.1. http://data.kew.org/sid/

[9] Sacande, M., Sanou, L. & Beentje, H. (2012). *Guide d'Identification des Arbres du Burkina Faso*. Royal Botanic Gardens, Kew.

[10] von Maydell, H.-J. (1990). *Trees and Shrubs of the Sahel: Their Characteristics and Uses*. Josef Margraf, Weikersheim, Germany.

[11] Watt, J. M. & Breyer-Brandwijk, M. G. (1962). *The Medicinal and Poisonous Plants of Southern and Eastern Africa*, 2nd edition. E. & S. Livingstone, Edinburgh & London.

[12] Zida, D., Tigabu, M., Sawadogo, L. & Oden, P. C. (2005). Germination requirements of seeds of four woody species from the Sudanian savanna in Burkina Faso, West Africa. *Seed Science and Technology* 33: 581–593.

Erythrina senegalensis DC.

TAXONOMY AND NOMENCLATURE

FAMILY. Leguminosae – Papilionoideae

SYNONYMS. *Duchassaingia senegalensis* (DC.) Hassk., *Erythrina guineensis* G.Don, *E. latifolia* Schumach. & Thonn.

VERNACULAR / COMMON NAMES. Coral tree (English); erythrine du Sénégal (French); ntenbilen, ntimini, santigui nianden (Bambara); ntenbilen (Jula); leru, ntenkisèdalilen, nti, ntijè (Malinke); kaferinye (Minyanka); kulintiga (Moore); mbototay, mototày (Peul); kafernanye (Senufo).

PLANT DESCRIPTION

LIFE FORM. Tree.

PLANT. 6–7(15) m tall (Figure 1), bole rarely straight, branches twisted forming irregular crown, bark rough, corky, cracked, armed with stout prickles. **LEAVES.** Alternate, trifoliate, glabrous, 12–30 cm long, terminal leaflet larger than two lateral. **FLOWERS.** Bright red, in erect terminal racemes (Figure 1) [1,9].

Figure 1 – *E. senegalensis* in flower (left) and inflorescence (above). (Photos: S. Sanogo, IER.)

FRUITS AND SEEDS

Fruits are irregular-shaped pods (Figure 2), containing 5–9 seeds, which are shiny, red, ovoid, 6–7 mm long (Figure 3) [6].

FLOWERING AND FRUITING. Trees are in leaf between October and April; flowering occurs between November and March; fruiting occurs from January to April [6].

Figure 2 – Ripe fruits and seed dispersal.
(Photo: S. Sanogo, IER.)

Figure 3 – External views and longitudinal section of the seed showing the embryonal axis and cotyledon.
(Photos: P. Gómez-Barreiro, RBG Kew.)

DISTRIBUTION

Throughout tropical West Africa, from Senegal and Mali to Chad, south to Cameroon [4].

HABITAT

Dry open savanna woodlands, burned savanna, coastal savanna (e.g. in Ghana), stream banks and roadsides, up to 1,200 m a.s.l. [3,4].

USES

Erythrina senegalensis is used widely in traditional medicine [3,7]. A root decoction is used in the treatment of onchocerciasis (river blindness), stomach pain, dysentery and kidney pain. Powdered fibre is used in the treatment of hepatitis, hepatobiliary disorders, icterus and cirrhosis of the liver. Macerated bark root fibres are used against vomiting [6]. The leaves are purgative. The bark from the trunk, macerated in water, is used for treating malaria, hepatobiliary disorders and amenorrhea. A decoction of the bark is used for treating fevers [3]. *Erythrina senegalensis* is an ornamental tree with attractive bright red flowers. It is often planted as a hedge and fence. Its wood is used as tinder to light fires [1].

KNOWN HAZARDS AND SAFETY

The bark is highly poisonous and has been used as an ordeal poison [14]. The raw seeds are poisonous [13]. The prickles on the trunk and branches are sharp.

CONSERVATION STATUS

Assessed as Least Concern (LC) according to IUCN Red List criteria [5,9]. Although widely exploited for traditional medicinal uses, the population dynamics of the species do not appear to be affected at present [3].

SEED CONSERVATION

HARVESTING. Fruits should be harvested from the trees when seed dispersal begins (Figure 2) [9].
PROCESSING AND HANDLING. Ripe seeds can be extracted easily from open pods.
STORAGE AND VIABILITY. Seeds of this species are reported to be orthodox [8]. Therefore, after appropriate drying, they can be stored at sub-zero temperatures for long-term conservation.

PROPAGATION

SEEDS

Dormancy and pre-treatments: Seeds should be scarified before planting. A germination rate of 85% can be achieved after treating with boiling water, followed by soaking in water for 24 hours [10]. The need for pre-treatment suggests the presence of physical dormancy, which has been reported for other *Erythrina* spp. [2]. However, this should be verified by further studies.
Germination, sowing and planting: Under laboratory controlled conditions, chipped seeds incubated at 21°C, with a photoperiod of 12 hours of light per day, reached 85% final germination [8]. In the nursery, seedlings reach 40–60 cm in four months [10]. When planted in 50 x 50 cm holes, and given 15 litres of water per plant every 15 days during the dry season, plants bear fruits after three years [11]. After four years, seedlings planted in Zégoua (annual rainfall 1,000–1,200 mm; Figure 4) and Kougué (annual rainfall 800–1,000 mm) had an average height of 200 cm and 265 cm, respectively [12].

VEGETATIVE PROPAGATION

Roots easily and grows rapidly from cuttings [10] and branch stakes [3].

PHYTOCHEMICAL ANALYSES

The bark of the trunk and roots were analysed by DMT/INRSP to determine the active pharmacological components.

Figure 4 – 18-month-old plant in the medicinal plants garden of the Association of Zégoua therapists. (Photo: S. Sanogo, IER.)

TRADE

Exploited by traditional healers, but no trade statistics are available.

Authors

Sidi Sanogo, Abdoul K. Sanogo, Bokary A. Kelly, Rokia Sanogo, Stéphane Rivière, Efisio Mattana and Tiziana Ulian.

References

[1] Arbonnier, M. (2004). *Trees, Shrubs and Lianas of West African Dry Zones*. CIRAD, France; Margraf Publishers, Weikersheim, Germany; and Muséum national d'histoire naturelle (MNHN), Paris, France.

[2] Baskin, C. C. & Baskin, J. M. (2014). *Seeds: Ecology, Biogeography, and Evolution of Dormancy and Germination*, 2nd edition. Academic Press, San Diego, USA.

[3] Burkill, H. M. (1995). *The Useful Plants of West Tropical Africa, Vol. 3: Families J–L*. 2nd edition. Royal Botanic Gardens, Kew.

[4] Conservatoire et Jardin botaniques & South African National Biodiversity Institute (SANBI) (2016). *African Plant Database*. Version 3.4.0. http://www.ville-ge.ch/musinfo/bd/cjb/africa/

[5] IUCN (2017). *The IUCN Red List of Threatened Species*. Version 2017-3. http://www.iucnredlist.org/

[6] Malgras, D. (1992). *Arbres et Arbustes Guérisseurs des Savanes Maliennes*. Agence de Cooperation Culturelle et Technique, Paris, France.

[7] Royal Botanic Gardens, Kew (1999–2015). *Survey of Economic Plants for Arid and Semi-Arid Lands (SEPASAL) Database*. http://apps.kew.org/sepasalweb/sepaweb

[8] Royal Botanic Gardens, Kew (2017). *Seed Information Database (SID)*. Version 7.1. http://data.kew.org/sid/

[9] Sacande, M., Sanou, L. & Beentje, H. (2012). *Guide d'Identification des Arbres du Burkina Faso*. Royal Botanic Gardens, Kew.

[10] Sanogo, A. K. (2000). *Techniques Simples de Production de Plants Forestiers*. Fiche technique, PRF/IER, Mali and Programme GDRN/Intercoopération, Switzerland.

[11] Sanogo, A. K., Sanogo, S. & Sidibé, S. I. (2011). *Conservation et Gestion Durable des Plantes Utiles aux Communautés Rurales*. Rapport de campagne 17, Commission Scientifique, IER, Bamako, Mali.

[12] Sanogo, A. K., Sanogo, S. & Sidibé, S. I. (2013). *Conservation et Gestion Durable des Plantes Utiles aux Communautés Rurales*. Rapport de campagne 19, Comité de Programme, IER, Bamako, Mali.

[13] von Maydell, H.-J. (1990). *Trees and Shrubs of the Sahel: Their Characteristics and Uses*. Josef Margraf, Weikersheim, Germany.

[14] Watt, J. M. & Breyer-Brandwijk, M. G. (1962). *The Medicinal and Poisonous Plants of Southern and Eastern Africa*, 2nd edition. E. & S. Livingstone, Edinburgh & London.

Evolvulus alsinoides (L.) L.

TAXONOMY AND NOMENCLATURE

FAMILY. Convolvulaceae

SYNONYMS. *Convolvulus alsinoides* L., *C. linifolius* L., *C. valerianoides* Blanco, *Evolvulus albiflorus* M.Martens & Galeotti, *E. azureus* Vahl ex Schumach. & Thonn., *E. chinensis* Choisy, *E. debilis* Kunth, *E. filiformis* Willd. ex Steud., *E. hirsutulus* Choisy, *E. pilosissimus* M.Martens & Galeotti, *E. pudicus* Hance, *E. pumilus* Span., *E. ramiflorus* Bojer ex Choisy, and many more (see [12]).

VERNACULAR / COMMON NAMES. Slender dwarf-morning-glory (English); koni ka koa (Bambara); lemlehy, ndottiyel (Peul).

Figure 1 – *E. alsinoides* with flowers at anthesis. (Photo: S. Sanogo, IER.)

PLANT DESCRIPTION

LIFE FORM. Annual or perennial herb.

PLANT. Spreading or ascending, stems <50 cm long (Figure 1), thinly or sometimes densely covered with somewhat long, spreading, silky hairs, with few or several stems. **LEAVES.** Alternate, entire, widely spaced, ovate-oblong to elliptic or lanceolate, subsessile, 5–45 x 1–15 mm, acute or rounded at both ends, distinctly mucronate. **FLOWERS.** Inflorescences axillary, 1 to few-flowered (Figure 1), corolla blue, rarely white, <8 mm long and wide [5].

FRUITS AND SEEDS

Fruits are globose capsules, 3–4 mm long, glabrous, 4-valved and 4-seeded. Seeds are brown to black, ovoid, 1.7 mm long, smooth, glabrous (Figure 2) [5].

FLOWERING AND FRUITING. In West Africa, flowering and fruiting occurs from November to April.

Figure 2 – External views of the seeds. (Photo: P. Gómez-Barreiro, RBG Kew.)

DISTRIBUTION

Widespread throughout tropical and subtropical regions of both hemispheres [12].

HABITAT

Woodlands, grasslands, savanna, thicket edges, roadsides and cultivated ground, on sandy soils [5] from sea level to 1,650 m a.s.l. [2].

USES

Used in traditional medicine by ethnic groups across Africa, India and the Philippines to treat fever, coughs, common cold, venereal diseases and depression. The leaves and roots are also used for digestive system disorders [11]. Investigations *in vivo* and *in vitro* have shown anti-stress, anti-anaemic, antimicrobial and gastro-protective properties [10]. A study on diabetic rats showed that an extract of the plant reduced the oxidative stress induced by streptozotocin and potentially increased insulin levels [3].

KNOWN HAZARDS AND SAFETY

No known hazards.

CONSERVATION STATUS

This widespread species is assessed as Least Concern (LC) according to IUCN Red List criteria in South Africa [7] and in parts of its range in Asia [6], but its status has yet to be confirmed in the IUCN Red List [4]. A variety previously assessed as Endangered (EN) in Japan (*E. alsinoides* var. *rotundifolius* Hayata ex Ooststr.) is no longer considered to be a distinct taxon [6,12].

SEED CONSERVATION

HARVESTING. Ripe infructescences can be collected directly from plants.

PROCESSING AND HANDLING. Seeds can be separated manually in the laboratory using sieves.

STORAGE AND VIABILITY. Seeds of this species are reported to be orthodox [8]. Therefore, after appropriate drying, they can be stored at sub-zero temperatures for long-term conservation.

PROPAGATION

SEEDS

Dormancy and pre-treatments: Convolvulaceae spp. are reported to have a physical component to seed dormancy [1]. Pre-treatments which allow water imbibition should therefore be carried out before sowing. Seeds stored at the MSB have been chipped with a scalpel before germination experiments [8].

Germination, sowing and planting: Under controlled laboratory conditions, chipped seeds reached high germination percentages (75–100%) when sown in the light at constant temperatures in the range 15–25°C [8].

VEGETATIVE PROPAGATION

No information available.

PHYTOCHEMICAL ANALYSES

Analyses carried out by DMT/INRSP revealed that the main phytochemical constituents are tannins, coumarins, flavonoids, sapononins, leucoanthocyanins, holosides, mucilage, triterpenes, steroids and alkaloids [9].

TRADE

No data available for Mali.

Authors

Sidi Sanogo, Abdoul K. Sanogo, Bokary A. Kelly, Stéphane Rivière, Efisio Mattana, Tiziana Ulian, Steve Davis and Rokia Sanogo.

References

[1] Baskin, C. C. & Baskin, J. M. (2014). *Seeds: Ecology, Biogeography, and Evolution of Dormancy and Germination*, 2nd edition. Academic Press, San Diego, USA.

[2] Germishuizen, G. & Meyer, N. L. (eds) (2003). *Plants of Southern Africa: An Annotated Checklist. Strelitzia* 14, National Botanical Institute, Pretoria, South Africa.

[3] Gomathi, D., Ravikumar, G., Kalaiselvi, M., Devaki, K. & Uma, C. (2013). Efficacy of *Evolvulus alsinoides* (L.) L. on insulin and antioxidants activity in pancreas of streptozotocin induced diabetic rats. *Journal of Diabetes & Metabolic Disorders* 12: 39.

[4] IUCN (2017). *The IUCN Red List of Threatened Species.* Version 2017-3. http://www.iucnredlist.org/

[5] Launert, E. (ed.) (1987). Convolvulaceae. In: *Flora Zambesiaca*, Vol. 8, part 1. Flora Zambesiaca Managing Committee, Royal Botanic Gardens, Kew.

[6] National Red List (2017). *Evolvulus alsinoides.* National Red List Network. http://www.nationalredlist.org/

[7] Raimondo, D., von Staden, L., Foden, W., Victor, J. E., Helme, N. A., Turner, R. C., Kamundi, D. A. & Manyama, P. A. (2009). *Red List of South African Plants. Strelitzia* 25, South African National Biodiversity Institute, Pretoria, South Africa.

[8] Royal Botanic Gardens, Kew (2017). *Seed Information Database (SID).* Version 7.1. http://data.kew.org/sid/

[9] Sanogo, A. K., Sanogo, S. & Sanogo, R. (2010). *Rapport Projet UPP. Atelier de Fin de Phase du Projet.* MGU–Useful Plants Project (UPP), Mali.

[10] Singh, A. (2008). Review of ethnomedicinal uses and pharmacology of *Evolvulus alsinoides* Linn. *Ethnobotanical Leaflets* 12: 734–740.

[11] Staugård, F. & Anderson, S. V. (1989). *Traditional Medicine in Botswana: Traditional Medicinal Plants.* Ipelegeng Publishers, Gaborone, Botswana.

[12] WCSP (2017). *World Checklist of Selected Plant Families.* Royal Botanic Gardens, Kew. http://wcsp.science.kew.org/

Flemingia faginea (Guill. & Perr.) Baker

TAXONOMY AND NOMENCLATURE

FAMILY. Leguminosae – Papilionoideae

SYNONYMS. *Moghania faginea* (Guill. & Perr.) Kuntze, *Rhynchosia faginea* Guill. & Perr.

VERNACULAR / COMMON NAMES. Sanfito (Manding).

PLANT DESCRIPTION

LIFE FORM. Shrub.

PLANT. Erect, bushy, 2–3 m tall (Figure 1), branching from base, branchlets chestnut brown or reddish tomentose [3]. **LEAVES.** Simple, alternate, rotund to cordate, c. 5 x 5 cm, with pubescent petioles. **FLOWERS** Fasciculate, at the level of leaf axils, sometimes accompanied by short racemes, corolla white or pink and purple-streaked, 7–8 mm long [2,5].

Figure 1 – *F. faginea* in full leaf on the bank of the Banifing River. (Photo: S. Sanogo, IER.)

FRUITS AND SEEDS

Fruits are pods 12 x 5 mm, with a hooked tip, and containing two large, round, and slightly mottled seeds (Figure 2) [5].

FLOWERING AND FRUITING: Seeds can be collected in February.

Figure 2 – External views of the seeds showing the mottled teguments and hilum region, and longitudinal and cross-sections showing the embryonal axis and cotyledon. (Photos: P. Gómez-Barreiro, RBG Kew.)

DISTRIBUTION

Tropical West Africa, from Senegal and Mali to Ghana [3].

HABITAT

Riverbanks in savanna, gallery forests, pond edges and temporarily flooded valleys [3,5].

USES

In traditional medicine, the bark is used for treating haemorrhoids and venereal diseases. A decoction of the leaves and bark is used against abortion, skin infections and child malnutrition. A decoction of the leaves or leaf powder is used to lower blood pressure and blood sugar levels. A decoction of the roots is used against scabies and itchy skin [2].

KNOWN HAZARDS AND SAFETY

No known hazards.

CONSERVATION STATUS

Yet to be assessed according to IUCN Red List criteria [4].

SEED CONSERVATION

HARVESTING. Fruits should be harvested when they start to open.

PROCESSING AND HANDLING. Ripe seeds can be easily separated from the open pods.

STORAGE AND VIABILITY. No information is available on seed storage behaviour.

Figure 3 – Harvesting leafy twigs of *F. faginea* for phytochemical analysis of extracts in the laboratory. (Photo: S. Sanogo, IER.)

PROPAGATION

SEEDS

Dormancy and pre-treatments: No information on seed dormancy is available. Seeds of the congeneric *Flemingia macrophylla* (Willd.) Merr. are reported to be physically dormant [1], suggesting that treatments to enhance seed imbibition should be tested for *F. faginea*.

Germination, sowing and planting: No information is available on optimum conditions for germination under laboratory controlled conditions.

VEGETATIVE PROPAGATION

No information available.

PHYTOCHEMICAL ANALYSES

Leaf extracts were analysed by DMT/INRSP (Figure 3). The main chemical groups detected were coumarins, flavonoids, tannins, reducing compounds, saccharides and holosides [6]. Five flavonoid 3-O-glycosides were isolated from the leaves, some of which were reported for the first time [7].

TRADE

No data available.

Authors

Sidi Sanogo, Abdoul K. Sanogo, Bokary A. Kelly, Rokia Sanogo, Stéphane Rivière and Efisio Mattana.

References

[1] Baskin, C. C. & Baskin, J. M. (2014). *Seeds: Ecology, Biogeography, and Evolution of Dormancy and Germination*, 2nd edition. Academic Press, San Diego, USA.

[2] Burkill, H. M. (1995). *The Useful Plants of West Tropical Africa. Vol. 3: Families J–L*, 2nd edition. Royal Botanic Gardens, Kew.

[3] Conservatoire et Jardin botaniques & South African National Biodiversity Institute (SANBI) (2016). *African Plant Database*. Version 3.4.0. http://www.ville-ge.ch/musinfo/bd/cjb/africa/

[4] IUCN (2017). *The IUCN Red List of Threatened Species.* Version 2017-3. http://www.iucnredlist.org/

[5] Mugnier, J. (2008). *Nouvelle Flore Illustrée du Sénégal et Régions Voisines. Les Dicotylédones*, Tome 1. Agroservices, France.

[6] Sanogo, A. K., Sanogo, S., Yossi, H., Diallo, D., Sanogo, R., Coulibaly, K. & Sidibé, S. I. (2010). *Conservation et Gestion Durable des Plantes Utiles aux Communautés*. Rapport de champagne, 16th session du comité de programme de l'Institut d'Economie Rurale (IER) du Mali.

[7] Soicke, H., Görler, K. & Waring, H. (1990). Flavonol glycosides from *Moghania faginea*. *Planta Medica* 56 (4): 410–412.

Gymnosporia senegalensis (Lam.) Loes.

TAXONOMY AND NOMENCLATURE

FAMILY. Celastraceae

SYNONYMS. *Catha senegalensis* (Lam.) G.Don, *Celastrus europaeus* Boiss., *C. saharae* Batt., *C. senegalensis* Lam., *Gymnosporia baumii* Loes., *G. benguelensis* Loes., *G. dinteri* Loes., *G. eremoecusa* Loes., *G. europaea* (Boiss.) Masf., *G. intermedia* Chiov., *G. saharae* (Batt.) Loes. ex Engl., *G. senegalensis* (Lam.) Loes. var. *senegalensis*, *Maytenus senegalensis* (Lam.) Exell, and many more (see [4]).

VERNACULAR / COMMON NAMES. Spike thorn (English); gnikele, koronkole, ngele, ntogoyo (Bambara); flakumo fogojoda (Bwa); kafu kwoni, soukoma (Minyanka, Senufo); gialgoti, gielgotel (Peul).

Figure 1 – General habit of *G. senegalensis*. (Photo: S. Sanogo, IER.)

Figure 2 – Flowering branches. (Photo: S. Sanogo, IER.)

PLANT DESCRIPTION

LIFE FORM. Shrub or tree.

PLANT. Semi-deciduous, dioecious, rarely monoecious (Figure 1), unarmed or with spines <4 cm long [11], branches reddish purple to reddish brown, unlined, ± flattened. **LEAVES.** Fasciculate or not, petiolate, 6 x c. 0·5–6.3 cm, oblong or rarely ovate to obovate or oblong-elliptic, pale green coriaceous or subcoriaceous, usually glaucous, margins usually minutely serrate. **FLOWERS.** Inflorescence cymose (Figure 2), dichasial at first, becoming monochasial, solitary and axillary or 1–6 on short axillary shoots, flowers scented, 3–60 per cyme, petals white or greenish white to pale yellow, sometimes tinged pink, 1–3.5 mm long [6].

FRUITS AND SEEDS

Fruits are pink to deep red capsules, 2–6 mm long, globose to pyriform, coriaceous and smooth, with 1–2 dark reddish brown, glossy seeds (Figure 3), with a fleshy smooth rose pink aril obliquely covering the lower 1/3–2/3 [6].

FLOWERING AND FRUITING. In West Africa, flowers in the dry season after coming into leaf [2].

Figure 3 – External views of the seeds showing the aril, and longitudinal and cross-sections showing the spatulate embryo and endosperm. (Photos: P. Gómez-Barreiro, RBG Kew.)

DISTRIBUTION

Widely distributed in sub-Saharan Africa, from Senegal to Eritrea, southwards through South Tropical Africa to South Africa; also found in parts of North Africa, Madagascar, and eastwards to India [6].

HABITAT

Deciduous woodland, bushland, scrub and wooded grassland, riverbanks and swamp margins, from sea level to 2,400 m a.s.l. [6].

USES

Gymnosporia senegalensis is used in traditional medicine for the treatment of numerous ailments, including respiratory diseases, inflammation, microbial infections and as a topical application for healing wounds [5]. The roots and bark are used, the latter for treating wounds, ulcers, boils and oedema [2]. Root bark extracts have been shown by laboratory studies to have inhibitory activity on *Escherichia coli*, *Staphylococcus aureus*, *Salmonella* spp. and *Candida albicans* [1]. Root extracts have been studied for their anti-inflammatory properties [12]. The young leaves and flowers are added to soup. The wood is used for making handles of spears and other tools, and is also used as fuelwood [2].

KNOWN HAZARDS AND SAFETY

The spines on the branches are sharp.

CONSERVATION STATUS

Assessed as Least Concern (LC) on the basis of its distribution and habitat in West Africa [11] and South Africa [10], but its status has yet to be confirmed in the IUCN Red List [7]. Other assessments have been made according to previous IUCN criteria for Egypt (Vulnerable [VU]) and for Bihar and Chhattisgarh (India) (Rare [R]) [9]. A subspecies previously assessed in Spain as being Near Threatened (NT) is no longer considered to be a distinct taxon [4,9]. In Mali, the use of *G. senegalensis* is not restricted by law despite overexploitation by traditional healers.

SEED CONSERVATION

HARVESTING. Ripe fruits can be collected directly from the trees.

PROCESSING AND HANDLING. Seeds can be extracted by hand from the dry capsules.

STORAGE AND VIABILITY. No information is available on seed storage behaviour.

PROPAGATION

SEEDS

Dormancy and pre-treatments: No information is available on seed dormancy. Congeneric species are reported to have physiological dormancy [3], highlighting the need for further studies. Removal by hand of the aril, followed by mechanical scarification with sandpaper, has been applied in an *in vitro* germination study [8].

Germination, sowing and planting: No information is available on optimum conditions for germination under laboratory controlled conditions or in the nursery.

VEGETATIVE PROPAGATION

A micropropagation protocol has been established using *in vitro* germinated 6-week-old seedlings as a source of explants [8].

PHYTOCHEMICAL ANALYSES

Analyses of leaves in decoction carried out by DMT/INRSP confirmed the use of the plant by therapists against malaria.

TRADE

Despite overexploitation of the roots by traditional healers in Mali, there are no trade statistics available.

Authors

Sidi Sanogo, Abdoul K. Sanogo, Bokary A. Kelly, Stéphane Rivière, Efisio Mattana, Steve Davis and Rokia Sanogo.

References

[1] Agban, A., Karou, D. S., Tchacondo, T., Batawila, K., Atchou, K., Gbeassor, M. & de Souza, C. (2012). Activité antimicrobienne de *Maytenus senegalensis* (Lam.) Exell (celastraceae) [*sic*], une plante de la médecine traditionnelle du Togo. *Journal de la Recherche Scientifique de l'Universite de Lome* 14 (1) (Series A): 63–68.

[2] Arbonnier, M. (2004). *Trees, Shrubs and Lianas of West African Dry Zones*. CIRAD, France; Margraf Publishers, Weikersheim, Germany; and Muséum national d'histoire naturelle (MNHN), Paris, France.

[3] Baskin, C. C. & Baskin, J. M. (2014). *Seeds: Ecology, Biogeography, and Evolution of Dormancy and Germination*, 2nd edition. Academic Press, San Diego, USA.

[4] Conservatoire et Jardin botaniques & South African National Biodiversity Institute (SANBI) (2016). *African Plant Database*. Version 3.4.0. http://www.ville-ge.ch/musinfo/bd/cjb/africa/

[5] da Silva, G., Serrano, R. & Silva, O. (2011). *Maytenus heterophylla* and *Maytenus senegalensis*, two traditional herbal medicines. *Journal of Natural Science, Biology and Medicine* 2 (1): 59–65.

[6] Exell, A. W., Fernandes, A. & Wild, H. (eds) (1966). Celastraceae. In: *Flora Zambesiaca*, Vol. 2, part 2. Flora Zambesiaca Managing Committee, Royal Botanic Gardens, Kew.

[7] IUCN (2017). *The IUCN Red List of Threatened Species*. Version 2017-3. http://www.iucnredlist.org/

[8] Matu, E. N., Lindsey, K. L. & van Staden, J. (2006). Micropropagation of *Maytenus senegalensis* (Lam.) Excell [*sic*]. *South African Journal of Botany* 72 (2006): 409–415.

[9] National Red List (2017). *Gymnosporia senegalensis*. National Red List Network. http://www.nationalredlist.org/

[10] Raimondo, D., von Staden, L., Foden, W., Victor, J. E., Helme, N. A., Turner, R. C., Kamundi, D. A. & Manyama, P. A. (2009). *Red List of South African Plants*. *Strelitzia* 25, South African National Biodiversity Institute, Pretoria, South Africa.

[11] Sacande, M., Sanou, L. & Beentje, H. (2012). *Guide d'Identification des Arbres du Burkina Faso*. Royal Botanic Gardens, Kew.

[12] Sosa, S., Morelli, C. F., Tubaro, A., Cairoli, P., Speranza, G. & Manitto, P. (2007). Anti-inflammatory activity of *Maytenus senegalensis* root extracts and of maytenoic acid. *Phytomedicine* 14 (2–3): 109–114.

Lawsonia inermis L.

TAXONOMY AND NOMENCLATURE

FAMILY. Lythraceae

SYNONYMS. *Lawsonia speciosa* L., *L. spinosa* L.

VERNACULAR / COMMON NAMES. Henna (English); henné, réséda de France (French); djabi (Bambara); pouddi (Peul).

Figure 1 – General habit of *L. inermis*. (Photo: S. Sanogo, IER.)

PLANT DESCRIPTION

LIFE FORM. Shrub or tree.

PLANT. Evergreen, 2–7 m tall (Figure 1), glabrous, branches often densely tangled, young branches 4-angular, becoming terete, rigid, often spinescent, spines <3.5 cm long [4].

LEAVES. Elliptic, obovate to oblanceolate, c. 1–6.5 x 0.5–2.5 cm, apex acute or mucronate, petiole short. **Flowers**. Panicles 5–25 cm long, flowers creamy white, scented, petals c. 2 mm long, slightly clawed [9].

FRUITS AND SEEDS

Spherical, indehiscent, glabrous capsules (Figure 2), with the calyx at the base and the persistent style at the top, c. 5 mm in diameter, pale brown when ripe, containing numerous pyramidal seeds (Figure 3) [1].

FLOWERING AND FRUITING. Flowers in the second half of the dry season and at the start of the rainy season in the Sikasso region of Mali.

Figure 2 – Ripening fruits. (Photo: S. Sanogo, IER.)

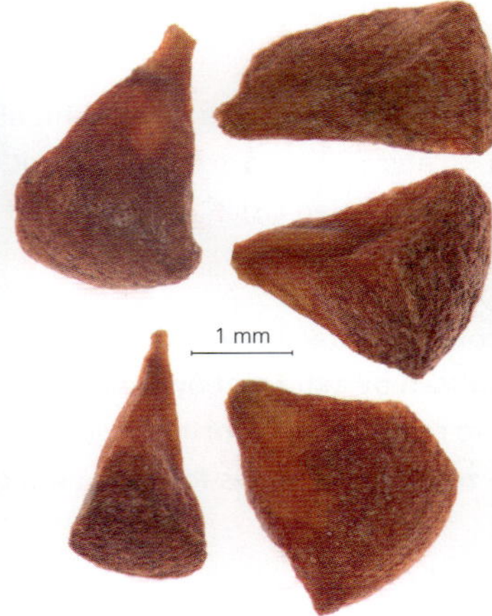

Figure 3 – External views of the seeds. (Photo: P. Gómez-Barreiro, RBG Kew.)

Figure 4 – Hand decorated with henna. (Photo: R. Dackouo, IER.)

DISTRIBUTION

Native to western tropical Asia, North Africa and probably the eastern coast of Africa; widely introduced, naturalised and cultivated throughout the tropics [9].

HABITAT

Temporarily flooded, rocky watercourses in dry areas, riverine thickets and forest, bushland, cliffs and rocky crevices, up to 1,400 m a.s.l. [4]. It is often found on alluvial soils along rivers and in villages [9], and also prefers sandy soils [1]. It is thought to have been cultivated in Sudano-Sahelian areas since prehistory [1].

USES

Henna is one of the world's oldest cosmetics. The dye from its leaves is used to decorate hands (Figure 4), nails, feet and hair, and is also used to colour leather and clothes. Written records of use date back more than 2,500 years: henna was used in Ancient Egypt [7] and is mentioned in the Bible. Subsequently, henna assumed great importance in Islam (it is used in many ceremonies, especially marriage), Hinduism and Buddhism. In traditional medicine, henna is used as a panacea in the treatment of many diseases [5,6]. The roots are used as a diuretic, to promote childbirth, and to treat gonorrhoea and bronchitis. The leaves have antibiotic properties and are used to treat leprosy and scurvy. A decoction of the leaves and roots is used to treat diarrhoea. A perfume is made from the flowers. The wood is used in India for making stakes and handles, and also provides firewood. Henna is used for soil conservation; and is widely planted as a living fence and as an ornamental [6]. The flowers attract bees [12].

KNOWN HAZARDS AND SAFETY

Generally considered to be safe but should not be used by those with glucose-6-phosphate dehydrogenase (G6PD) deficiency [14]. Allergic reactions are rare. However, adulterants added to henna paste, or products sold as substitutes for henna that contain para-phenylene diamine (PPD) (e.g. black henna), can sometimes cause severe allergic reactions [11]. Care should be taken when handling the plant as the spines on the stems are sharp.

CONSERVATION STATUS

This widespread species is not considered to be either rare or threatened. It has been assessed as Least Concern (LC) in West Africa [17] and Sri Lanka according to IUCN Red List criteria [10], but its status has yet to be confirmed in the IUCN Red List [8].

SEED CONSERVATION

HARVESTING. Ripe fruits can be collected directly from the plants.
PROCESSING AND HANDLING. Seeds can be extracted by hand from the dry capsules.
STORAGE AND VIABILITY. Seeds are reported to be orthodox [16]. Therefore, after appropriate drying, they can be stored at sub-zero temperatures for long-term conservation.

PROPAGATION

SEEDS

Dormancy and pre-treatments: Seeds of this species are reported to be physiologically dormant [2]. Soaking the seeds in water for 1–3 days promotes germination [3]. Seed lots stored at the MSB were incubated without any pre-treatments, or by adding potassium nitrate (KNO_3) to the germination substrate. High germination percentages were achieved in both cases [16].
Germination, sowing and planting: Under laboratory controlled conditions, untreated and KNO_3-treated seeds stored at the MSB reached high germination percentages when incubated in the light at constant temperatures of 20–26°C or under alternating temperature regimes of 15/25°C and 20/35°C [16]. In the nursery, germination takes 2–3 weeks after sowing, with a germination rate of up to 70% [12].

VEGETATIVE PROPAGATION

Easy to propagate from cuttings [12]. Micropropagation is also reported to be successful [13]; a protocol for mass propagation has been developed [15].

TRADE

No trade statistics are available for Mali. Large quantities of henna are produced at home or for local markets, so accurate estimates of international production are also unavailable.

Authors

Sidi Sanogo, Abdoul K. Sanogo, Bokary A. Kelly, Stéphane Rivière, Efisio Mattana, Steve Davis and Tiziana Ulian.

References

[1] Arbonnier, M. (2004). *Trees, Shrubs and Lianas of West African Dry Zones*. CIRAD, France; Margraf Publishers, Weikersheim, Germany; and Muséum national d'histoire naturelle (MNHN), Paris, France.

[2] Baskin, C. C. & Baskin, J. M. (2014). *Seeds: Ecology, Biogeography, and Evolution of Dormancy and Germination*, 2nd edition. Academic Press, San Diego, USA.

[3] Centre National de Semences Forestières (CNFS) (2012–2015). *Catalogue des Semences Forestières (2012–2015)*. CNFS, Ouagadougou, Burkina Faso.

[4] Conservatoire et Jardin botaniques & South African National Biodiversity Institute (SANBI) (2016). *African Plant Database*. Version 3.4.0. http://www.ville-ge.ch/musinfo/bd/cjb/africa/

[5] Forgues, M. & Bailleul, C. (2009). *Richesses Médicinales du Bénin, Burkina Faso, Mali, Sénégal, Togo... Pays de la Zone Sahélo-Soudano-Guinéenne*. Editions Donniya, Bamako, Mali.

[6] Getachew, A. & Suzanne, T. L. (2005). *Lawsonia inermis* L. In: *PROTA (Plant Resources of Tropical Africa/Ressources Végétales de l'Afrique Tropicale)*. Edited by P. C. M. Jansen & D. Cardon, Wageningen, The Netherlands. http://www.prota4u.org/

[7] Hepper, F. N. (1990). *Pharaoh's Flowers: The Botanical Treasures of Tutankhamun*. Royal Botanic Gardens, Kew/HMSO, London.

[8] IUCN (2017). *The IUCN Red List of Threatened Species*. Version 2017-3. http://www.iucnredlist.org/

[9] Launert, E. (ed.) (1978). Lythraceae. In: *Flora Zambesiaca*, Vol. 4. Flora Zambesiaca Managing Committee, Royal Botanic Gardens, Kew.

[10] Ministry of Environment (MOE) (2012). *The National Red List 2012 of Sri Lanka; Conservation Status of the Fauna and Flora*. Ministry of Environment, Colombo, Sri Lanka.

[11] NHS Choices (2015). *The Dangers of Black Henna*. National Health Service, UK. https://www.nhs.uk/Livewell/skin/Pages/black-henna-tattoo-ppd.aspx

[12] Orwa, C., Mutua, A., Kindt, R., Jamnadass, R. & Simons, A. (2009). *Agroforestree Database: A Tree Reference and Selection Guide*. Version 4.0. World Agroforestry Centre, Kenya. http://www.worldagroforestry.org/output/agroforestree-database

[13] Rajkumar, M. H., Sringeswara, A. N. & Rajanna, M. D. (2011). *Ex-situ* conservation of medicinal plants at University of Agricultural Sciences, Bangalore, Karnataka. *Recent Research in Science and Technology* 3 (4): 21–27.

[14] Raupp, P., Ali Hassan, J., Varughese, M. & Kristiansson, B. (2001). Henna causes life threatening haemolysis in glucose-6-phosphate dehydrogenase deficiency. *Archives of Disease in Childhood* 85 (5): 411–412.

[15] Rout, G. R., Das, G., Samantaray, S. & Das, P. (2001). *In vitro* micropropagation of *Lawsonia inermis* (Lythraceae). *Rev. Biol. Trop*. 49 (3–4): 957–963.

[16] Royal Botanic Gardens, Kew (2017). *Seed Information Database (SID)*. Version 7.1. http://data.kew.org/sid/

[17] Sacande, M., Sanou, L. & Beentje, H. (2012). *Guide d'Identification des Arbres du Burkina Faso*. Royal Botanic Gardens, Kew.

Lippia multiflora Moldenke

TAXONOMY AND NOMENCLATURE

FAMILY. Verbenaceae

SYNONYMS: None.

VERNACULAR/COMMON NAMES: Bush tea, healer herb (English); thé de Gambie (French); kankaliba (Bambara, Jula).

PLANT DESCRIPTION

LIFE FORM. Shrub or tree.

PLANT. Robust woody perennial, <2–4 m tall (Figure 1), stems ridged. **LEAVES.** Oblong-lanceolate or elliptic, 3–4 verticillate, c. 5–17 x 1–5.5 cm, bluish green, apex acute, margins finely toothed, whitish pubescence below. **FLOWERS.** Aromatic, small (2 mm long), whitish with yellow centres, borne on terminal inflorescences <20 cm long, 10 cm wide, with numerous spikes, and cone-shaped heads [12].

Figure 1 – Leafy branch of *L. multiflora* in full bloom. (Photo: S. Sanogo, IER.)

FRUITS AND SEEDS

Fruits are mericarps derived from dry schizocarps which split when mature (Figure 2).

FLOWERING AND FRUITING. Flowers from September to November; fruits mature in January [6].

Figure 2 – External views of the mericarps. (Photos: P. Gómez-Barreiro, RBG Kew.)

DISTRIBUTION

Tropical West Africa, from Gambia to Sudan, south to Angola [13].

HABITAT

Savanna, wooded savanna, and transitional areas with coastal savanna [1]; also on disturbed ground and previously cultivated sites [4].

USES

Widely used in traditional medicine as a panacea to treat many kinds of ailments [4]. The aromatic, thyme-scented leaves are used, fresh or dried, to make a fragrant infusion which is drunk as a popular tea substitute, or they are eaten or added to soup as a flavouring [4]. *Lippia multiflora* is also used to treat gastrointestinal disorders, diarrhoea and malaria [5]. The essential oil has antifungal properties and has been shown to significantly reduce contamination by the seed-borne fungus *Phoma sorghina* (Sacc.) Boerema, Dorenb. & Kesteren, on *Sorghum bicolor* (L.) Moench [3]. *Lippia multiflora* is often grown in villages for home use.

KNOWN HAZARDS AND SAFETY

Generally regarded as safe. A clinical toxicity study by the World Health Organization (WHO)/ Collaborating Centre for Scientific Research into Plant Medicine (CSRPM) in Mali, showed there were no side-effects or toxicity in a trial of 50 people who had consumed *Lippia* tea for over 25 years. No adverse effects are known for the essential oil [7].

CONSERVATION STATUS

Widespread and common; yet to be assessed according to IUCN Red List criteria [8].

SEED CONSERVATION

HARVESTING. Fruits can be collected directly from the plants.

PROCESSING AND HANDLING. Fruits can be separated from other plant parts by sieving.

STORAGE AND VIABILITY. Seeds are reported to be orthodox [11]. Therefore, after appropriate drying, they can be stored at sub-zero temperatures for long-term conservation.

PROPAGATION

SEEDS

Dormancy and pre-treatments: Seeds of this species are suspected to be non-dormant because applied treatments (gibberellic acid [GA_3] and potassium nitrate [KNO_3]) did not improve seed germination compared to untreated seeds. However, further studies are needed to exclude any type of physiological dormancy [9].

Germination, sowing and planting: Seeds stored at the MSB and incubated in a range of constant temperatures (5–35°C) achieved maximum germination (c. 53%) at 35°C, and <20% germination at temperatures ≤25°C [9].

VEGETATIVE PROPAGATION

Can be propagated by cuttings; the use of growth regulators improves rooting [2,10].

TRADE

Dried and packaged leaves of *L. multiflora* (Figure 3) are sold at local markets in Côte d'Ivoire or exported to Mali, Burkina Faso and Niger [10].

Figure 3 – Dried and packaged leaves of *L. multiflora* (left); small bags of dried leaves cost around 0.15 Euro. (Photos: R. Dackouo, IER.)

Authors

Sidi Sanogo, Abdoul K. Sanogo, Bokary A. Kelly, Stéphane Rivière and Efisio Mattana.

References

[1] Acquaye, D., Smith, M,, Letchamo, W. & Simon, J. (2001). *Lippia Tea*. Centre for New Use Agriculture and Natural Products, Rutgers University, New Jersey, USA.

[2] Ameyaw, Y. (2009). A growth regulator for the propagation of *Lippia multiflora* Moldenke, a herbal for the management of mild hypertension in Ghana. *Journal of Medicinal Plants Research* 3 (9): 681–685.

[3] Bonzi, S., Somda, I., Sereme, P. & Adam, T. (2013). Efficacy of essential oils of *Cymbopogon citratus* (D.C.) [*sic*] Stapf, *Lippia multiflora* Moldenke and hot water in the control of seed-borne fungi *Phoma sorghina* and their effects on *Sorghum bicolor* (L.) Moench seed germination and plant development in Burkina Faso. *Net Journal of Agricultural Science* 1 (4): 111–115.

[4] Burkill, H. M. (2000). *The Useful Plants of West Tropical Africa. Vol. 5*: *Families S–Z*, 2nd edition. Royal Botanic Gardens, Kew.

[5] Ekissi, A. C., Konan, A. G., Yao-Kouame, A., Bonfoh, B. & Kati-Coulibaly, S. (2011). Evaluation of the chemical constituents of savannah tea (*Lippia multiflora*) leaves. *Journal of Applied Biosciences* 42: 2854–2858.

[6] Etou-Ossibi, A. W, Dimo, T., Elion-Itou, R. D. G., Nsondé-Ntandou, G. F., Nzonzi, J., Bilanda, D. C., Ouamba, J. M. & Abeena, A. A. (2012). Effets de l'extrait aqueux de *Lippia multiflora* Moldenke sur l'hypertension artérielle induite par le DOCA-sel chez le rat. *Phytothérapie* 10 (6): 363–368.

[7] Folashade, K. O. & Omoregie, E. H. (2012). Essential oil of *Lippia multiflora* Moldenke: a review. *Journal of Applied Pharmaceutical Science* 2 (1): 15–23.

[8] IUCN (2017). *The IUCN Red List of Threatened Species.* Version 2017-3. http://www.iucnredlist.org/

[9] Mattana, E., Sacande, M., Sanogo, A. K., Lira, R., Gómez-Barreiro, P., Rogledi, M. & Ulian, T. (2017). Thermal requirements for seed germination of underutilized *Lippia* species. *South African Journal of Botany* 109: 223–230.

[10] N'Guessan, K. A. & Yao-Kouame, A. (2010). Filière de commercialisation et usages des feuilles de *Lippia multiflora* en Côte d'Ivoire. *Journal of Applied Biosciences* 29: 1743–1752.

[11] Royal Botanic Gardens, Kew (2017). *Seed Information Database (SID).* Version 7.1. http://data.kew.org/sid/

[12] Verdcourt, B. (1992). Verbenaceae. In: *Flora of Tropical East Africa*. Edited by R.M. Polhill. A.A.Balkema, Rotterdam, The Netherlands.

[13] WCSP (2017). *World Checklist of Selected Plant Families.* Royal Botanic Gardens, Kew. http://wcsp.science.kew.org/

Nauclea latifolia Sm.

TAXONOMY AND NOMENCLATURE

FAMILY. Rubiaceae

SYNONYMS. *Nauclea esculenta* (Afzel. ex Sabine) Merr., *Sarcocephalus esculentus* Afzel. ex Sabine, *S. latifolius* (Sm.) E.A.Bruce

VERNACULAR / COMMON NAMES. African peach (English); pêcher africain (French); balimafila-toro, bari, baro (Bambara); pembé (Dogon); bati (Malinke); baure, donnoso, dundakè (Peul); tinierekassan (Senufo).

PLANT DESCRIPTION

LIFE FORM. Shrub or tree.

PLANT. Straggling, 2–9 m tall, bark grey or brown, fissured, crown spreading, branches entangled (Figure 1). **LEAVES.** Simple, opposite 10–22 x 7–15 cm, apex shortly acuminate, base cuneate to cordate. **FLOWERS.** Inflorescences are dense spheres (Figure 2), 4–5 cm in diameter, with many sweet-scented, white, yellowish or pinkish tubular flowers, corolla tube 7–10 mm, lobes c. 2 mm [4].

Figure 1 – General habit of *N. latifolia*. (Photo: S. Sanogo, IER.)

FRUITS AND SEEDS

The fruits are fleshy berries, irregularly globose (Figure 4), warty, 3–8 cm in diameter, red when ripe, containing numerous seeds (Figure 3) embedded in pink flesh, with a strawberry-like scent [2].

FLOWERING AND FRUITING. In the Sudan region, flowering occurs between April and June; fruits ripen between July and September [7].

Figure 2 – Inflorescence. (Photo: S. Sanogo, IER.)

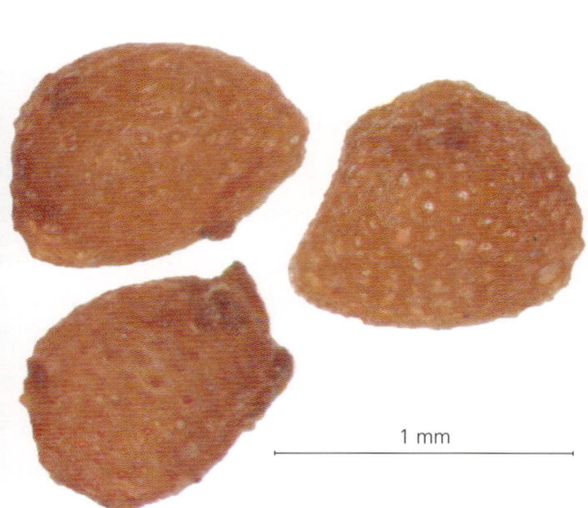

Figure 3 – External views of the seeds. (Photo: P. Gómez-Barreiro, RBG Kew.)

Figure 4 – Ripe fruit. (Photo: S. Sanogo, IER.)

DISTRIBUTION

Tropical West Africa, east to Ethiopia and East Africa, and south to Angola [16].

HABITAT

Fringing forests, lowland savanna and around ponds, on moist and fairly well-drained soils [2], at 900–1,500 m a.s.l. [5].

USES

This species has many uses in traditional medicine. The bark has antipyretic properties and is used for treating malaria and other infections [6]. The bark is also eaten as a remedy for constipation, and to treat kidney failure and gonorrhoea [10]. Aqueous extracts of the leaves have significant anthelmintic activity [1]. Previous reports that the roots contain a compound identical to that of a synthetic analgesic drug have been shown to be erroneous — the presence of the compound in the roots having arisen from cross-contamination [9]. The wood is used as fuel, and the roots and bark provide a yellow dye. *Nauclea latifolia* can be used for soil stabilisation, shade and as a windbreak. The flowers provide nectar and pollen for bees [12].

KNOWN HAZARDS AND SAFETY

The root, root bark and young leaves can be emetic and cause diarrhoea if eaten in excess [6]. An alkaloid in the leaves shows a marked and cumulative cardio-inhibiting action, resulting in headaches and nausea [11]. In laboratory tests, the aqueous extract of stem bark was shown to have allergenic properties and caused inflammatory reactions in rats and mice [8].

CONSERVATION STATUS

Assessed as Least Concern (LC) in Burkina Faso on the basis of its distribution and habitat [14], but its status has yet to be confirmed in the IUCN Red List [7]. In Mali, the use of *N. latifolia* is not restricted by law despite overexploitation by traditional healers.

SEED CONSERVATION

HARVESTING: The indehiscent fruits can be collected directly from the plants when ripe.
PROCESSING AND HANDLING: Seed can be extracted from the fruits and depulped under running water.
STORAGE AND VIABILITY: Seeds are reported to be orthodox [13]. Therefore, after appropriate drying, they can be stored at sub-zero temperatures for long-term conservation.

PROPAGATION

SEEDS

Dormancy and pre-treatments: Seeds of this species are reported to be non-dormant [3]. However, seed germination can be improved after the fruits have passed through the digestive tract of animals [15].
Germination, sowing and planting: Studies carried out in Uganda reported that seeds of this species require light to germinate. A total germination rate of 60% was obtained 28 days after sowing at temperatures of 20–35°C [15].

VEGETATIVE PROPAGATION

Can be propagated by cuttings [12]; also regenerates from suckers [15].

TRADE

Despite overexploitation of the bark by therapists and herbalists, no reliable statistics are available on the trade of products derived from this species.

Authors

Sidi Sanogo, Abdoul K. Sanogo, Bokary A. Kelly, Rokia Sanogo, Stéphane Rivière, Steve Davis and Efisio Mattana.

References

[1] Ademola, I. O., Fagbemi, B. O. & Idowu, S. O. (2007). Anthemintic efficacy of *Nauclea latifolia* extract against gastrointestinal nematodes of sheep: *in vitro* and *in vivo* studies. *African Journal of Traditional, Complementary and Alternative Medicines* 4 (2): 148–156.

[2] Arbonnier, M. (2004). *Trees, Shrubs and Lianas of West African Dry Zones*. CIRAD, France; Margraf Publishers, Weikersheim, Germany; and Muséum national d'histoire naturelle (MNHN), Paris, France.

[3] Baskin, C. C. & Baskin, J. M. (2014). *Seeds: Ecology, Biogeography, and Evolution of Dormancy and Germination*, 2nd edition. Academic Press, San Diego, USA.

[4] Beentje, H. J. (1994). *Kenya Trees, Shrubs and Lianas*. National Museums of Kenya, Nairobi, Kenya.

[5] Bridson, D. & Verdcourt, B. (1988). Rubiaceae (Part 2). In: *Flora of Tropical East Africa*. Edited by R. M. Polhill. A.A. Balkema, Rotterdam, The Netherlands.

[6] Burkill, H. M. (1997). *The Useful Plants of West Tropical Africa. Vol. 4: Families M–R*, 2nd edition. Royal Botanic Gardens, Kew.

[7] IUCN (2017). *The IUCN Red List of Threatened Species*. Version 2017-3. http://www.iucnredlist.org/

[8] Kouadio, J. H., Bleyere, M. H., Kone, M. & Dano, S. D. (2014). Acute and sub-acute toxicity of aqueous extract of *Nauclea latifolia* in Swiss mice and in OFA rats. *Tropical Journal of Pharmaceutical Research* 13 (1). http://dx.doi.org/10.4314/tjpr.v13i1.16

[9] Kusari, S., Tatsimo, S. J. N., Zühlke, S., Talontsi, F. M., Kouam, S. F. & Spiteller, M. (2014). Tramadol — a true natural product? *Angewandt Chemie International Edition* 53 (45): 12073–12076.

[10] Mugnier, J. (2008). *Nouvelle Flore Illustrée du Sénégal et Régions Voisines. Les Dicotylédones*, Tome 1. Agroservices, France.

[11] Oliver-Bever, B. (1983). Medicinal plants in tropical west Africa II. Plants acting on the nervous system. *Journal of Ethnopharmacology* 7 (1): 1–93.

[12] Orwa, C., Mutua, A., Kindt, R., Jamnadass, R. & Simons, A. (2009). *Agroforestree Database: A Tree Reference and Selection Guide*. Version 4.0. World Agroforestry Centre, Kenya. http://www.worldagroforestry.org/output/agroforestree-database

[13] Royal Botanic Gardens, Kew (2017). *Seed Information Database (SID)*. Version 7.1. http://data.kew.org/sid/

[14] Sacande, M., Sanou, L. & Beentje, H. (2012). *Guide d'Identification des Arbres du Burkina Faso*. Royal Botanic Gardens, Kew.

[15] Stangeland, T., Tabuti, J. R. S. & Lye, K. A. (2007). The influence of light and temperature on the germination of two Ugandan medicinal trees. *African Journal of Ecology* 46 (4): 565–571.

[16] WCSP (2017). *World Checklist of Selected Plant Families*. Royal Botanic Gardens, Kew. http://wcsp.science.kew.org/

Securidaca longepedunculata Fresen.

TAXONOMY AND NOMENCLATURE

FAMILY. Polygalaceae

SYNONYMS. *Elsota longipedunculata* (Fresen.) Kuntze

VERNACULAR / COMMON NAMES. Violet tree (English); arbre violette (French); diro, joro, joti (Bambara); so hwinu, so 'onu (Bwa); kièfrèke, mankana (Malinke); dugupya, jege, juri (Minyanka); pelga (Moore); alale (Peul); fyeme, fyire (Senufo).

PLANT DESCRIPTION

LIFE FORM. Shrub or tree.

PLANT. Deciduous or semi-deciduous, <6 m tall (Figure 1), crown open, branches drooping, spindly [1], sometimes spiny [5]; bark light yellow to beige. **LEAVES.** Alternate, spirally arranged, 2–9 x 1–3 cm, very variable in shape, linear or elliptical, finely hairy at first, becoming hairless, petiole 4–8 mm [1]. **FLOWERS.** Pink, purple or violet, sometimes variegated with white, asymmetric (Figure 1), sweet-scented [6], in terminal or lateral racemes, 5–8 cm long [1].

Figure 1 – General habit of *S. longepedunculata* and branch with flowers. (Photos: S. Sanogo, IER.)

FRUITS AND SEEDS

Fruits (Figures 2&3) 3–5 x 0.8–2 cm, with an oblong or elliptic somewhat obliquely curved wing (Figure 3), sometimes with a second rudimentary wing. Nut containing the seed 8–10 mm in diameter, wrinkled or smooth (Figure 3) [6].

FLOWERING AND FRUITING. In West Africa, flowers from February to May; fruits from March to June, fruits sometimes remaining on the tree until the next flowering season [15,17]. Plants are in leaf from March to November.

Figure 2 – Dry fruits. (Photo: S. Sanogo, IER.)

Figure 3 – External views of the fruits and longitudinal section of the seed showing the endosperm and embryo. (Photo: P. Gómez-Barreiro, RBG Kew.)

DISTRIBUTION

Widespread throughout tropical and subtropical Africa, from Mauritania to Sudan, south to South Africa [5,6].

HABITAT

Found in many types of woodland and bushland, forest edges and secondary vegetation, from near sea level to 1,600 m a.s.l. [3,5,6].

USES

A multipurpose species with many uses in traditional medicine [4,11,13]. The roots are used to treat rheumatism, labour pains, snake bites, tuberculosis, meningitis and ulcers [13,19]. A decoction of the roots is used for treating epilepsy, as an antiseptic for treating wounds and injuries, and as a taeniafuge and purgative [1,9]. The roots are also used to treat internal and external parasites of livestock. The bark of the trunk (Figure 4) is ground to a powder and used against filariasis. Leafy

Figure 4 – Damage to the trunk caused by bark harvesting. (Photo: A. K. Sanogo, IER.)

twigs are chopped and eaten to delay the effects of snake bites [9]. The bark has been used as an ingredient of arrow poison [21].The bark fibres are used to make ropes and coarse cloth. The wood is used as firewood [9]. The flowers are frequented by bees for honey production [10]. The roots, root bark and stem bark are used for both pre- and post-harvest protection of grain from predation by insects [8,10].

KNOWN HAZARDS AND SAFETY

All parts of the plant are highly poisonous. The roots contain methyl salicylate; both the roots and seed oil are known to have caused human fatalities [21]. Care should be taken when using the plant medicinally, when handling the plant, or when preparing and applying products such as insecticides which have been derived from the plant.

CONSERVATION STATUS

Assessed as Least Concern (LC) in West Africa [15] and South Africa [12] according to IUCN Red List criteria, on the basis of its distribution and habitat. However, its status has yet to be confirmed in the IUCN Red List [7].

SEED CONSERVATION

HARVESTING. Fruits can be picked directly from the trees.
PROCESSING AND HANDLING. Seeds can be damaged when extracting them from the winged fruits. Therefore, the conservation unit is the whole fruit. The wing can be cut using scissors.
STORAGE AND VIABILITY. Seeds are reported to be orthodox [14]. Therefore, after appropriate drying, they can be stored at sub-zero temperatures for long-term conservation.

PROPAGATION

SEEDS

Dormancy and pre-treatments: Seeds of this species are reported to be either non-dormant or physiologically dormant [2], highlighting the need for further studies.
Germination, sowing and planting: Seed lots stored at the MSB reached high germination percentages (c. 80%) when untreated seeds were incubated at 25°C with 8 hours of light per day [14]. In the nursery, seeds sown in pots germinate better than those sown in seedbeds [20]. Seedlings have a very slow growth rate, reaching 15–20 cm in five months [16]. Seedlings should not be planted in periodically flooded soils; sandy loam soils are preferred [20]. The average height of seedlings after four years was 75 cm on sandy soil in Kougué, compared to 39 cm on lateritic soil in Zégoua (southern Mali) [18].

VEGETATIVE PROPAGATION

Can be propagated by root cuttings [10].

TRADE

No data are available for Mali.

Authors

Sidi Sanogo, Abdoul K. Sanogo, Bokary A. Kelly, Stéphane Rivière, Steve Davis and Efisio Mattana.

References

[1] Arbonnier, M. (2004). *Trees, Shrubs and Lianas of West African Dry Zones*. CIRAD, France; Margraf Publishers, Weikersheim, Germany; and Muséum national d'histoire naturelle (MNHN), Paris, France.

[2] Baskin, C. C. & Baskin, J. M. (2014). *Seeds: Ecology, Biogeography, and Evolution of Dormancy and Germination*, 2nd edition. Academic Press, San Diego, USA.

[3] Beentje, H. J. (1994). *Kenya Trees, Shrubs and Lianas*. National Museums of Kenya, Nairobi, Kenya.

[4] Burkill, H. M. (1997). *The Useful Plants of West Tropical Africa. Vol. 4: Families M–R*, 2nd edition. Royal Botanic Gardens, Kew.

[5] Conservatoire et Jardin botaniques & South African National Biodiversity Institute (SANBI) (2016). *African Plant Database*. Version 3.4.0. http://www.ville-ge.ch/musinfo/bd/cjb/africa/

[6] Exell, A. W. & Wild, H. (eds) (1960). Polygalaceae. In: *Flora Zambesiaca*, Vol. 1, part 1. Flora Zambesiaca Managing Committee, Royal Botanic Gardens, Kew.

[7] IUCN (2017). *The IUCN Red List of Threatened Species.* Version 2017-3. http://www.iucnredlist.org/

[8] Jayasekera, T. K., Stevenson, P. C., Hall, D. R. & Belmain, S. R. (2005). Effect of volatile constituents from *Securidaca longepedunculata* on insect pests of stored grain. *Journal of Chemical Ecology* 31 (2): 303–313.

[9] Malgras, D. (1992). *Arbres et Arbustes Guérisseurs des Savanes Maliennes.* Agence de Cooperation Culturelle et Technique, Paris, France.

[10] Natural Resources Institute (2017). *Securidaca longepedunculata*. Optimising Pesticidal Plants: Technology Information, Outreach and Networks (OPTIONs) project. Natural Resources Institute (NRI), University of Greenwich, UK. http://projects.nri.org/options/background/plants-database/26-securidaca-longepedunculata

[11] Organisation Ouest Africaine de la Santé (OOAS) (2013). *Pharmacopée Afrique de l'Ouest*. OOAS/West African Health Organization (WAHO), Bobo Dioulasso, Burkina Faso.

[12] Raimondo, D., von Staden, L., Foden, W., Victor, J. E., Helme, N. A., Turner, R. C., Kamundi, D. A. & Manyama, P. A. (2009). *Red List of South African Plants. Strelitzia* 25, South African National Biodiversity Institute, Pretoria, South Africa.

[13] Royal Botanic Gardens, Kew (1999–2015). *Survey of Economic Plants for Arid and Semi-Arid Lands (SEPASAL) Database*. http://apps.kew.org/sepasalweb/sepaweb

[14] Royal Botanic Gardens, Kew (2017). *Seed Information Database (SID).* Version 7.1. http://data.kew.org/sid/

[15] Sacande, M., Sanou, L. & Beentje, H. (2012). *Guide d'Identification des Arbres du Burkina Faso*. Royal Botanic Gardens, Kew.

[16] Sanogo, A. K. (2000). *Techniques Simples de Production de Plants Forestiers*. Fiche technique, PRF/IER, Mali and Programme GDRN/Intercoopération, Switzerland.

[17] Sanogo A. K., Sanogo, S., Kelly, B. A. & Diallo O. (2009). *Mise au Point de Techniques Améliorées d'Exploitation des Plantes Médicinales dans les Terroirs: cas de* Nauclea latifolia *Sm.*, Securidaca longepedunculata *Fres. et* Trichilia emetica *Vahl.* Rapport Final de recherche, 15th Commission Scientifique, IER, Mali.

[18] Sanogo, A. K., Sanogo, S. & Sidibé, S. I. (2013). *Conservation et Gestion Durable des Plantes Utiles aux Communautés Rurales*. Rapport de Campagne 19, Comité de Programme, IER, Mali.

[19] Sanogo, A. K., Sanogo, S., Sidibé, S. I., Senou, O. & Dembélé, M. (2007). *Identification d'Espèces de Plantes Utiles aux Communautés Rurales*. Rapport d'enquête, IER, DMT and Royal Botanic Gardens, Kew.

[20] Sanogo, S., Sanogo, A. K., Senou, O., Kouyaté, A. M. & Yossi, H. (2005). *Réintroduction d'Espèces Forestières Utiles en Voie de Disparition*. Rapport de Recherche Final, 11th session du Comité de Programme, IER, Mali.

[21] Watt, J. M. & Breyer-Brandwijk, M. G. (1962). *The Medicinal and Poisonous Plants of Southern and Eastern Africa*, 2nd edition. E. & S. Livingstone, Edinburgh & London.

Stylosanthes erecta P.Beauv.

TAXONOMY AND NOMENCLATURE

FAMILY. Leguminosae – Papilionoideae

SYNONYMS. *Ononis coriifolia* Reichb. ex Guill. & Perr., *Stylosanthes guineensis* Schumach. & Thonn.

VERNACULAR / COMMON NAMES. Nigerian stylo (English); diofaga (Bambara); mbono-muso (Mandinka, Gambia); bala korana, nbonomuso (Mandinka, Senegal); lèddèl, lékon fero (Peul).

PLANT DESCRIPTION

LIFE FORM. Perennial herb or subshrub.

PLANT. Prostrate or erect, stems <1.5 m long (Figure 1); much-branched, glabrescent or pubescent in narrow longitudinal lines alternating in position between nodes. **LEAVES.** Alternate, trifoliate, elliptic to lanceolate, 0.5–3 cm long, 1.5–7 mm wide, pubescent or mostly glabrous with bristles, stipules 1–1.5 cm long. **FLOWERS.** Clustered in short terminal spikes, rachis 1–6(14) cm long, standard yellow or orange (Figure 1) with orange red mark at base, 6–8 x 5–6 mm, wings yellow, sometimes darker at the base, keel white or yellow [5].

Figure 1 – Branch with flower. (Photo: S. Sanogo, IER.)

FRUITS AND SEEDS

Pods 4–7 mm long, 1–2-jointed, oblong, compressed, with a circinnate beak <2 mm long [5] (Figure 2). Seeds approximately ovoid or irregularly oblong and compressed, 2 x 1.5 x 1 mm (Figure 2) [5].

FLOWERING AND FRUITING. In West Africa, flowers and fruits from November to June [1].

Figure 2 – External views of the pods and longitudinal section showing the seed.
(Photos: P. Gómez-Barreiro, RBG Kew.)

DISTRIBUTION

Tropical Africa from Senegal to Tanzania, south to Angola; also found in Madagascar [4].

HABITAT

Sandy soils and roadsides; in much of its range, it is found along the coast, in coastal meadows and marshes, from sea level to 500 m a.s.l. [4].

USES

The root is used in traditional medicine as a stimulant and is an ingredient of aphrodisiac prescriptions. An infusion of the plant is taken to treat coughs and common cold. The plant is used as a fish poison and has superstitious values, the root being commonly worn as a charm. The plant is grazed by livestock and it has been planted in *Eucalyptus* plantations to feed animals [1,3].

KNOWN HAZARDS AND SAFETY

No known hazards.

CONSERVATION STATUS

Assessed as Not Threatened (nt) according to previous IUCN criteria [6], but its current status has yet to be confirmed in the IUCN Red List [7].

SEED CONSERVATION

HARVESTING. As with other *Stylosanthes* spp., seeds should be harvested at the end of the rainy season [8]; the whole plant is mown at fruit maturity (Figure 3).

PROCESSING AND HANDLING. The seeds are extracted by pounding in a mortar, then winnowing outdoors [10].

STORAGE AND VIABILITY. Seeds are reported to be orthodox [9]. Therefore, after appropriate drying, they can be stored at sub-zero temperatures for long-term conservation.

PROPAGATION

SEEDS

Dormancy and pre-treatments: Seeds of *Stylosanthes* spp. are reported to have physical dormancy; seeds have been reported to germinate after fire [2]. Seeds should be scarified before incubation [9].

Germination, sowing and planting: Experiments carried out under laboratory controlled conditions on seed lots stored at the MSB revealed high germination (80%) for chipped seeds incubated at constant 25°C in the light [9].

Figure 3 – Drying mown plants before threshing to extract the seeds. (Photo: A. K. Sanogo, IER.)

Figure 4 – Leafy branches collected for preparing extracts for phytochemical analysis. (Photo: A. K. Sanogo, IER.)

VEGETATIVE PROPAGATION

No information available.

PHYTOCHEMICAL ANALYSES

Phytochemical analyses of extracts of leafy branches were conducted as part of the UPP following requests from therapists (Figure 4). Results revealed the extracts have analgesic properties.

TRADE

No data available.

Authors

Sidi Sanogo, Abdoul K. Sanogo, Bokary A. Kelly, Rokia Sanogo, Stéphane Rivière, Steve Davis and Efisio Mattana.

References

[1] Adjanohoun, E. J. & Aké Assi, L. (1979). *Contribution au Recensement des Plantes Médicinales de Côte d'Ivoire*. Centre National de Floristique, Abidjan, Côte d'Ivoire.

[2] Baskin, C. C. & Baskin, J. M. (2014). *Seeds: Ecology, Biogeography, and Evolution of Dormancy and Germination*, 2nd edition. Academic Press, San Diego, USA.

[3] Burkill, H. M. (1995). *The Useful Plants of West Tropical Africa. Vol. 3: Families J–L*, 2nd edition. Royal Botanic Gardens, Kew.

[4] Conservatoire et Jardin botaniques & South African National Biodiversity Institute (SANBI) (2016). *African Plant Database*. Version 3.4.0. http://www.ville-ge.ch/musinfo/bd/cjb/africa/

[5] Gillett, J. B., Polhill, R. M. & Verdcourt, B. (1971). Leguminosae Subfamily Papilionoideae. In: *Flora of Tropical East Africa*. Edited by E. Milne-Redhead & R. M. Polhill. Crown Agents for Oversea Governments and Administrations, London.

[6] ILDIS (2006–2013). *World Database of Legumes*. International Legume Database and Information Service. http://www.legumes-online.net/ildis/aweb/database.htm

[7] IUCN (2017). *The IUCN Red List of Threatened Species.* Version 2017-3. http://www.iucnredlist.org/

[8] Mouity, J. A. (1999). *Diversité Biologique des Fabaceae et Leur Interêt Ethnobotanique dans la Province du Bazega*. Mémoire de fin d'Etudes présenté en vue de l'obtention du Diplôme d'Ingénieur du Développement Rural, Université Polytechnique de Bobo Dioulasso, Burkina Faso.

[9] Royal Botanic Gardens, Kew (2017). *Seed Information Database (SID).* Version 7.1. http://data.kew.org/sid/

[10] Sanogo, A. K. & Sanogo, S. (2000). *Techniques de Récolte de Semences Forestiers*. Fiche technique, PRF/IER, Mali and Programme GDRN/Intercoopération, Switzerland.

Vachellia nilotica (L.) P.J.H.Hurter & Mabb.

TAXONOMY AND NOMENCLATURE

FAMILY. Leguminosae – Caesalpinioideae

SYNONYMS. *Acacia arabica* (Lam.) Willd., *A. nilotica* (L.) Willd. ex Delile, *Mimosa nilotica* L.

VERNACULAR / COMMON NAMES. Babul, Egyptian thorn (English); gommier rouge (French); baanan, bagana (Bambara); bakani (Bobo); aboro (Bwa); gavdi (Fula); bagana jiri, bwana (Malinke); paanan, pakhi (Minyanka, Senufo); pêg-nanga (Moore).

PLANT DESCRIPTION

LIFE FORM. Tree.

PLANT. Evergreen or deciduous, <25 m high (Figure 1); root system deep and extensive, trunk <1 m in diameter; crown flat or umbrella-shaped, bark blackish or dark brown, rough and longitudinally fissured, branches with paired stipular spines, 1–5 cm long. **LEAVES.** Alternate, bipinnate, bluish, with 3–6 pairs of pinnae and 10–30 pairs of elliptical or narrowly oblong leaflets, 1.5–7 x 0.5–2 mm [3,7]. **FLOWERS.** Bisexual or male, bright yellow, in axillary, globose heads of up to 50 flowers, 6–15 mm in diameter, sweet-scented (Figure 2) [5,7].

Figure 1 – General habit of *V. nilotica*. (Photo: S. Sanogo, IER.)

Figure 2 – Flowering branches. (Photo: S. Sanogo, IER.)

FRUITS AND SEEDS

Pods (Figure 3) are variable in shape, indehiscent, (4)8–17(24) x 1.3–2.2 cm, straight or curved, glabrous to grey velvety, ± turgid. Seeds are deep blackish brown, 7–9 x 6–7 mm, smooth, subcircular, compressed with an areole of 6–7 x 4.5–5 mm (Figure 4) [5].

FLOWERING AND FRUITING. In West Africa, flowering begins in the early rainy seasons; fruits ripen from January to March [14].

Figure 3 – Fruiting branches.
(Photo: S. Sanogo, IER.)

Figure 4 – External views of the seeds showing the areole surrounded by the pleurogram and longitudinal section showing the embryonic axis and the cotyledon. (Photos: P. Gómez-Barreiro, RBG Kew.)

DISTRIBUTION

Drylands of tropical Africa from Senegal to Sudan and Egypt, southwards to eastern Africa, Mozambique and South Africa, and west to the Arabian Peninsula and India [3,7]. Introduced and naturalised in many areas of the tropics and subtropics [7] and can be invasive (e.g. in parts of Australia).

HABITAT

Wooded grassland, savanna and dry scrub forests, tolerating drought and waterlogged conditions for several months depending on subspecies [7,12]. In West Africa, the species occurs mainly in Sahelian and Sahelo-Sudanese regions. In some places of Sudan, it prefers heavy soils [10]. Elsewhere, it occurs on a variety of soils, from deep sandy soils to loam, shallow granite or clay soils, and seasonally flooded alluvium [7].

USES

Vachellia nilotica (more widely known as *Acacia nilotica*) [9] has a wide range of uses [7,12]. In traditional medicine, powdered root bark is used to treat toothache and sores in the digestive tract. It is also used in the treatment of diarrhoea and dysentery [3,10]. Powdered bark of the trunk is used to treat haemorrhoids, gastrointestinal pains, dysentery, stomatitis and gingivitis. Decoctions of leafy branches are used as a mouthwash to treat tooth decay. The leaves are used to treat cataracts. Macerated pods are used to prepare beverages for treating dysentery [10]. The fruits have antidiarrhoeal, antitussive, anti-inflammatory, analgesic, diuretic and tonic properties [11]. The wood is used in building construction, and for making canoes and tool handles, and makes good firewood and charcoal [3]. Bark fibre is used for making ropes and ties for the construction of sheds, barns and huts [17]. The bark and fruits are highly valued as a source of tannins [6]. The gum is of lower quality than that obtained from *Senegalia senegal* (L.) Britton [syn. *Acacia senegal* (L.) Willd.] [17]. The flowers attract bees, which provide a good honey. The tree is often planted for shade and as a windbreak. Its extensive roots fix nitrogen, making it suitable for soil improvement and reafforestation programmes [6,7].

KNOWN HAZARDS AND SAFETY

The spines on the stems are sharp.

CONSERVATION STATUS

This widespread species is assessed as Least Concern (LC) according to IUCN Red List criteria [14], but its status has yet to be confirmed in the IUCN Red List [8].

SEED CONSERVATION

HARVESTING. Fruits can be collected directly from the trees from January to March.

PROCESSING AND HANDLING. Seeds can be extracted by hand-shelling or by threshing the pods in a bag followed by winnowing.

STORAGE AND VIABILITY. Seeds are reported to be orthodox [13]. Therefore, after appropriate drying, they can be stored at sub-zero temperatures for long-term conservation. When kept in hermetic storage at 10°C with 4.5–9% moisture content, seeds can maintain viability for several years [1]. Seeds stored in plastic bottles at room temperature remain viable for five years [15].

PROPAGATION

SEEDS

Dormancy and pre-treatments: Seeds of this species are reported to be physically dormant [4]. Therefore, teguments should be scarified to allow seed imbibition. Seeds can be mechanically scarified by chipping with a scalpel or by filing [13]. They can also be chemically scarified by soaking them in sulphuric acid (90% H_2SO_4) for 30 minutes, followed by extensive washing under running water and then soaking in water for 24 hours before sowing; this results in a final germination of 80–95% in one week [15]. Alternatively, seeds can be treated by pouring hot water (80–90°C) over the seeds, then cooling in cold water for 72 hours, or by placing the seeds in boiling water for three minutes and then soaking in warm water for 24 hours. Both pre-treatments result in a final germination of 70–85% in 15 days [16].

Germination, sowing and planting: Under laboratory controlled conditions, pre-treated seeds can reach high germination (≥90%) when incubated at constant temperatures of 21–26°C [13]. In the nursery, one or two pre-treated seeds per hole can be sown in equal parts by volume of potting soil, sand and compost, at a depth of 1 cm. The best plants are obtained by using plastic pots, 14 cm wide and 30 cm deep, to allow for the rapid development of the root system [16]. Young plants should be shaded during the hottest parts of the day during the first month and watered at least once per day [18]. Plants should be 30–50 cm tall after four months from sowing, at which point they are ready to be planted in the field. Before planting, digging around the roots once per month and a reduction in watering for two months will help plants to adapt to field conditions. Planting holes (Figure 5) should be 50 x 50 x 50 cm at a minimum distance of 5 x 5 m, and half refilled a month before planting to retain the maximum amount of rainwater. In the Sikasso region of Mali, planting should be carried out in late July or early August [19].

VEGETATIVE PROPAGATION

Experiments carried out in Sudan indicate that stem cuttings can be rooted when treated with indole-3-butyric acid [2].

Figure 5 – 18-month-old plant in a farm in Bankass. (Photo: R. Dackouo, IER.)

TRADE

No data are available for Mali. However, production and trade in tannin, timber and other products derived from *V. nilotica* are important locally and nationally in Africa and in India [7].

Authors

Sidi Sanogo, Abdoul K. Sanogo, Bokary A. Kelly, Rokia Sanogo, Stéphane Rivière, Efisio Mattana, Steve Davis and Tiziana Ulian.

References

[1] Albrecht, J. (1993). *Tree Seed Handbook of Kenya*. GTZ Forestry Seed Centre, Muguga, Kenya.

[2] Ali, Y. H. & El-Tigani, S. (2007). A note on the propagation of *Acacia nilotica* subspecies *nilotca* [*sic*] Brenan by stem cuttings. *Sudan Silva* 13 (1): 90–95.

[3] Arbonnier, M. (2004). *Trees, Shrubs and Lianas of West African Dry Zones*. CIRAD, France; Margraf Publishers, Weikersheim, Germany; and Muséum national d'histoire naturelle (MNHN), Paris, France.

[4] Baskin, C. C. & Baskin, J. M. (2014). *Seeds: Ecology, Biogeography, and Evolution of Dormancy and Germination*, 2nd edition. Academic Press, San Diego, USA.

[5] Brenan, J. P. M. (ed.) (1970). Leguminosae (Mimosoideae). In: *Flora Zambesiaca*, Vol. 3, part 1. Flora Zambesiaca Managing Committee, Royal Botanic Gardens, Kew.

[6] Burkill, H. M. (1995). *The Useful Plants of West Tropical Africa, Vol. 3: Families J–L*, 2nd edition. Royal Botanic Gardens, Kew.

[7] Fagg, C. W. & Mugedo, J. Z. A. (2005). *Acacia nilotica* (L.) Willd. ex Delile. In: *PROTA (Plant Resources of Tropical Africa/Ressources Végétales de l'Afrique Tropicale)*. Edited by P. C. M. Jansen & D. Cardon. Wageningen, The Netherlands. http://www.prota4u.org/

[8] IUCN (2017). *The IUCN Red List of Threatened Species*. Version 2017-3. http://www.iucnredlist.org/

[9] Kyalangalilwa, B., Boatwright, J. S., Daru, B. H., Maurin, O. & Van der Bank, M. (2013). Phylogenetic position and revised classification of *Acacia* s.l. (Fabaceae: Mimosoideae) in Africa, including new combinations in *Vachellia* and *Senegalia*. *Botanical Journal of the Linnean Society* 172 (4): 500–523.

[10] Malgras, D. (1992). *Arbres et Arbustes Guérisseurs des Savanes Maliennes*. Agence de Cooperation Culturelle et Technique, Paris, France.

[11] Organisation Ouest Africaine de la Santé (OOAS) (2013). *Pharmacopée Afrique de l'Ouest*. OOAS/ West African Health Organization (WAHO), Bobo-Dioulasso, Burkina Faso.

[12] Royal Botanic Gardens, Kew (1999–2015). *Survey of Economic Plants for Arid and Semi-Arid Lands (SEPASAL) Database*. http://apps.kew.org/sepasalweb/sepaweb

[13] Royal Botanic Gardens, Kew (2017). *Seed Information Database (SID)*. Version 7.1. http://data.kew.org/sid/

[14] Sacande, M., Sanou, L. & Beentje, H. (2012). *Guide d'Identification des Arbres du Burkina Faso*. Royal Botanic Gardens, Kew.

[15] Sanogo, A. K. (2000). *Récolte de Semences Forestières*. Fiche technique, PFR/CRRA/IER.

[16] Sanogo, A. K. (2000). *Techniques Simples de Production de Plants Forestiers*. Fiche technique, PRF/ IER, Mali and Programme GDRN/Intercoopération, Switzerland.

[17] Sanogo, A. K. (2002). *Problématique de l'Exploitation des Plantes Médicinales. Etude des Modes d'Exploitation Traditionnelle et Leurs Contraintes sur la Régénération des Plantes dans les Terroirs. Cas de la Préfecture de Kadiolo au Mali*. Mémoire de fin d'étude; Ecole Nationale d'Economie Appliquée (ENEA), Dakar, Senegal.

[18] Sanogo, A. K., Sanogo, S. & Kelly, B. A. (2007). *Plantation d'Espèces Médicinales et Test d'Exploitation d'Organes Végétaux dans les Parcelles des Tradithérapeutes de Kadiolo*. Final Report, CRU/ CRRA-SIK/IER.

[19] Sanogo, A. K., Sanogo, S. & Sidibé, S. I. (2012). *Conservation et Gestion Durable des Plantes Utiles aux Communautés Rurales*. Rapport de campagne, Comité de Programme 2013, IER, Mali.

Vitex madiensis Oliv.

TAXONOMY AND NOMENCLATURE

FAMILY. Lamiaceae

SYNONYMS. None.

VERNACULAR / COMMON NAMES. Dankele, koro (Bambara); dankelekele (Malinke); korole (Senufo).

Figure 1 – General habit of *V. madiensis*. (Photo: S. Sanogo, IER.)

PLANT DESCRIPTION

LIFE FORM. Shrub or tree.

PLANT. Up to 8 m tall (Figure 1), or producing annual stems of 0.3–1.5 m tall from a massive underground woody rootstock; bark fissured, older stems dark red brown, yellow and grey; young stems shortly tomentose with reddish brown buds [13]. **LEAVES.** Aromatic when crushed, often 3-whorled, (3)5(6)-foliate, often drying yellow green; petioles 7–16 cm long, leaflets 6–17 x 4–9 cm, elliptic, narrow to broadly obovate or oblanceolate. **FLOWERS.** Axillary inflorescences of sturdy, usually few-flowered compound dichasia, 5–19(30) cm long, tomentose; corolla 8–12 mm long, tube greenish cream or greenish mauve, 5–7 mm long, curved, corolla lobes pale mauve to brilliant purple, c. 4 mm [13].

FRUITS AND SEEDS

The fruits are globose, glabrous drupes (Figure 2), c. 2.5 cm long, surrounded at the base by the persistent calyx, widened into a cupule, blackish or purplish black when ripe, and enclosing a hard stone (Figure 3) embedded in a thin pulp [1].

FLOWERING AND FRUITING. In West Africa, flowers early in the rainy season; fruits mature from August to September [11].

Figure 2 – Ripe fruits. (Photo: S. Sanogo, IER.)

Figure 3 – External views, and longitudinal and transverse sections of the woody endocarps showing the seeds. (Photos: P. Gómez-Barreiro, RBG Kew.)

Figure 4 – Collecting roots of *V. madiensis*. (Photo: S. Sanogo, IER.)

DISTRIBUTION

Tropical Africa, from Senegal to Somalia and East Africa, south to Angola and Zimbabwe [18].

HABITAT

Woody savanna, mainly on rocky soils [1], and in miombo and mixed deciduous woodlands in the southern parts of its range [17].

USES

One of the main agroforestry species in Central and West Africa with many uses, including for food and medicine [3]. The fruit pulp is eaten fresh, dried or cooked [1]. A decoction of the leaves or leafy stems is used to treat headache, muscle aches and joint pains [5,9]. The leaves are used in the treatment of epilepsy [5,6]. A decoction of the roots (Figure 4) is used for treating abdominal pain and colic [4,9]. The essential oil has potential use in the cosmetic industry [12].

KNOWN HAZARDS AND SAFETY

No known hazards.

CONSERVATION STATUS

Assessed as Least Concern (LC) on the basis of its distribution and habitat [15], but its status has yet to be confirmed in the IUCN Red List [8].

SEED CONSERVATION

HARVESTING. The fruits are harvested when they become black [10].
PROCESSING AND HANDLING. After harvesting, the fruit can be pulped by lightly pounding in a mortar and then washed under running water [16].
STORAGE AND VIABILITY. No data are available on the physiological responses of seeds to desiccation. Congeneric species are reported to be either orthodox or recalcitrant [14], highlighting the need for further research.

PROPAGATION

SEEDS

Dormancy and pre-treatments: No information available on seed dormancy. However, physiological dormancy (PD) has been reported for many *Vitex* spp. [2]. Scarification, by soaking in concentrated sulphuric acid at 98% for one hour and then soaking in hot water (100°C) for 12 hours, is reported to improve seed germination [11], suggesting that PD could be present. Research is needed to confirm this.

Germination, sowing and planting: No information is available on the optimum germination requirements under laboratory controlled conditions and in the nursery.

VEGETATIVE PROPAGATION

Regenerates from coppice shoots; can be propagated by cuttings and air layering [7,11,12].

PHYTOCHEMICAL ANALYSES

Phytochemical analyses of leaf extracts undertaken during the UPP identified anti-analgesic and anti-pyretic properties supporting medicinal usage.

TRADE

The fruits (usually harvested by women) are edible and traded in local and regional markets [1,10].

Authors

Sidi Sanogo, Abdoul K. Sanogo, Rokia Sanogo, Bokary A. Kelly, Stéphane Rivière and Efisio Mattana.

References

[1] Arbonnier, M. (2004). *Trees, Shrubs and Lianas of West African Dry Zones*. CIRAD, France; Margraf Publishers, Weikersheim, Germany; and Muséum national d'histoire naturelle (MNHN), Paris, France.

[2] Baskin, C. C. & Baskin, J. M. (2014). *Seeds: Ecology, Biogeography, and Evolution of Dormancy and Germination*, 2nd edition. Academic Press, San Diego, USA.

[3] Baumer, M. (1995). *Arbres, Arbustes et Arbrisseaux Nourriciers en Afrique Occidentale*. Enda-Editions, Dakar, Senegal.

[4] Bossard, E. (1996). *La Médecine Traditionnelle au Centre et à l'Ouest de l'Angola*. Ministério da Ciencia e da Tecnología, Lisbon, Portugal.

[5] Claudie, H. (1979). *Phytothérapie et Médecine Familiale chez les Gbaya-Kara (République Centrafricaine).* Thèse de doctorat, Université de Paris, France.

[6] Diafouka, A. J. P. (1997). *Analyse des Usages des Plantes Médicinales dans 4 Régions de Congo-Brazzaville.* Thèse de doctorat, Université Libre de Bruxelles, Belgium.

[7] Food and Agriculture Organization of the United Nations (FAO) (1997). *Aménagement des Forêts Naturelles des Zones Tropicales Sèches.* Food and Agriculture Organization of the United Nations (FAO), Rome, Italy.

[8] IUCN (2017). *The IUCN Red List of Threatened Species.* Version 2017-3. http://www.iucnredlist.org/

[9] Malgras, D. (1992). *Arbres et Arbustes Guérisseurs des Savanes Maliennes*. Agence de Cooperation Culturelle et Technique, Paris, France.

[10] Mapongmetsem, P. M. (2005). *Phénologie et Apport au Sol des Substances Biogènes par la Litière des Fruitiers Sauvages dans les Savanes Soudano-Guinéennes*. Thèse de Doctorat d'Etat, Université de Yaoundé, Cameroon.

[11] Mapongmetsem, P. M. (2006). Domestication of *Vitex madiensis* in the Adamawa Highlands of Cameroon: phenology and propagation. *Akdeniz Üniversitesi Ziraat Fakültesi Dergisi* 19 (2): 269–278.

[12] Mapongmetsem, P. M., Ngassoum, M. B., Yonkeu, S., Lamaty, G. & Menut, C. (2004). Domestication of *Vitex madiensis* Oliv.: essential oil chemical composition and vegetative propagation by cuttings. In: *IIIème Sympo. Inter. Plantes Aromatiques et Médicinales.* Brazil.

[13] Pope, G. V. & Martins, E. S. (eds) (2005). Verbenaceae. In: *Flora Zambesiaca*, Vol. 8, part 7. Flora Zambesiaca Managing Committee, Royal Botanic Gardens, Kew.

[14] Royal Botanic Gardens, Kew (2017). *Seed Information Database (SID).* Version 7.1. http://data.kew.org/sid/

[15] Sacande, M., Sanou, L. & Beentje, H. (2012). *Guide d'Identification des Arbres du Burkina Faso.* Royal Botanic Gardens, Kew.

[16] Sanogo, A. K. & Sanogo, S. (2000). *Récolte de Semences Forestières*. Fiche technique, PFR/CRRA/IER.

[17] Smith, P. & Allen, Q. (2004). *Field Guide to the Trees and Shrubs of the Miombo Woodlands*. Royal Botanic Gardens, Kew.

[18] WCSP (2017). *World Checklist of Selected Plant Families.* Royal Botanic Gardens, Kew. http://wcsp.science.kew.org/

Zanthoxylum zanthoxyloides (Lam.) Zepern. & Timler

TAXONOMY AND NOMENCLATURE

FAMILY. Rutaceae

SYNONYMS. *Fagara senegalensis* (DC.) A.Chev., *F. xanthoxyloides* Lam., *Zanthoxylum senegalense* DC.

VERNACULAR / COMMON NAMES. Candlewood, Senegal prickly-ash, toothache bark (English); fagara jaune (French); huo, n'dé, uo (Bambara); barkelé, bule barkelé (Fula-Pulaar); uo (Mandinka, Maninka); wuho (Xaasongaxango).

PLANT DESCRIPTION

LIFE FORM. Tree or shrub.

PLANT. Up to 12(16) m tall, trunk grey, bark finely fissured longitudinally, with large woody thorns <1.2 cm long (Figure 1) on branches, stems, petioles; hooked thorns sometimes on undersides of leaves. **LEAVES.** Hairless, glossy, brittle, obovate or elliptic, 5–10 x 2–4 cm [1], with a citronella-like smell when crushed (Figure 1). **FLOWERS.** White or greenish, in dense narrow panicles 5–25 cm long [1,5].

Figure 1 – Branches with fruits and trunk with fissured bark and woody thorns. (Photos: S. Sanogo, IER.)

FRUITS AND SEEDS

Fruits are ovoid follicles, 5–6 mm in diameter, brown, with glandular dots, dehiscent and single-seeded. Seeds are black to bluish, shiny (Figure 2), and long persistent in the fruit [6].

FLOWERING AND FRUITING. Often flowers twice a year, during the first part of the dry season and during the rainy season [1,6].

Figure 2 – External view of the seeds. (Photo: P. Gómez-Barreiro, RBG Kew.)

DISTRIBUTION

Tropical West Africa, from Senegal east to Cameroon [5,6].

HABITAT

Savanna, thickets, dry and transitional forests, forest patches on dry ground, secondary dry forests, coastal dunes and thickets; locally abundant in coastal areas [5].

USES

The roots, stem bark and leaves are commonly used in traditional medicine. They are considered antiseptic and analgesic. Root or stem bark macerations, decoctions or infusions are widely taken to treat malaria, fever, sickle cell anaemia, tuberculosis, paralysis, oedema and general body weakness. They are also widely taken to treat intestinal problems, intestinal worms, gonorrhoea and urethritis, as an emmenagogue or a stimulant, and to treat migraine and neuralgia during childbirth. The roots are externally applied to ulcers, swellings, haemorrhoids, abscesses, snake bites, yaws, wounds, leprosy and syphilitic sores, as well as to treat rheumatic and arthritic pain and hernias [6]. The wood is used for making torches. The timber is yellow, very hard and termite-resistant and used for building purposes, including use as poles and posts. It also makes good firewood. The roots, young shoots and twigs are commonly used as chewing sticks. The bark or young branches are resinous, making them suitable for ceremonial torches. The spines are thrown into a fire to give off a scented smoke. The leaves, which smell like citronella, and the seeds, which taste strongly of cinnamon or pepper, are commonly used to season food. Necklaces are made from the seeds [2]. *Zanthoxylum zanthoxyloides* has numerous magico-religious uses, including protection against spirits. It also serves as a fetish plant [6].

KNOWN HAZARDS AND SAFETY

The thorns on the stems, leaf stalks and undersides of the leaves are sharp [1].

CONSERVATION STATUS

Assessed as Least Concern (LC) on the basis of its distribution and habitat [7], but its status has yet to be confirmed in the IUCN Red List [4]. In Mali, harvesting of roots and stem bark by herbalists and traditional healers seriously threatens the survival of the species in and around villages.

SEED CONSERVATION

HARVESTING. Fruits can be collected from November to February [8].

PROCESSING AND HANDLING. Seeds are extracted from the fruits, and washed and dried in the shade [8], then separated manually from any other plant material by paper rolling or with aspirators.

STORAGE AND VIABILITY. Initial seed viability is high as assessed by X-ray analysis, with c. 100% of filled seeds. Seed storage behaviour is unknown due to unresolved problems of seed development and dormancy. However, seeds maintained their initial poor viability of 2% germination when dried from 20% to 3% moisture content [8].

Figure 3 – Plant propagation at the Sikasso nursery. (Photo: A. K. Sanogo, IER.)

PROPAGATION

SEEDS

Dormancy and pre-treatments: Seeds are reported to have physiological dormancy [8]. Different pre-treatments have been tested at the MSB to break dormancy, including the use of gibberellic acid in the germination medium and heat shock treatments. However, Sanon *et al.* [8] found an increase in germination rate from 2% to 37% after nine months of storage at 4°C with 9% moisture content, suggesting that removal of dormancy may have been occurred.

Germination, sowing and planting: No information is available on optimum conditions for germination, including incubation temperature and photoperiod, because of unsolved problems of seed dormancy. However, plant propagation has started at the Sikasso nursery (Figure 3).

VEGETATIVE PROPAGATION

Root induction was carried out successfully using a new culture medium (called ZZ) with 100% rooting success [3].

TRADE

Roots, leaves and stems are commonly sold in markets in Mali, Côte d'Ivoire, Burkina Faso, Ghana and Nigeria [6].

Authors

Sidi Sanogo, Abdoul K. Sanogo, Bokary A. Kelly, Stéphane Rivière, Efisio Mattana, Tiziana Ulian and Rokia Sanogo.

References

[1] Arbonnier, M. (2004). *Trees, Shrubs and Lianas of West African Dry Zones*. CIRAD, France; Margraf Publishers, Weikersheim, Germany; and Muséum national d'histoire naturelle (MNHN), Paris, France.

[2] Burkill, H. M. (1997). *The Useful Plants of West Tropical Africa. Vol. 4: Families M–R*, 2nd edition. Royal Botanic Gardens, Kew.

[3] Etsè, K. D., Aïdam, A. V., de Souza, C., Crèche, J. & Lanoue, A. (2011). *In vitro* propagation of *Zanthoxylum zanthoxyloides* Lam., an endangered African medicinal plant. *Acta Botanica Gallica* 158 (1): 47–55.

[4] IUCN (2017). *The IUCN Red List of Threatened Species.* Version 2017-3. http://www.iucnredlist.org/

[5] Lebrun, J.-P. & Stork, A. L. (2010). *Tropical African Flowering Plants, Ecology and Distribution*. Vol. 5 – Buxaceae – Simaroubaceae. Conservatoire et Jardin Botaniques de la Ville de Genève, Switzerland.

[6] Matu, E. N. (2011). *Zanthoxylum zanthoxyloides* (Lam.) Zepern. & Timler. In: *PROTA (Plant Resources of Tropical Africa/Ressources Végétales de* l'*Afrique Tropicale)*. Edited by G. H. Schmelzer & A. Gurib-Fakim. Wageningen, The Netherlands. http://www.prota4u.org/

[7] Sacande, M., Sanou, L. & Beentje, H. (2012). *Guide d'Identification des Arbres du Burkina Faso*. Royal Botanic Gardens, Kew.

[8] Sanon, M. D., Gaméné, C. S., Sacandé, M. & Neya, O. (2004). Desiccation and storage of *Kigelia africana, Lophira lanceolata, Parinari curatellifolia* and *Zanthoxylum zanthoxyloides* seeds from Burkina Faso. In: *Comparative Storage Biology of Tropical Seeds*. Edited by M. Sacandé, D. Jøker, M. E. Dulloo & K. A. Thomsen. International Plant Genetic Resources Institute, Rome, Italy. pp. 16–23.

Nursery at LNBG in Mbombela (formerly Nelspruit).
(Photo: T. Ulian, RBG Kew.)

SOUTH AFRICA

Adenium swazicum Stapf

TAXONOMY AND NOMENCLATURE

FAMILY. Apocynaceae

SYNONYMS. *Adenium boehmianum* var. *swazicum* (Stapf) G.D.Rowley, *A. obesum* subsp. *swazicum* (Stapf) G.D.Rowley

VERNACULAR / COMMON NAMES. Sabi star, summer impala lily, summer-flowering impala lily (English); sisila semphala (Swati).

PLANT DESCRIPTION

LIFE FORM. Shrub.

PLANT. Caudiciform succulent, <0.7 m; stems branching, white or grey, from a large underground tuber <1 m in diameter; sap a clear, poisonous latex. **LEAVES.** Deciduous, simple, subsessile, lamina oblong to narrowly oblong, long tapering, 4–13 x 0.5–3 cm, smooth, folded and borne in terminal clusters, often folded (Figure 1), midrib prominent beneath. **FLOWERS.** Corolla lobes c. 1.5–2.5 (–3.5) x 1–2 cm, pink to white, crimson, mauve to deep reddish purple, with darker throats [1,6,10] (Figure 1).

Figure 1 – An adult plant in flower (left) and details of leaves and fruit follicles (right).
(Photos: T. Ulian, RBG Kew and W. Froneman, LNBG.)

FRUITS AND SEEDS

The seeds are in paired follicles which are c. 15 cm long and 1–2 cm wide. At maturity, the follicles are pinkish brown and covered in dense hair, and split vertically to release parachuted pale brown, glabrous seeds, 1.2–1.4 x 0.3 cm (Figure 2) [6].

FLOWERING AND FRUITING. Flowers from late November to the end of April [10]; follicles split open between September and November to release the seeds [10].

Figure 2 – Open follicles with seeds.
(Photo: W. Froneman, LNBG.)

DISTRIBUTION

South Africa (Mpumalanga and KwaZulu-Natal), Mozambique and Swaziland [6,8,13].

HABITAT

Open woodland on sand and often brackish soil at 300–400 m [6] and in the Lowveld Bushveld on turf soils. In South Africa, it is found on deep clay soils mainly derived from basalt and rhyolite [3].

USES

Extracts from *Adenium* spp. are extensively used as arrow poison [11]. *Adenium swazicum* is harvested for medicinal and ornamental or horticultural uses [9,10]. Tubers are soaked or boiled and the extract is taken orally to cure stomach ailments [9]. The harvesting method is destructive as the tubers are uprooted, killing the plant.

KNOWN HAZARDS AND SAFETY

As with other members of the genus, *A. swazicum* contains cardiac glycosides and is highly toxic to both humans and livestock [11,12].

CONSERVATION STATUS

Assessed as Critically Endangered (CR) in South Africa according to IUCN Red List criteria [3], Endangered (EN) in Swaziland [2], and Lower Risk-Least Concern (LR-lc) in Mozambique [5], but yet to be confirmed in the IUCN Red List [4]. The main threats are unsustainable harvesting of tubers for traditional medicine, uprooting of wild plants for horticulture, and habitat loss as a result of sugarcane cultivation and cattle grazing [10]. In South Africa and Swaziland, this species is protected by legislation [2,10].

SEED CONSERVATION

HARVESTING. Seeds can be harvested from open follicles after they burst open in November.
PROCESSING AND HANDLING. Seeds are easily cleaned by manually removing the parachutes.
STORAGE AND VIABILITY. No data are available on seed storage behaviour or viability [7]. However, high germination rates in garden experiments suggest that seed viability at dispersal is high.

PROPAGATION

SEEDS

Dormancy and pre-treatments: No pre-treatments are required.

Germination, sowing and planting: Germination tests carried out at LNBG showed that germination percentages of 80–85% can be achieved with optimum soil temperatures of 15–20°C. In the nursery, seeds should be sown in a mixture of equal parts by volume of river sand and compost or kraal manure. Containers should have sufficient drainage holes. Seeds should be sown 3 cm apart and 2 mm deep and watered thoroughly. Germination starts after 10–15 days if the seeds are sown during the summer months (September to January). Seedlings should be watered only once a week because the plants are sensitive to overwatering. During the winter months (May to August), the seedlings should be watered once every two weeks. Seedlings should be transplanted in individual pots or bags when they are 10 cm high 10 (Figures 3&4).

Figure 3 – *A. swazicum* seedlings. (Photo: W. Froneman, LNBG.)

Figure 4 – *A. swazicum* plant multiplication at LNBG. (Photo: K. van der Walt, LNBG.)

VEGETATIVE PROPAGATION

Can be propagated by cuttings and grafting. Cuttings should be 10–22 cm long and taken in late spring or early summer [1,10]. The base of the cuttings can be treated with hormone rooting powder; the cuttings are then planted directly into clean river sand or a potting mixture with at least 50% of sand [10], or they can be left to dry for about one week before planting [1]. Cuttings made from branches thinner than 1 cm at LNBG did not form roots, and rotted within two weeks [10]. *Adenium swazicum* can also be grafted using *A. obesum* (Forssk.) Roem. & Schult. as the stock plant [10].

TRADE

Plant material is collected when needed by local people and is not available in the muthi (traditional medicine) markets in Mpumalanga. However, destructive harvesting for medicinal and cultural uses was recorded as the biggest threat in a survey made between 2009 and 2011 [9].

Authors

Willem Froneman, Efisio Mattana, Karin van der Walt, Steve Davis and Tiziana Ulian.

References

[1] Bester, S.P. (2007). *Adenium* Roem. & Schult. South African National Biodiversity Institute (SANBI), South Africa. http://pza.sanbi.org/

[2] Dlamini, T. S. & Dlamini, G. M. (2002). Swaziland. In: *Southern African Plant Red Data Lists*. Edited by J. S. Golding. Southern African Botanical Diversity Network Report No. 14, SABONET, Pretoria, South Africa. pp. 121–134.

[3] Hurter, P. J. H., Lötter, M., Krynauw, S., Burrows, J. E. & Victor, J. E. (2009). *Adenium swazicum* Stapf. In: *National Assessment*: *Red List of South African Plants*. Version 2017.1. http://redlist.sanbi.org/

[4] IUCN (2017). *The IUCN Red List of Threatened Species.* Version 2017-3. http://www.iucnredlist.org/

[5] Izidine, S. & Bandeira, S. O. (2002). Mozambique. In: *Southern African Plant Red Data Lists*. Edited by J. S. Golding. Southern African Botanical Diversity Network Report No. 14, SABONET, Pretoria, South Africa. pp. 43–60.

[6] Launert, E. (ed.) (1985). Apocynaceae. In: *Flora Zambesiaca*, Vol. 7, part 2. Flora Zambesiaca Managing Committee, Royal Botanic Gardens, Kew.

[7] Royal Botanic Gardens, Kew (2017). *Seed Information Database (SID).* Version 7.1. http://data.kew.org/sid/

[8] Schmidt, E., Lötter, M. & McCleland, W. (2002). *Trees and Shrubs of Mpumalanga and Kruger National Park*. Jacana Media, Johannesburg, South Africa.

[9] Van der Walt, K. (2010). The critically endangered succulent *Adenium swazicum*. *Aloe* 47 (2): 4–7.

[10] Van der Walt, K. (2015). *Population Biology and Ecology of the Critically Endangered Succulent* Adenium swazicum. Master's thesis, University of the Witwatersrand, Johannesburg, South Africa.

[11] Van Wyk, B.-E., Van Heerden, F. & Van Oudtshoorn, B. (2002). *Poisonous Plants of South Africa*. Briza Publications, Pretoria, South Africa.

[12] Watt, J. M. & Breyer-Brandwijk, M. G. (1962). *The Medicinal and Poisonous Plants of Southern and Eastern Africa*, 2nd edition. E. & S. Livingstone, Edinburgh & London.

[13] WCSP (2017). *World Checklist of Selected Plant Families.* Royal Botanic Gardens, Kew. http://wcsp.science.kew.org/

Aloe arborescens Mill.

TAXONOMY AND NOMENCLATURE

FAMILY. Xanthorrhoeaceae

SYNONYMS. *Aloe arborea* Medik. *A. frutescens* Salm-Dyck, *A. fruticosa* Lam., *A. fulgens* Tod., *A. natalensis* J.M.Wood & M.S.Evans, *A. perfoliatum* Meyen, *A. principis* (Haw.) Stearn, *A. salm-dyckiana* Schult. & Schult.f., *A. sigmoidea* Baker

VERNACULAR / COMMON NAMES. Candelabra aloe, krantz aloe (English); kransaalwyn (Afrikaans); iNhlaba-encane, mangana leykulu (Swati).

PLANT DESCRIPTION

LIFE FORM. Shrub.

PLANT. Succulent, much-branched at or near base, <5 m tall (Figure 1); stems covered with persistent dried leaf remains. **LEAVES.** In dense rosettes at apex of branches, spreading and becoming recurved, lanceolate-attenuate, greyish green to bright green, without spots, 50–60 cm long, 5–7 cm wide at base; margins armed with firm forward-pointing, yellowish, deltoid teeth, 3–5 mm long, 10–15 mm apart. **FLOWERS.** Inflorescence erect, to 90 cm tall, simple or rarely with one short branch, with few sterile bracts below the raceme, which is conical, very densely flowered; buds covered by the bracts; flowers bright orange scarlet, or yellow [8,12].

Figure 1 – *A. arborescens* habit. (Photo: W. Froneman, LNBG.)

FRUITS AND SEEDS

The fruits are capsules, 17 x 7 mm, turning brown when ripe. Mature seeds are black, oblong-ovoid, with very narrow whitish wings, 1.75 x 7 mm (Figure 2).

FLOWERING AND FRUITING. Flowers during the winter months (i.e. May to July) [4]; capsules ripen from August to September.

Figure 2 – External views of the seeds and longitudinal section showing the fully developed linear embryo. (Photo: P. Gómez-Barreiro, RBG Kew.)

DISTRIBUTION

South Tropical Africa (Malawi, Mozambique and Zimbabwe) to South Africa, Botswana and Swaziland. In South Africa, it occurs in the Western Cape and along the coast and eastern mountains to the Northern Provinces, occurring in the Cape, Free State, KwaZulu-Natal, Mpumalanga and Limpopo [8,12,14]. It is introduced and naturalised in parts of North Africa, the Mediterranean region, Australia, Mexico and elsewhere [14].

HABITAT

Mountainous areas, sometimes forming dense bushes, or on exposed ridges and precipitous or overhanging walls of rock (i.e. 'krantzes') [12], in montane grasslands, in open evergreen forest and in coastal forest, from sea level to 2,800 m a.s.l. [4,8].

USES

In traditional medicine, leaf decoctions are used in childbirth, whereas cold water leaf infusions are used for stomach ache [4,5]. Leaf infusions are used to minimise high blood pressure. Leaf and root infusions are used to purify the blood; the leaves have purgative properties [4]. The leaf gel is used to treat skin rashes, burns and wounds [1,3]. The plant is often planted as a living hedge and for ornamental purposes [1,4].

KNOWN HAZARDS AND SAFETY

The leaf gel can occasionally cause allergic contact dermatitis [10]. Ingestion of large amounts of the leaf juice is purgative [11]. The teeth on the leaf margins are sharp.

CONSERVATION STATUS

A widespread and common species assessed as Least Concern (LC) in South Africa according to IUCN Red List criteria [13], but Vulnerable (VU) in Malawi [7]. Its status has yet to be confirmed in the IUCN Red List [6].

SEED CONSERVATION

HARVESTING. The seed capsules are ready for harvesting from August to September.

PROCESSING AND HANDLING. Black (mature) seeds can be easily shaken out of the capsules.

STORAGE AND VIABILITY. X-ray analysis carried out at the MSB showed c. 80–100% filled seeds. Seeds of this species are reported to be orthodox [9] and, therefore, after appropriate drying, they can be stored at sub-zero temperatures for long-term conservation.

PROPAGATION

SEEDS

Dormancy and pre-treatments: No pre-treatments are required.

Germination, sowing and planting: Germination tests carried out at LNBG showed that the optimum soil temperatures for germination are 20–25°C, with a final germination of c. 90%. Seeds should be sown in a mixture of equal parts by volume of washed river sand, compost and loamy soil, in seed trays or containers that are at least 10 cm deep. Seeds should be covered with 1 cm of substrate. Germination starts after 10–15 days if seeds are sown during the summer months (September to January). Seedlings should be watered twice a week in summer and once a week in winter (May to August). The seedlings should be transplanted after a year in the seed trays (Figure 3). Experiments carried out on seeds banked for long-term conservation at the MSB revealed final germination percentages of c. 80–85% for seeds re-moisturised at 95% RH and incubated for 1 day at 20°C, and then at constant temperatures of 10–15°C with 8 hours of light per day [9].

Figure 3 – *A. arborescens* seedlings.
(Photo: W. Froneman, LNBG.)

Figure 4 – Rooted cutting.
(Photo: W. Froneman, LNBG.)

VEGETATIVE PROPAGATION

Aloe arborescens grows easily from stem cuttings taken in summer [4,12] and truncheons [12]. Cuttings should be 15 x 1 cm and can be dipped in rooting powder before planting in coarse river sand in a mist bed or container. If available, automatic mist sprayers should be set at one minute of water application every 10 minutes. If planted in a container, the cuttings should be watered twice a week. Rooted cuttings can be transplanted after three months (Figure 4). Alternatively, cuttings can be allowed to dry before planting them in well-drained soil or sand in their permanent positions [4].

TRADE

Internationally traded and known in commerce as 'Japan aloe' [12]. International trade of plant material is regulated by CITES (listed on Appendix II) [2]. Plant material is available in muthi markets and in traditional Indian pharmacies in Mbombela (formerly Nelspruit).

Authors

Willem Froneman, Steve Davis and Efisio Mattana.

References

[1] Bosch, C. H. (2006). *Aloe arborescens* Mill. In: *PROTA (Plant Resources of Tropical Africa/Ressources Végétales de l'Afrique Tropicale).* Edited by G. H. Schmelzer & A. Gurib-Fakim. Wageningen, The Netherlands. http://www.prota4u.org/

[2] CITES (2017). *Convention on International Trade in Endangered Species of Wild Fauna and Flora*. https://cites.org/eng/app/index.php

[3] Grace, O. M., Simmonds, M. S. J., Smith, G. F. & Van Wyk, A. E. (2009). Documented utility and biocultural value of *Aloe* L. (Asphodelaceae): a review. *Economic Botany* 63 (2): 167–178.

[4] Hankey, A. & Notten, A. (2004). *Aloe arborescens* Mill. South African National Biodiversity Institute (SANBI), South Africa. http://pza.sanbi.org/

[5] Hutchings, A., Scott, A. H., Lewis, G. & Cunningham, A. B. (1996). *Zulu Medicinal Plants: An Inventory*. University of Natal Press, Pietermaritzburg, South Africa.

[6] IUCN (2017). *The IUCN Red List of Threatened Species.* Version 2017-3. http://www.iucnredlist.org/

[7] Msekandiana, G. & Mlangeni, E. (2002). Malawi. In: *Southern African Plant Red Data Lists.* Edited by J. S. Golding. Southern African Botanical Diversity Network Report No. 14, SABONET, Pretoria, South Africa. pp. 31–42.

[8] Pope, G. V. (ed.) (2001). Aloaceae. In: *Flora Zambesiaca*, Vol. 12, part 3. Flora Zambesiaca Managing Committee, Royal Botanic Gardens, Kew.

[9] Royal Botanic Gardens, Kew (2017). *Seed Information Database (SID).* Version 7.1. http://data.kew.org/sid/

[10] Shoji, A. (1982). Contact dermatitis to *Aloe arborescens. Contact Dermatitis* 8 (3): 164–167.

[11] Spoerke, D. G. & Smolinske, S. C. (1990). *Toxicity of Houseplants*. CRC Press, Boca Raton, Florida, USA.

[12] Van Wyk, B.-E. & Smith, G. F. (2014). *Guide to the Aloes of South Africa*, 3rd edition. Briza Publications, Pretoria, South Africa.

[13] von Staden, L. (2009). *Aloe arborescens* Mill. In: *National Assessment: Red List of South African Plants*. Version 2017.1. http://redlist.sanbi.org/

[14] WCSP (2017). *World Checklist of Selected Plant Families.* Royal Botanic Gardens, Kew. http://wcsp.science.kew.org/

Argyrolobium tomentosum (Andrews) Druce

TAXONOMY AND NOMENCLATURE

FAMILY. Leguminosae – Papilionoideae

SYNONYMS. *Argyrolobium andrewsianum* (E.Mey.) Steud., *A. shirense* Taub., *A. stuhlmannii* Taub., *Cytisus tomentosus* Andrews

VERNACULAR / COMMON NAMES. Velvet yellow bush pea (English); umadlozana (Swati).

PLANT DESCRIPTION

LIFE FORM. Shrub.

PLANT. Erect or scrambling, <2 m tall, evergreen; stems <7 mm in diameter at base; branches thinly to densely appressed or spreading, fulvous to greyish pubescent, sometimes the hairs grouped, glabrescent. **LEAVES.** 3-foliate; leaflets 10–50 x 6.25 mm, elliptic-lanceolate to elliptic, sparsely woolly above, more densely so beneath, with hairy margins, stipulate; terminal leaflets 2–7 x c. 1–3.5 cm. **FLOWERS.** Inflorescence leaf-opposed, pedunculate, racemose, sometimes long and lax, sometimes contracted and sub-umbelliform (8–32-flowered); flowers yellow, turning orange red with age, hairy outside [6,7] (Figure 1).

Figure 1 – *A. tomentosum* in flower. (Photo: W. Froneman, LNBG.)

FRUITS AND SEEDS

The seeds are produced in 3–6 cm long, oblong, slightly curved pods containing 12–18 seeds. The seeds are brown to black and smooth [7] (Figure 2).

FLOWERING AND FRUITING. Flowers from September to May, when ripe pods are also available.

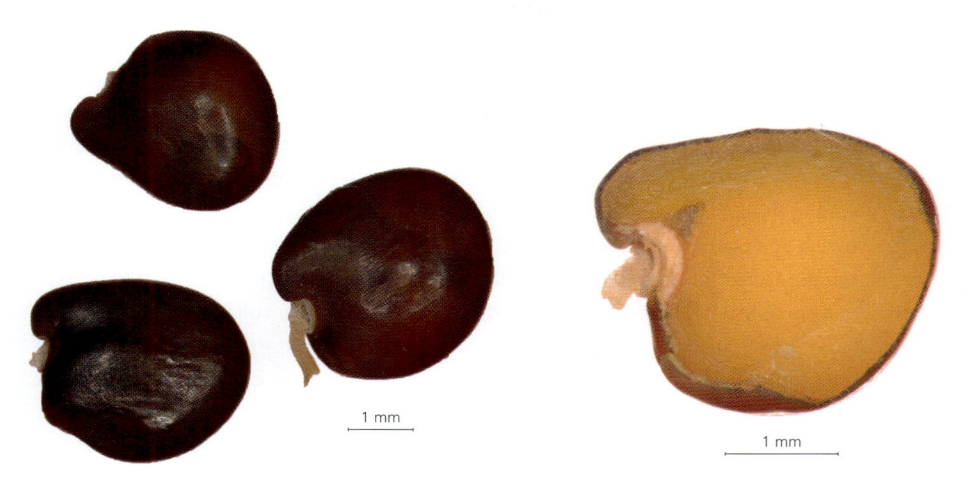

Figure 2 – External views of the seeds and longitudinal section showing the embryo.
(Photo: P. Gómez-Barreiro, RBG Kew.)

DISTRIBUTION

South Africa (Eastern Cape, KwaZulu-Natal, Limpopo, Mpumalanga and Western Cape), Mozambique, Zimbabwe, Zambia, Malawi, Tanzania, Uganda and Democratic Republic of Congo [3,6,7].

HABITAT

Brachystegia woodlands on rocky slopes, gully forests, margins of evergreen and swamp forests, moist grasslands on river and dambo margins, rarely growing in water, at altitudes of 600–2,400 m a.s.l. [1,7].

USES

In traditional medicine, a root infusion is taken as an emetic to sharpen the divining powers of sangomas (spiritual healers) [2,5]. Powdered leaves are mixed with a lotion and applied to sick people to prevent the spread of illness and 'bad spirits' to other people. The roots are used as a wash 'to clean' sick people. Powdered leaves are used to treat poisoned feet. Sometimes planted as a hedge or screen, or as an ornamental [5].

KNOWN HAZARDS AND SAFETY

Argyrolobium spp. are poisonous [10].

CONSERVATION STATUS

This widespread species is assessed as Least Concern (LC) in South Africa according to IUCN Red List criteria [8], but its status has yet to be confirmed in the IUCN Red List [4].

SEED CONSERVATION

HARVESTING. The pods are ready to be harvested from November.

PROCESSING AND HANDLING. Seeds can be extracted by manually opening the pods.

STORAGE AND VIABILITY. Seeds of this species are reported to be orthodox [9]. X-ray analysis carried out on the seed lot stored at the MSB showed 100% filled seeds.

PROPAGATION

SEEDS

Dormancy and pre-treatments: Before sowing, seeds can be left overnight in water to soften the hard tegument. In the laboratory, for experiments under controlled conditions, seeds can be scarified by chipping with a scalpel [9].

Germination, sowing and planting: Germination tests carried out at LNBG showed that the optimum soil temperatures for germination are 20–25°C, with final germination of 60–70%. Seeds should be sown 4 cm apart and 1 cm deep in a mixture of equal parts by volume of washed river sand, compost and loamy soil, in containers at least 10 cm deep. Seeds start to germinate after 15–20 days during the summer months (September to January), although germination is sporadic.

Figure 3 – *A. tomentosum* seedlings. (Photo: W. Froneman, LNBG.)

Seedlings should be watered twice a week in summer and once a week in winter (May to August) and transplanted when 10 cm high (Figure 3). Germination tests carried out on seeds stored at the MSB revealed a high germination percentage (100%) for scarified seeds incubated at constant temperatures of 20°C and 25°C with 8 hours of light per day [9]. *Argyrolobium tomentosum* grows best in loamy soil of pH 6–7, with moderate amounts of water and an optimum growing temperature of 25°C. It does not survive severe frost [5].

VEGETATIVE PROPAGATION

No protocols available.

TRADE

Plant material is available in muthi markets and in traditional Indian pharmacies in Mbombela (formerly Nelspruit).

Authors

Willem Froneman, Efisio Mattana, Lucy Shai and Jabulani Mahlangu.

References

[1] Conservatoire et Jardin botaniques & South African National Biodiversity Institute (SANBI) (2016). *African Plant Database*. Version 3.4.0. http://www.ville-ge.ch/musinfo/bd/cjb/africa/

[2] Hutchings, A., Scott, A. H., Lewis, G. & Cunningham, A. B. (1996). *Zulu Medicinal Plants: An Inventory*. University of Natal Press, Pietermaritzburg, South Africa.

[3] Hyde, M. A., Wursten, B. T., Ballings, P. & Coates Palgrave, M. (2017). *Flora of Zimbabwe*. www.zimbabweflora.co.zw

[4] IUCN (2017). *The IUCN Red List of Threatened Species*. Version 2017-3. http://www.iucnredlist.org/

[5] Mufhati, B. (2016). *Argyrolobium tomentosum* (Andrew) [*sic*] Druce. South African National Biodiversity Institute (SANBI), South Africa. http://pza.sanbi.org/

[6] Pooley, E. (1998). *A Field Guide to Wild Flowers KwaZulu-Natal and the Eastern Region*. Natal Flora Publications Trust, Durban, South Africa.

[7] Pope, G. V., Polhill, R. M. & Martins, E. S. (eds) (2003). Leguminosae (Papilionoideae). In: *Flora Zambesiaca*, Vol. 3, part 7. Flora Zambesiaca Managing Committee, Royal Botanic Gardens, Kew.

[8] Raimondo, D., von Staden, L., Foden, W., Victor, J. E., Helme, N. A., Turner, R. C., Kamundi, D. A. & Manyama, P. A. (2009). *Red List of South African Plants. Strelitzia* 25, South African National Biodiversity Institute, Pretoria, South Africa.

[9] Royal Botanic Gardens, Kew (2017). *Seed Information Database (SID)*. Version 7.1. http://data.kew.org/sid/

[10] Van Wyk, B.-E., Van Heerden, F. & Van Oudtshoorn, B. (2002). *Poisonous Plants of South Africa*. Briza Publications, Pretoria, South Africa.

Artemisia afra Jacq. ex Willd.

TAXONOMY AND NOMENCLATURE

FAMILY. Compositae

SYNONYMS. *Artemisia tenuifolium* Gaterau, *A. altaica* Desf., *A. balsamita* Willd., *A. grandiflora* Fisch. ex Herder, *A. pallida* Salisb., *A. pontica* Burm.f., *A. pseudopontica* Schur, *A. tenuifolia* Moench

VERNACULAR / COMMON NAMES. African wormwood, wild wormwood (English); wilde-als, wildeals (Afrikaans); mhlonyane, umhlonyane (Swati).

PLANT DESCRIPTION

LIFE FORM. Shrub.

PLANT. Highly aromatic, multi-stemmed, bushy, slightly untidy, <2.5 m tall, stems woody at base. **LEAVES.** Finely divided, fern-like in appearance, upper surface dark green, lower surface covered with hairs giving the shrub a silver grey colour (Figure 1). **FLOWERS.** Creamy yellow or brown, small, in pendulous heads (capitula) c. 3–5 mm in diameter; inflorescences axillary or terminal with many heads [1,9].

Figure 1 – *A. afra* leaves. (Photo: W. Froneman, LNBG.)

FRUITS AND SEEDS

The fruits are slightly 3-angled, curved cypselas, c. 1 mm long (Figure 2), covered with a silvery-white coating [4].

FLOWERING AND FRUITING. Flowers from January to June [6], mainly in late summer (March to May) in South Africa [8]. Fruits start to ripen in May or June.

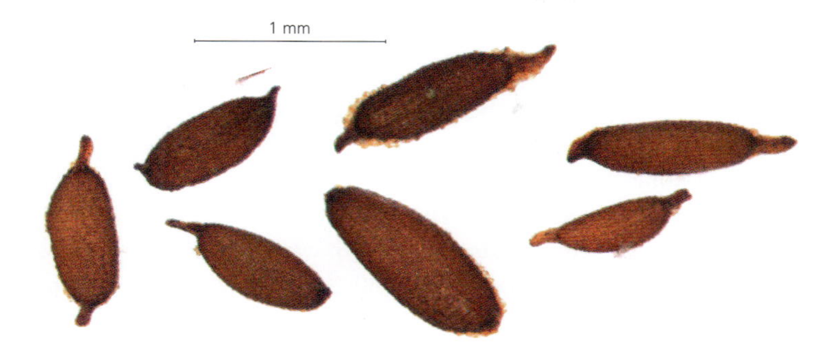

Figure 2 – External views of the cypselas. (Photo: P. Gómez-Barreiro, RBG Kew.)

DISTRIBUTION

Widely distributed in South Africa, from Gauteng and Limpopo, along the eastern parts of South Africa to the Western Cape [4,9]; also found in Swaziland, Lesotho, Angola, Botswana, Namibia and Zimbabwe, and in East Africa (Kenya, Tanzania, Uganda and Ethiopia) [6].

HABITAT

Grows in full sun in open woodland and forest margins, grassland, on damp slopes and along streams, at altitudes of 20–2,440 m a.s.l. [4]. Colonises burnt areas, sometimes forming pure stands [1].

USES

Used by people of many different cultures for treating a variety of ailments, including coughs, common cold, influenza, fever, loss of appetite, colic, headache, malaria and intestinal worms [2,8,9]. Leaf infusions are taken as teas or used as enemas or emetics for fertility complaints. The roots, stems, and leaves are used as enemas, poultices, infusions, body washes and lotions, and are smoked, snuffed or drunk as a tea. The oil acts as a local anaesthetic for rheumatism, neuralgia and arthritis [11]. An infusion of the leaves is used to treat body odour. *Artemisia afra* is also used as an insect repellent [8].

KNOWN HAZARDS AND SAFETY

High doses and chronic use could be harmful as the plant contains thujone, which can be toxic and hallucinogenic [9]. The essential oil can cause severe poisoning [2]. The plant is also poisonous to livestock. Cattle fed with *A. afra* suffered pulmonary oedema, liver changes and blood nephritis [10].

CONSERVATION STATUS

A widespread species assessed as Least Concern (LC) in South Africa according to IUCN Red List criteria [5], but its status has yet to be confirmed in the IUCN Red List [3].

SEED CONSERVATION

HARVESTING. The cypselas are ready to be harvested from June to September.

PROCESSING AND HANDLING. Cypselas can be sorted by hand.

STORAGE AND VIABILITY. Seeds of this species are reported to be orthodox [7]. X-ray analysis carried out on the seed lot stored at the MSB showed a high percentage of filled seeds (up to 90–100%).

PROPAGATION

SEEDS

Dormancy and pre-treatments: No pre-treatments required.

Germination, sowing and planting: Germination tests carried out at LNBG showed that the optimum soil temperatures for germination are between 20–25°C, with a final germination of c. 90%. Seeds should be sown on a mixture of equal parts by volume of washed river sand, compost and loamy soil, in a container at least 10 cm deep and covered with 5 mm of substrate. Seeds start to germinate after 10–15 days if planted during the summer months (i.e. September to January). Seeds and seedlings should be watered twice a week in summer and once a week in winter (i.e. May to August). Germination tests carried out on seeds stored at the MSB highlighted high germination percentages (95–100%) for scarified seeds incubated at constant temperatures in the range 15–30°C with 8 hours of light per day [7].

VEGETATIVE PROPAGATION

This species is normally propagated by hardwood cuttings taken during summer (i.e. from October to March). Cuttings should be 10 cm long and treated with a hormone powder to stimulate rooting, inserted into a mixture of washed river sand and potting compost (in a ratio of 1:2 by volume), and placed under a mist sprayer system or watered twice a week. Rooted cuttings are ready for transplanting after three months (Figure 3). Established plants can be propagated by division [8].

Figure 3 – *A. afra* rooted cutting. (Photo: W. Froneman, LNBG.)

TRADE

Plant material is available in muthi markets and in traditional Indian pharmacies in Mbombela (formerly Nelspruit).

Authors

Willem Froneman, Efisio Mattana, Lucy Shai, Steve Davis and Jabulani Mahlangu.

References

[1] Beentje, H. J. (1994). *Kenya Trees, Shrubs and Lianas*. National Museums of Kenya, Nairobi, Kenya.

[2] Hutchings, A., Scott, A. H., Lewis, G. & Cunningham, A. B. (1996). *Zulu Medicinal Plants: An Inventory*. University of Natal Press, Pietermaritzburg, South Africa.

[3] IUCN (2017). *The IUCN Red List of Threatened Species.* Version 2017-3. http://www.iucnredlist.org/

[4] Liu, N. Q., Van der Kooy, F. & Verpoorte, R. (2009). *Artemisia afra*: a potential flagship for African medicinal plants? *South African Journal of Botany* 75 (2): 185–195.

[5] Raimondo, D., von Staden, L., Foden, W., Victor, J. E., Helme, N. A., Turner, R. C., Kamundi, D. A. & Manyama, P. A. (2009). *Red List of South African Plants. Strelitzia* 25, South African National Biodiversity Institute, Pretoria, South Africa.

[6] Royal Botanic Gardens, Kew (1999–2015). *Survey of Economic Plants for Arid and Semi-Arid Lands (SEPASAL) Database*. http://apps.kew.org/sepasalweb/sepaweb

[7] Royal Botanic Gardens, Kew (2017). *Seed Information Database (SID).* Version 7.1. http://data.kew.org/sid/

[8] Van der Walt, L. (2004). *Artemisia afra* Jacq. ex Willd. South African National Biodiversity Institute (SANBI), South Africa. http://pza.sanbi.org/

[9] Van Wyk, B.-E., Van Oudtshoorn, B. & Gericke, N. (2009). *Medicinal Plants of South Africa*, 2nd edition. Briza Publications, Pretoria, South Africa.

[10] Von Koenen, E. (2001). *Medicinal, Poisonous, and Edible Plants in Namibia*. Edition Namibia, Vol. 4. Klaus Hess Publishers, Windhoek, Namibia.

[11] Watt, J. M. & Breyer-Brandwijk, M. G. (1962). *The Medicinal and Poisonous Plants of Southern and Eastern Africa*, 2nd edition. E. & S. Livingstone, Edinburgh & London.

Berchemia zeyheri (Sond.) Grubov

TAXONOMY AND NOMENCLATURE

FAMILY. Rhamnaceae

SYNONYMS. *Phyllogeiton zeyheri* Suess., *Rhamnus zeyheri* Sond.

VERNACULAR / COMMON NAMES. Red ivory (English); rooi-ivoor (Afrikaans); umNeyi Nyiri (Swati).

Figure 1 – *B. zeyheri* showing the green fruits and bark. (Photos: W. Froneman, LNBG.)

PLANT DESCRIPTION

LIFE FORM. Tree or sometimes a shrub.

PLANT. Evergreen or semi-deciduous, <6–8(15) m tall [2,5]; main stem pale green when young, young branchlets puberulous; bark dark grey, roughly fissured (Figure 1). **LEAVES.** Usually opposite to subopposite, entire, elliptic to ovate, glossy dark green above, paler below, with prominent midrib beneath, lamina 12–40(60) x 6–25 mm, rounded or cordate at base (Figure 1). **FLOWERS.** Small (2 mm in diameter in bud), star-like, held on 1 cm long pedicels, in clusters of 1–4(6), axillary, petals 2 mm long, yellowish green [2,5,8].

FRUITS AND SEEDS

The fruits are fleshy drupes, ± ovate, 6–9 x 3 mm, yellow to red when ripe, with several separate single-seeded pyrenes. Pyrenes brown and ± ovate (Figures 2&3) [7,8].

FLOWERING AND FRUITING. Flowers from September to December; fruits from January to April [5].

Figure 2 – External views of the dry drupes. (Photo: P. Gómez-Barreiro, RBG Kew.)

DISTRIBUTION

South Africa (KwaZulu-Natal and the Northern Provinces), Botswana, Swaziland, Zimbabwe and Mozambique [2,8].

HABITAT

Open woodland, rocky ridges, old termite mounds and riverbanks [5,8] at 60–1,980 m a.s.l [1]. Drought resistant, but does not tolerate frost [5].

USES

The fruits are edible and can be eaten fresh or after storage (Figure 3). Bark infusions are used in traditional medicine as enemas for back pains and for treating rectal ulceration in children [3]. The wood is durable and used to make furniture, wooden bowls, walking sticks, ornaments and curios. Fibres and woven materials are dyed with an extract made of the bark to give them a purplish colour [8].

KNOWN HAZARDS AND SAFETY

The sap can cause dermatitis.

Figure 3 – *B. zeyheri* fresh drupes and pyrenes. (Photo: W. Froneman, LNBG.)

CONSERVATION STATUS

Assessed as Least Concern (LC) in South Africa according to IUCN Red List criteria [6], but its status has yet to be confirmed in the IUCN Red List [4].

SEED CONSERVATION

HARVESTING. The fleshy drupes are ready for harvesting from December.

PROCESSING AND HANDLING. Seeds (pyrenes) can be easily extracted and cleaned by washing the drupes in water to separate them from the fleshy pulp.

STORAGE AND VIABILITY. No data on seed desiccation sensitivity are available. However, the related *B. discolor* (Klotzsch) Hemsl. is reported to have desiccation-tolerant seeds [7]. X-ray analysis carried out on the seed lot stored at the MSB showed a high percentage of filled seeds (85–90%).

PROPAGATION

SEEDS

Dormancy and pre-treatments: No pre-treatments are required when sowing in the nursery. However, for germination experiments under laboratory controlled conditions, seeds should be sterilised by immersing in a solution of 10% commercial bleach for 5 minutes, then soaked in water for one day at 20°C to remove the fruit pulp, and then scarified by removing the covering structure before sowing [7].

Germination, sowing and planting: Germination tests carried out at LNBG showed that the optimum soil temperatures for germination are 20–25°C, with final germination of 50–60%. Seeds should be sown in a mixture of equal parts by volume of washed river sand and loamy soil, in a container at least 10 cm deep. Seeds should be sown 2 cm apart and 1 cm deep. Seeds and seedlings should be watered twice a week in summer (but only once every two weeks in winter). Seeds germinate after 5–9 days [5] if sown during the summer (i.e. September to January). Seedlings should be transplanted when they are 10 cm tall. Germination tests carried out on seeds stored at the MSB revealed a high germination percentage (85%) for seeds pre-treated as above and incubated at 20°C with 8 hours of light per day [7].

VEGETATIVE PROPAGATION

No protocols available.

TRADE

The fruits are an important source of income for local communities and are sold in rural markets. As well as being available in muthi markets, plant material is also available in traditional Indian pharmacies in Mbombela (formerly Nelspruit). Craftwork and ornaments made from the wood are available in formal and informal curio markets in Southern Africa [9].

Authors

Willem Froneman, Efisio Mattana, Vusi Lukhele and Tiziana Ulian.

References

[1] Conservatoire et Jardin botaniques & South African National Biodiversity Institute (SANBI) (2016). *African Plant Database*. Version 3.4.0. http://www.ville-ge.ch/musinfo/bd/cjb/africa/

[2] Exell, A. W., Fernandes, A. & Wild, H. (eds) (1966). Rhamnaceae. In: *Flora Zambesiaca*, Vol. 2, part 2. Flora Zambesiaca Managing Committee, Royal Botanic Gardens, Kew.

[3] Hutchings, A., Scott, A. H., Lewis, G. & Cunningham, A. B. (1996). *Zulu Medicinal Plants: An Inventory*. University of Natal Press, Pietermaritzburg, South Africa.

[4] IUCN (2017). *The IUCN Red List of Threatened Species.* Version 2017-3. http://www.iucnredlist.org/

[5] Ndou, A. P. (2005). *Berchemia zeyheri* (Sond.) Grubov. South African National Biodiversity Institute (SANBI), South Africa. http://pza.sanbi.org/

[6] Raimondo, D., von Staden, L., Foden, W., Victor, J. E., Helme, N. A., Turner, R. C., Kamundi, D. A. & Manyama, P. A. (2009). *Red List of South African Plants. Strelitzia* 25, South African National Biodiversity Institute, Pretoria, South Africa.

[7] Royal Botanic Gardens, Kew (2017). *Seed Information Database (SID).* Version 7.1. http://data.kew.org/sid/

[8] Schmidt, E., Lötter, M. & McCleland, W. (2002). *Trees and Shrubs of Mpumalanga and Kruger National Park*. Jacana Media, Johannesburg, South Africa.

[9] Van Wyk, B.-E. & Gericke, N. (2000). *People's Plants: A Guide to Useful Plants of Southern Africa.* Briza Publications, Pretoria, South Africa.

Coddia rudis (E.Mey. ex Harv.) Verdc.

TAXONOMY AND NOMENCLATURE

FAMILY. Rubiaceae

SYNONYMS. *Heinsia capensis* H.Buek ex Harv., *Lachnosiphonium rude* (E.Mey. ex Harv.) J.G.García, *L. rude* var. *parvifolium* Yamam., *Randia rudis* E.Mey. ex Harv., *Xeromphis rudis* (E.Mey. ex Harv.) Codd

VERNACULAR / COMMON NAMES. Small bone apple (English); kleinbeenappel (Afrikaans); siKhwakhwane (Swati).

PLANT DESCRIPTION

LIFE FORM. Shrub or tree.

PLANT. Evergreen, 1–4 m tall, much-branched with arching branches, grey when young.

LEAVES. Opposite, broadly ovate, usually 2 cm wide, 1.5 cm long [7], acuminate, contracted into the petiole, shortly pubescent on midrib above and with hairy domatia in leaf axils beneath.

FLOWERS. Fragrant, white fading to yellowish or cream, grouped in axillary clusters, corolla tube 5–7 mm long; lobes 3–6 x 1.7–5.5 mm (Figure 1) [3,6].

Figure 1 – *C. rudis* flowers (Photo: A. Notten, SANBI) and fruits (Photo: J. Nichols, conservation consultant.)

FRUITS AND SEEDS

Fruits are 5–7 mm long and wide, ellipsoid to subglobose, glabrous, or slightly to densely minutely bristly or pilose [3], crowned with the remains of the calyx [6] (Figure 1). Seeds are pale chestnut coloured (Figure 2).

FLOWERING AND FRUITING. Flowers from August to January; fruits available from January to June.

Figure 2 – External views of the seeds and longitudinal section showing the endosperm and the spatulate embryo. (Photos: P. Gómez-Barreiro, RBG Kew.)

DISTRIBUTION

South Africa (Mpumalanga, KwaZulu-Natal and Eastern Cape), Swaziland, Mozambique and Zimbabwe [3,6].

HABITAT

Woodlands including *Brachystegia*, riverine and other fringing vegetation, in thickets and rocky areas [6], at altitudes from sea level to 1,290 m a.s.l. [3].

USES

The fruits are edible [7]. The roots are used in traditional medicine for treating fevers and as an emetic. Pounded root decoctions are used to treat impotence [1]. Also planted as an ornamental and as a hedge [7].

KNOWN HAZARDS AND SAFETY

No known hazards.

CONSERVATION STATUS

Assessed as Least Concern (LC) in South Africa according to IUCN Red List criteria [4], but its status has yet to be confirmed in the IUCN Red List [2].

SEED CONSERVATION

HARVESTING. Fruits can be picked directly from the plants.

PROCESSING AND HANDLING. Seeds can be separated from the fruits by gently rubbing them (rubber gloves can be worn), and then using sieves to separate them from other plant material.

STORAGE AND VIABILITY. X-ray analysis carried out on the seed lot stored at the MSB showed a high percentage of filled seeds (75%). No information is available on desiccation tolerance [5]. However, the high germination percentage achieved on the seed lot stored for long-term conservation at the MSB suggests that seeds of this species are unlikely to be desiccation sensitive.

PROPAGATION

SEEDS

Dormancy and pre-treatments: No pre-treatments are required.

Germination, sowing and planting: Germination tests carried out at LNBG showed that 15–20% of seeds germinated when sown in a soil temperature of 20–25°C. Seeds were sown in a mixture of equal parts by volume of washed river sand, compost and loamy soil in a container 10 cm deep. Seeds were sown 1 cm apart and 1 cm deep. Seeds and seedlings (Figure 3) were watered twice a week in summer, and the seedlings once a week in winter (i.e. from May to August). Seeds sown in September started to germinate after 30–40 days. Seedlings were transplanted when they were 10 cm high. Germination tests carried out on seeds stored at the MSB revealed a high germination percentage (88%) for seeds incubated at 25°C with 8 hours of light per day.

VEGETATIVE PROPAGATION

No protocols available.

Figure 3 – *C. rudis* seedlings. (Photo: W. Froneman, LNBG.)

TRADE

Plant material is available in muthi markets and in traditional Indian pharmacies in Mbombela (formerly Nelspruit).

Authors

Willem Froneman, Efisio Mattana, Lucy Shai and Jabulani Mahlangu.

References

[1] Hutchings, A., Scott, A. H., Lewis, G. & Cunningham, A. B. (1996). *Zulu Medicinal Plants: An Inventory*. University of Natal Press, Pietermaritzburg, South Africa.

[2] IUCN (2017). *The IUCN Red List of Threatened Species.* Version 2017-3. http://www.iucnredlist.org/

[3] Pope, G. V. (ed.) (2003). Rubiaceae (subfamily Cinchonoideae). In: *Flora Zambesiaca*, Vol. 5, part 3. Flora Zambesiaca Managing Committee, Royal Botanic Gardens, Kew.

[4] Raimondo, D., von Staden, L., Foden, W., Victor, J. E., Helme, N. A., Turner, R. C., Kamundi, D. A. & Manyama, P. A. (2009). *Red List of South African Plants. Strelitzia* 25, South African National Biodiversity Institute, Pretoria, South Africa.

[5] Royal Botanic Gardens, Kew (2017). *Seed Information Database (SID).* Version 7.1. http://data.kew.org/sid/

[6] Schmidt, E., Lötter, M. & McCleland, W. (2002). *Trees and Shrubs of Mpumalanga and Kruger National Park*. Jacana Media, Johannesburg, South Africa.

[7] Sepheka, W. K. (2012). *Coddia rudis* (E.Mey. ex Harv.) Verdc. South African National Biodiversity Institute (SANBI), South Africa. http://pza.sanbi.org/

Cordyla africana Lour.

TAXONOMY AND NOMENCLATURE

FAMILY. Leguminosae – Papilionoideae

SYNONYMS. None.

VERNACULAR / COMMON NAMES. Wild mango (English); wildemango (Afrikaans); xivuvule (Shangaan); umbubuli (Swati).

Figure 1 – *C. africana* bark (Photo: W. Froneman, LNBG) and branch with flowers. (Photo: SANBI.)

PLANT DESCRIPTION

LIFE FORM. Tree.

PLANT. Deciduous, to 25 m (but can reach 40 m) [1], with longitudinally fissured brown or grey bark (Figure 1) and a spreading crown. Stems greyish brown and rough when old; young branches glabrous or minutely pubescent. **LEAVES.** Compound, imparipinnate, with 11–28 pairs of leaflets which are elliptic to oblong, rounded to emarginate at apex, rounded to truncate at base, each 1–5 x 0.6–2.4 cm. **FLOWERS.** Green with protruding, showy, yellow orange or yellow stamens, in clusters below the leaves (Figure 1) [1,11,12].

FRUITS AND SEEDS

Fruits are up to 8 x 6 cm, yellow or orange yellow when ripe (Figure 2). Seeds are oblong and embedded in yellow sticky flesh (Figure 2) [12].

FLOWERING AND FRUITING. Flowers from August to October; fruits from December to February [11].

Figure 2 – *C. africana* seeds and fruits. (Photo: W. Froneman, LNBG.)

DISTRIBUTION

South Africa (KwaZulu-Natal and Mpumalanga), Swaziland, Mozambique, northwards to coastal districts of Tanzania and Kenya [1,11,12].

HABITAT

Riverbanks in Bushveld [11], and in evergreen riverine forest or miombo woodland on escarpment slopes or alluvial plains, from sea level to 1,000 m a.s.l. [12].

USES

The fruits are edible and rich in vitamin C [13]. The bark is used in traditional medicine as an emetic [4]. The wood is used for buildings, and the trunks are hollowed out to make drums [7]. *Cordyla africana* is planted as an ornamental tree; flowering specimens are spectacular [6,11].

KNOWN HAZARDS AND SAFETY

The bark is purgative and emetic [8].

CONSERVATION STATUS

A widespread species assessed according to IUCN Red List criteria as Least Concern (LC) in South Africa [9], Swaziland [3] and Zambia [2], but its status has yet to be confirmed in the IUCN Red List [5].

SEED CONSERVATION

HARVESTING. Fruits are yellow when ripe (Figure 2) and are available from December to February. The fruits usually drop from the tree before they are fully ripe [7].

PROCESSING AND HANDLING. Fruits can be opened manually and seeds removed from the fleshy pulp [11].

STORAGE AND VIABILITY. No information available on seed desiccation tolerance. However, seeds of the congeneric *C. pinnata* (A.Rich.) Milne-Redh. are reported to be recalcitrant [10].

PROPAGATION

SEEDS

Dormancy and pre-treatments: No pre-treatments are required.

Germination, sowing and planting: In the wild, seeds often germinate while still inside the fruit [7]. Germination tests carried out at LNBG showed that the optimum soil temperatures for germination are 20–25°C, with a final germination value of 98%. Seeds should be sown in a mixture of equal parts by volume of washed river sand, compost and loamy soil in a container that is at least 10 cm deep. Seeds should be sown 6 cm apart and 2 cm deep. Seeds and seedlings should be watered twice a week in summer and once a week in winter (i.e. from May to August). Seeds start to germinate after 10–15 days if sown in the summer (i.e. from September to January). Seedlings should be transplanted when they are 10 cm high.

VEGETATIVE PROPAGATION

No protocols available.

TRADE

Plant material is available in muthi markets and in traditional Indian pharmacies in Mbombela (formerly Nelspruit).

Authors

Willem Froneman, Steve Davis and Efisio Mattana.

References

[1] Beentje, H. J. (1994). *Kenya Trees, Shrubs and Lianas*. National Museums of Kenya, Nairobi, Kenya.

[2] Bingham, M. G. & Smith, P. P. (2002). Zambia. In: *Southern African Plant Red Data Lists*. Edited by J. S. Golding. Southern African Botanical Diversity Network Report No. 14, SABONET, Pretoria, South Africa. pp. 135–156.

[3] Dlamini, T. S. & Dlamini, G. M. (2002). Swaziland. In: *Southern African Plant Red Data Lists*. Edited by J. S. Golding. Southern African Botanical Diversity Network Report No. 14, SABONET, Pretoria, South Africa. pp. 121–134.

[4] Hutchings, A., Scott, A. H., Lewis, G. & Cunningham, A. B. (1996). *Zulu Medicinal Plants: An Inventory*. University of Natal Press, Pietermaritzburg, South Africa.

[5] IUCN (2017). *The IUCN Red List of Threatened Species*. Version 2017-3. http://www.iucnredlist.org/

[6] Lemmens, R. H. M. J. & Nyunaï, N. (2011). *Cordyla africana* Lour. In: *PROTA (Plant Resources of Tropical Africa/Ressources Végétales de l'Afrique Tropicale)*. Edited by R. H. M. J. Lemmens, D. Louppe & A. A. Oteng-Amoako. Wageningen, The Netherlands. http://www.prota4u.org/

[7] Meyer, J. (2006). *Cordyla africana* Lour. South African National Biodiversity Institute (SANBI), South Africa. http://pza.sanbi.org/

[8] Quattrocchi, U. (2012). *CRC World Dictionary of Medicinal and Poisonous Plants: Common Names, Scientific Names, Eponyms, Synonyms, and Etymology, Vol. II: C–D*. CRC Press, Boca Raton, Florida, USA.

[9] Raimondo, D., von Staden, L., Foden, W., Victor, J. E., Helme, N. A., Turner, R. C., Kamundi, D. A. & Manyama, P. A. (2009). *Red List of South African Plants. Strelitzia* 25, South African National Biodiversity Institute, Pretoria, South Africa.

[10] Royal Botanic Gardens, Kew (2017). *Seed Information Database (SID)*. Version 7.1. http://data.kew.org/sid/

[11] Schmidt, E., Lötter, M. & McCleland, W. (2002). *Trees and Shrubs of Mpumalanga and Kruger National Park*. Jacana Media, Johannesburg, South Africa.

[12] Timberlake, J. R., Pope, G. V., Polhill, R. M. & Martins, E. S. (eds) (2007). Leguminosae (Papilionoideae). In: *Flora Zambesiaca*, Vol. 3, part 3. Flora Zambesiaca Managing Committee, Royal Botanic Gardens, Kew.

[13] Watt, J. M. & Breyer-Brandwijk, M. G. (1962). *The Medicinal and Poisonous Plants of Southern and Eastern Africa*, 2nd edition. E. & S. Livingstone, Edinburgh & London.

Dioscorea strydomiana Wilkin

TAXONOMY AND NOMENCLATURE

FAMILY. Dioscoreaceae

SYNONYMS. None.

VERNACULAR / COMMON NAMES. Ebutsini wild yam, Strydom's yam (English); inyawolendlovu (Swati).

PLANT DESCRIPTION

LIFE FORM. Shrub-like caudex geophyte.

PLANT. Dioecious, <1.5 m high, with annual, non-twining stems from a caudiciform woody tuber which above ground is up to c. 1 x 1 m, with an outer corky layer divided by furrows into projections; stems one to several per shoot-bearing apex per growing season. **LEAVES.** Alternate, pale green on both surfaces, mostly narrowly ovate to elliptic (Figure 1). **FLOWERS.** Inflorescences are one per axil, simple racemose, erect to ascending, with flowers held erect to ascending at anthesis, few per inflorescence towards shoot apices, especially in female plants [5].

Figure 1 – Adult plant and female flowers.
(Photos: J. Burrows, Buffelskloof Herbarium, Lydenburg, South Africa.)

FRUITS AND SEEDS

The fruits are elliptic capsules, with floral components still persistent at capsule apex until dehiscence, when it opens at c. 1/4 to 1/3 of its depth [5]. Seeds are winged at apex only or with a narrow wing on the side and sometimes at the base of the seed, pale to dark brown, smooth to the naked eye but roughened to weakly striate under a dissecting microscope [5] (Figure 2).

FLOWERING AND FRUITING. Flowers from September to October; capsules ripen from March to May.

Figure 2 – External and internal (longitudinal section) views of the seeds. (Photos: P. Gómez-Barreiro, RBG Kew.)

DISTRIBUTION

Endemic to Mpumalanga, South Africa [5].

HABITAT

Open woodland with a grass-rich understorey on steep, rocky, SE- to SSE-facing slopes on dolerite, with quartzite intrusions at 1,100–1,500 m a.s.l. [5].

USES

Dioscorea strydomiana is used locally with another species of *Dioscorea* to treat cancer, but its actual efficacy is unknown [5]. Steroidal compound levels are known to be very high in the related species *D. elephantipes* (L'Hér.) Engl. and *D. sylvatica* Eckl. It is therefore likely that *D. strydomiana* may also contain similar anti-inflammatory compounds that could be useful in treating conditions such as arthritis or for wound healing [5]. Stems and tubers are also reported to treat water retention (Traditional Knowledge reported by Mpumalanga Parks and Tourism Agency, in [5]).

KNOWN HAZARDS AND SAFETY

No known hazards.

Figure 3 – Plant damaged by unsustainable harvesting of the tuber. (Photo: J. Burrows, Buffelskloof Herbarium, Lydenburg, South Africa.)

CONSERVATION STATUS

This rare species has been assessed as Critically Endangered (CR) according to IUCN Red List criteria [1,4]. The main threat is unsustainable harvesting of tubers for traditional medicine (Figure 3). Mining is also a potential threat [1,4].

SEED CONSERVATION

HARVESTING. Fruit capsules burst open between April and May. Seeds can be harvested from open capsules.

PROCESSING AND HANDLING. Seeds can be easily cleaned by separating them from fruit debris using aspirators or sieves.

STORAGE AND VIABILITY. Initial seed viability may vary among seed lots as assessed by X-ray analysis, with up to 30% of seeds found to be empty. Seeds of many *Dioscorea* spp. (including the related *D. sylvatica*) are reported to have orthodox seeds, but no data are available for *D. strydomiana* [3]. However, germination tests carried out at the MSB on *D. strydomiana* seeds stored for more than one month at –20°C showed high germination percentages (>80%), suggesting that seeds of this species are unlikely to be recalcitrant.

PROPAGATION

SEEDS

Dormancy and pre-treatments: No pre-treatments are required.

Germination, sowing and planting: Seeds of D. *strydomiana* reached maximum germination at 15°C (c. 47%), which decreased slightly to c. 37% at 25°C and was completely inhibited at 35°C, with a base temperature for germination of 9.3°C [2]. A mixture of river sand, compost/kraal manure and loamy soil in a ratio of 1:1:3 by volume should be used. Each seed should be planted in its own container, which should be at least 10 cm deep and 10 cm wide (10 cm commercial plant pots or 500 ml plastic cool drink bottles with drainage holes are suitable). Seeds should be covered with 15 mm of substrate and watered thoroughly. Seedlings (Figure 4)

Figure 4 – Seedlings propagated at LNBG. (Photo: E. Mattana, RBG Kew.)

should be watered only once a week because the plants are sensitive to overwatering. During the winter months (May to August), seedlings should be watered once every two weeks only. Seeds start to germinate after 22–26 days if sown during summer months (September to January). *Dioscorea* seedlings are sensitive to transplanting, so seedlings should remain in the same container for at least four years.

VEGETATIVE PROPAGATION

No protocols available.

TRADE

Plant material is collected from wild populations when needed by local people and is not available in muthi markets in Mpumalanga.

Authors

Efisio Mattana, Paul Wilkin, Vusi Lukhele, Pablo Gómez-Barreiro and Tiziana Ulian.

References

[1] IUCN (2017). *The IUCN Red List of Threatened Species.* Version 2017-3. http://www.iucnredlist.org/

[2] Mattana, E., Gomez Barreiro, P., Lötter, M., Hankey, A. J., Froneman, W., Mamatsharaga, A., Wilkin, P. & Ulian, T. (2018). Morphological and functional seed traits of the wild medicinal plant *Dioscorea strydomiana*, the most threatened yam in the world. *Plant Biology* in press. doi:10.1111/plb.12887

[3] Royal Botanic Gardens, Kew (2017). *Seed Information Database (SID).* Version 7.1. http://data.kew.org/sid/

[4] von Staden, L., Victor, J. E., Raimondo, D. & Hurter, P. J. H. (2011). *Dioscorea strydomiana* Wilkin. In: *National Assessment: Red List of South African Plants.* Version 2017.1. http://redlist.sanbi.org/

[5] Wilkin, P., Burrows, J., Burrows, S., Muthama Muasya, A. & van Wyk, E. (2010). A critically endangered new species of yam (*Dioscorea strydomiana* Wilkin, Dioscoreaceae) from Mpumalanga, South Africa. *Kew Bulletin* 65 (3): 421–433.

Drimia delagoensis (Baker) Jessop

TAXONOMY AND NOMENCLATURE

FAMILY. Asparagaceae

SYNONYMS. *Sekanama delagoensis* (Baker) Speta, *Urginea delagoensis* Baker, *U. lydenburgensis* R.A.Dyer

VERNACULAR / COMMON NAMES. Gifbol (Afrikaans); uMahlanganisa isiklenama (Swati).

PLANT DESCRIPTION

LIFE FORM. Perennial herb.

PLANT. A succulent geophyte, <45 cm tall [7]; bulb solitary or in tight clusters protruding above the soil surface (Figure 1), coppery green, up to 6 cm in diameter. **LEAVES.** Grey green, rounded, cylindric, 50 x 1 cm, fleshy to tough rubbery, tapering to a slender point. **FLOWERS.** Inflorescence c. 30 cm long, with solitary or slightly grouped flowers which are 8 mm in diameter, creamy to pale brown or mauve, with or without a central green stripe which is purple beneath [1,7,9].

Figure 1 – *D. delagoensis* bulb and flowers. (Photos: SANBI.)

FRUITS AND SEEDS

The fruits are capsules containing black seeds that are flattened (Figure 2).

FLOWERING AND FRUITING. Flowers from August to September; capsules ripen from October to January [9].

Figure 2 – External view, and longitudinal and cross-sections of the seeds showing the linear fully developed embryo. (Photos: P. Gómez-Barreiro, RBG Kew.)

DISTRIBUTION

South Africa (KwaZulu-Natal and Mpumalanga), Swaziland and Mozambique [4].

HABITAT

Open woodland on rocky areas [9].

USES

The bulb is used in traditional medicine to ward off evil spirits and as a blood purifier [6]. *Drimia delagoensis* is also used for treating skin infections, although its efficacy has yet to be evaluated [2,5]. Used in traditional medicine for making splints for bone fractures, and worn as amulets [7].

KNOWN HAZARDS AND SAFETY

No known hazards.

CONSERVATION STATUS

Assessed as Least Concern (LC) in South Africa according to IUCN Red List criteria [8], but its status has yet to be confirmed in the IUCN Red List [3].

SEED CONSERVATION

HARVESTING. Fruit capsules burst open between October and November. Seeds can be harvested manually from open capsules.

PROCESSING AND HANDLING. Seeds are easily cleaned by separating them from fruit debris using aspirators or sieves.

STORAGE AND VIABILITY. No information is available on seed desiccation tolerance. However, seeds of the congeneric *D. elata* Jacq. and *D. maritima* (L.) Stearn are reported to be recalcitrant [10].

PROPAGATION

SEEDS

Dormancy and pre-treatments: No pre-treatments are required for seed propagation in nurseries. However, the seed dormancy breaking and germination requirements of this species are not well-investigated. Further studies under controlled conditions are therefore needed to understand the key factors that enhance seed germination.

Germination, sowing and planting: Germination tests carried out at LNBG showed that optimum germination temperatures are in the range 20–28°C, with a final germination value of 94%. Seeds should be sown in a container filled with a mixture of equal parts by volume of washed river sand, compost and good soil. Seeds should be spaced evenly across the container and covered with c. 2 mm of substrate. The container should be put in a shady place and watered twice a week in summer (i.e. from October to March). Seedlings should be transplanted when the bulbs are 5 cm in diameter.

VEGETATIVE PROPAGATION

An experiment was carried at LNBG to propagate *D. delagoensis* using bulb scales. After removing the scales from the bulb, the bulb scales were treated with a hormone rooting powder, planted 5 cm deep in coarse river sand (Figure 3) and placed in a mist bed equipped

Figure 3 – *D. delagoensis* bulb scales sown for an experimental trial, and close-up of one scale rooting. (Photos: W. Froneman, LNBG.)

with automatic sprayers. The mist sprayers were set at one minute water application every 10 minutes. Roots started to develop after three months (Figure 3). However, this method is very slow and unreliable.

TRADE

Plant material is only collected when needed by local people and is not available in muthi markets in Mpumalanga.

Authors

Willem Froneman, Efisio Mattana, Vusi Lukhele and Tiziana Ulian.

References

[1] Conservatoire et Jardin botaniques & South African National Biodiversity Institute (SANBI) (2016). *African Plant Database*. Version 3.4.0. http://www.ville-ge.ch/musinfo/bd/cjb/africa/

[2] De Wet, H., Nciki, S. & Van Vuuren, S. F. (2013). Medicinal plants used for the treatment of various skin disorders by a rural community in northern Maputaland, South Africa. *Journal of Ethnobiology and Ethnomedicine* 2013; 9:51.

[3] IUCN (2017). *The IUCN Red List of Threatened Species*. Version 2017-3. http://www.iucnredlist.org/

[4] Jessop, J. P. (1977). Studies in the bulbous Liliaceae in South Africa: 7. The taxonomy of *Drimia* and certain allied genera. *Journal of South African Botany* 43: 265–319.

[5] Nciki, S. (2015). *Validating the Traditional Use of Medicinal Plants in Maputaland to Treat Skin Diseases*. Master's thesis, University of the Witwatersrand, Johannesburg, South Africa.

[6] Ndawonde, B. G., Zobolo, A. M., Dlamini, E. T. & Siebert, S. J. (2007). A survey of plants sold by traders at Zululand muthi markets, with a view to selecting popular plant species for propagation in communal gardens. *African Journal of Range & Forage Science* 24 (2): 103–107.

[7] Pooley, E. (1998). *A Field Guide to Wild Flowers: KwaZulu-Natal and the Eastern Region*. Natal Flora Publications Trust, Durban, South Africa.

[8] Raimondo, D., von Staden, L., Foden, W., Victor, J. E., Helme, N. A., Turner, R. C., Kamundi, D. A. & Manyama, P. A. (2009). *Red List of South African Plants. Strelitzia* 25, South African National Biodiversity Institute, Pretoria, South Africa.

[9] Retief, E. & Herman, P. P. J. (1997). *Plants of the Northern Provinces of South Africa: Keys and Diagnostic Characters. Strelitzia* 6, National Botanical Institute, Pretoria, South Africa.

[10] Royal Botanic Gardens, Kew (2017). *Seed Information Database (SID)*. Version 7.1. http://data.kew.org/sid/

Eucomis autumnalis (Mill.) Chitt.

TAXONOMY AND NOMENCLATURE

FAMILY. Asparagaceae

SYNONYMS. *Fritillaria autumnalis* Mill.

VERNACULAR / COMMON NAMES. Pineapple flower (English); wilde pynappel, wildepynappel (Afrikaans); mathunga (Swati).

PLANT DESCRIPTION

LIFE FORM. Perennial herb.

PLANT. Deciduous summer-growing geophyte, <55 cm tall; bulb <10 cm in diameter.

LEAVES. Leaves of basal rosette wavy, keeled, with margins tightly scalloped or toothed (Figure 1), lanceolate or ovate, 15–55 x 4–13 cm. **FLOWERS.** Inflorescence a cylindric raceme, c. 11 cm long, with a stout stalk; flowers sweet-scented, white to pale yellow green [1,5]; the inflorescence resembles a pineapple, with a rosette of leaves above the flowers (Figure 1) [5,9].

Figure 1 – *E. autumnalis* inflorescence. (Photo: LNBG.)

FRUITS AND SEEDS

The fruits are trilocular capsules; seeds are round and black (Figure 2).

FLOWERING AND FRUITING. Flowers appear from December to April [5]; fruits ripen from February to July.

Figure 2 – External view and longitudinal section of the seeds, showing the linear fully developed embryo. (Photos: P. Gómez-Barreiro, RBG Kew.)

DISTRIBUTION

South Africa (Cape Provinces, KwaZulu–Natal, Free State and Northern Provinces), Lesotho, Swaziland, Botswana, Zimbabwe and Malawi [5,10].

HABITAT

Open woodland and forest margins, grasslands and marshes, often among rocks [1,5], from the coast to 2,450 m a.s.l. [11].

USES

In traditional medicine, the bulbs are used to treat urinary diseases, and used in small quantities as emetics and enemas for treating fevers. They are also used in infusions taken during pregnancy to facilitate delivery [2]. Decoctions of the bulb in water or milk are used to assist in post-operative recovery, to help heal fractures and as an enema to treat back pain. Decoctions are also used for stomach aches, fevers, colic, flatulence, hangovers and syphilis [9]. *Eucomis autumnalis* contains some steroidal triterpenoids that are known to be beneficial in wound therapy. The plant is often grown as an ornamental [5]. The inflorescence is used as a cut flower [4].

KNOWN HAZARDS AND SAFETY

The bulb is toxic if eaten [6].

CONSERVATION STATUS

Although currently assessed as Least Concern (LC) in South Africa according to IUCN Red List criteria, *E. autumnalis* has suffered some large declines in population numbers. The main threat is over-harvesting of bulbs for the medicinal plant trade [11]. The species has been assessed as Vulnerable (VU) in Lesotho [8], but its status has yet to be confirmed in the IUCN Red List [3].

SEED CONSERVATION

HARVESTING. Fruit capsules burst open between February and July. Seeds can be manually collected from open capsules.

PROCESSING AND HANDLING. Seeds can be easily cleaned by separating them from other plant material using aspirators or sieves.

STORAGE AND VIABILITY. Seeds of this species are reported to be orthodox [7]. X-ray analysis carried out on the seed lot stored at the MSB showed a high percentage of filled seeds (80–100%).

PROPAGATION

SEEDS

Dormancy and pre-treatments: Before sowing in the nursery, the hard tegument of the seeds should be cracked. However, high germination percentages were obtained for untreated seeds sown in the laboratory under controlled conditions.

Germination, sowing and planting: Germination tests carried out on cracked seeds at LNBG showed that the optimum soil temperature range for germination is 20–28°C, with a final germination value of 78%. Seeds should be sown in a container filled with a mixture of equal parts by volume of washed river sand, compost and good soil. Seeds should be spaced evenly and covered with c. 2 mm of substrate. The container should be put in a shady place and watered twice a week in summer (i.e. October to March), and only once a week after the seeds have germinated to prevent root and bulb rot. Seedlings should be transplanted when the bulbs are 5 cm in diameter. Germination tests carried out on seeds stored at the MSB highlighted a high germination percentage (85–95%) for seeds incubated in a constant temperature range of 15–25°C with 8 hours of light per day [7].

VEGETATIVE PROPAGATION

Can be propagated from offsets (best removed when the plant is dormant in late autumn and winter and planted in the spring) [4] or leaf cuttings. Leaves should be cut into 5 cm long sections and planted in a container filled with a mixture of potting soil and washed river sand in a ratio of 2:1 by volume (Figure 3), and kept in a humid environment. Tiny bulbs develop after 1–3 months (Figure 3).

Figure 3 – *E. autumnalis* leaf cuttings.
(Photos: W. Froneman, LNBG.)

TRADE

Plant material is not available in muthi markets in Mbombela (formerly Nelspruit), although large volumes of bulbs are reported as being sold for medicine in muthi markets elsewhere in South Africa [11].

Authors

Willem Froneman, Efisio Mattana, Vusi Lukhele, Steve Davis and Tiziana Ulian.

References

[1] Conservatoire et Jardin botaniques & South African National Biodiversity Institute (SANBI) (2016). *African Plant Database*. Version 3.4.0. http://www.ville-ge.ch/musinfo/bd/cjb/africa/

[2] Hutchings, A., Scott, A. H., Lewis, G. & Cunningham, A. B. (1996). *Zulu Medicinal Plants: An Inventory*. University of Natal Press, Pietermaritzburg, South Africa.

[3] IUCN (2017). *The IUCN Red List of Threatened Species.* Version 2017-3. http://www.iucnredlist.org/

[4] Notten, A. (2002). *Eucomis autumnalis* (Mill.) Chitt. South African National Biodiversity Institute (SANBI), South Africa. http://pza.sanbi.org/

[5] Pooley, E. (1998). *A Field Guide to Wild Flowers KwaZulu-Natal and the Eastern Region*. Natal Flora Publications Trust, Durban, South Africa.

[6] Quattrocchi, U. (2012). *CRC World Dictionary of Medicinal and Poisonous Plants: Common Names, Scientific Names, Eponyms, Synonyms, and Etymology, Vol. III.* CRC Press, Boca Raton, Florida, USA.

[7] Royal Botanic Gardens, Kew (2017). *Seed Information Database (SID).* Version 7.1. http://data.kew.org/sid/

[8] Talukdar, S. (2002). Lesotho. In: *Southern African Plant Red Data Lists*. Edited by J. S. Golding. Southern African Botanical Diversity Network Report No. 14, SABONET, Pretoria, South Africa. pp. 21–30.

[9] Van Wyk, B.-E., Van Oudtshoorn, B. & Gericke, N. (2009). *Medicinal Plants of South Africa*, 2nd edition. Briza Publications, Pretoria, South Africa.

[10] WCSP (2017). *World Checklist of Selected Plant Families.* Royal Botanic Gardens, Kew. http://wcsp.science.kew.org/

[11] Williams, V. L., Raimondo, D., Crouch, N. R., Cunningham, A. B., Scott-Shaw, C. R., Lötter, M. & Ngwenya, A. M. (2016). *Eucomis autumnalis* (Mill.) Chitt. In: *National Assessment: Red List of South African Plants*. Version 2017.1. http://redlist.sanbi.org/

Gerbera ambigua (Cass.) Sch.Bip.

TAXONOMY AND NOMENCLATURE

FAMILY. Compositae

SYNONYMS. *Gerbera coriacea* (DC.) Sch.Bip., *G. discolor* Harv., *G. discolor* Sond., *G. elegans* Muschl., *G. flava* R.E.Fr., *G. kraussii* Sch.Bip., *G. lynchii* Dummer, *G. nervosa* Sond., *G. randii* S.Moore, *G. welwitschii* S.Moore, *Lasiopus ambiguus* Cass.

VERNACULAR / COMMON NAMES. Pink and white gerbera (English); botterblom (Afrikaans); vembana (Swati).

PLANT DESCRIPTION

LIFE FORM. Perennial herb.

PLANT. Acaulous, <50 cm tall [1]; rootstock thickened, woody; root-crowns densely silky-lanate.

LEAVES. Radical, ascending or spreading, very variable (oblong lanceolate, elliptic or oblong to ovate), lamina 20 x 6.5(9) cm but usually smaller, light to dark green above, white or lemon tomentose beneath, leaf margins sometimes with obscure teeth. **FLOWERS.** Flower stalks emerge from the crown and bear a single inflorescence, 3–5 cm in diameter; ray florets white, pink, or yellow; disc florets black, yellow, reddish or white (Figure 1) [5,7].

Figure 1 – *G. ambigua* flower. (Photo: W. Froneman, LNBG.)

FRUITS AND SEEDS

The cypselas have silky hairs attached aiding wind dispersal of the seed (Figure 2).

FLOWERING AND FRUITING. Flowers throughout the year but mainly from September to December. Seeds are available throughout the flowering season.

Figure 2 – External views and longitudinal section of the cypselas. (Photos: P. Gómez-Barreiro, RBG Kew.)

DISTRIBUTION

South Africa (Eastern Cape, Free State, Gauteng, KwaZulu-Natal, Limpopo, Mpumalanga and North West), Lesotho, Swaziland, and from Angola to Tanzania [5,6].

HABITAT

Open woodland and savanna, damp and submontane grasslands (sometimes on termite mounds) [5,7] and in areas regularly burnt [5], at 200–2,000 m a.s.l. [1].

USES

Leaf infusions are used in traditional medicine to treat tapeworm and stomach ache; root infusions are used to treat coughs [2] and the roots are used as good luck. *Gerbera ambigua* is grown as an ornamental garden plant [4].

KNOWN HAZARDS AND SAFETY

No known hazards.

CONSERVATION STATUS

Assessed as Least Concern (LC) in South Africa according to IUCN Red List criteria [6], but its status has yet to be confirmed in the IUCN Red List [3].

SEED CONSERVATION

HARVESTING. Seeds are available throughout the year.

PROCESSING AND HANDLING. Ripe seeds can be easily cleaned by separating them from other plant material using aspirators or sieves.

STORAGE AND VIABILITY. No information is available on seed storage behaviour. However, no *Gerbera* spp. are reported to have desiccation-sensitive seeds [8]. X-ray analysis carried out on the seed lot stored at the MSB showed a high percentage of filled seeds (85%).

PROPAGATION

SEEDS

Dormancy and pre-treatments: No pre-treatments are required.

Germination, sowing and planting: Germination tests carried out at LNBG showed that the optimum soil temperatures for germination are in the 20–28°C range, with a final germination of c. 80%. Seeds should be sown in a container filled with washed river sand and compost in a ratio

Figure 3 – *G. ambigua* seedlings. (Photo: W. Froneman, LNBG.)

of 2:1 by volume. Seeds should be spaced evenly at a depth of c. 2 mm. The container should be kept in a shady place and watered once a week in summer (i.e. October to March) and once every two weeks in winter. Germination takes place within 10–15 days. Seedlings (Figure 3) should be pricked out when they are large enough to handle [4].

VEGETATIVE PROPAGATION

Large clumps of mature plants can be lifted in winter and divided [4].

TRADE

Plant material is available in muthi markets in Mpumalanga and in traditional Indian pharmacies in Mbombela (formerly Nelspruit).

Authors

Willem Froneman, Efisio Mattana and Lucy Shai.

References

[1] Conservatoire et Jardin botaniques & South African National Biodiversity Institute (SANBI) (2016). *African Plant Database*. Version 3.4.0. http://www.ville-ge.ch/musinfo/bd/cjb/africa/

[2] Hutchings, A., Scott, A. H., Lewis, G. & Cunningham, A. B. (1996). *Zulu Medicinal Plants: An Inventory*. University of Natal Press, Pietermaritzburg, South Africa.

[3] IUCN (2017). *The IUCN Red List of Threatened Species*. Version 2017-3. http://www.iucnredlist.org/

[4] Johnson, I. (2006). *Gerbera ambigua* (Cass.) Sch. Bip. South African National Biodiversity Institute (SANBI), South Africa. http://pza.sanbi.org/

[5] Pope, G. V. (ed.) (1992). Compositae. In: *Flora Zambesiaca*, Vol. 6, part 1. Flora Zambesiaca Managing Committee, Royal Botanic Gardens, Kew.

[6] Raimondo, D., von Staden, L., Foden, W., Victor, J. E., Helme, N. A., Turner, R. C., Kamundi, D. A. & Manyama, P. A. (2009). *Red List of South African Plants*. *Strelitzia* 25, South African National Biodiversity Institute, Pretoria, South Africa.

[7] Retief, E. & Herman, P. P. J. (1997). *Plants of the Northern Provinces of South Africa: Keys and Diagnostic Characters*. *Strelitzia* 6, National Botanical Institute, Pretoria, South Africa.

[8] Royal Botanic Gardens, Kew (2017). *Seed Information Database (SID)*. Version 7.1. http://data.kew.org/sid/

Gomphocarpus physocarpus E.Mey.

TAXONOMY AND NOMENCLATURE

FAMILY. Apocynaceae

SYNONYMS. *Asclepias brasiliensis* (E.Fourn.) Schltr., *A. physocarpa* (E.Mey.) Schltr., *Gomphocarpus brasiliensis* E.Fourn.

VERNACULAR / COMMON NAMES. Balloon cottonbush, hairy balls, milkwood (English); balbossie (Afrikaans); uMtsemuliso (Swati).

PLANT DESCRIPTION

LIFE FORM. Annual or perennial herb or shrub.

PLANT. Up to 2.5 m tall; stems usually solitary, branching near the top, pale yellowish green, hollow, with spreading hairs. **LEAVES.** Light green, opposite, narrowly oblong, 4–12 x 0.5–2 cm (Figure 1) [1]. **FLOWERS.** White to cream, with pale pink to purple corona, in pendulous clusters; lobes bend strongly backwards, 7–9 x 5–6 mm [5] (Figure 1).

Figure 1 – *G. physocarpus* in flower and fruit. (Photos: W. Froneman, LNBG.)

FRUITS AND SEEDS

Fruits are solitary (Figures 1&2), inflated and roundish, 3.8–7 cm, and covered by soft spines [5]. Seeds are small, ellipsoid to oblong, with silky hairs (Figures 2&3).

FLOWERING AND FRUITING. Flowers throughout the year but mainly from November to April. Seeds can be collected from July to August.

Figure 2 – Fruit with ripe seeds. (Photo: W. Froneman, LNBG.)

Figure 3 – External views and longitudinal and cross-sections of the seeds. (Photos: P. Gómez-Barreiro, RBG Kew.)

DISTRIBUTION

South Africa (Cape Provinces, KwaZulu-Natal, Mpumalanga, Gauteng and Limpopo), Swaziland and Mozambique; introduced to other parts of Africa [1].

HABITAT

Grassland and Bushveld, seasonally wet pastures and floodplains, disturbed areas and roadsides [1,5], from sea level to 1,000 m a.s.l. [1].

USES

Widely used in traditional medicine. Dried, ground leaves are taken as snuff for treating headaches. The roots are used for treating stomach ache. Milk infusions of the leaves are used sparingly as enemas for newborn babies. Stripped green bark is tied around the waist of newborn babies with apparent urinary problems [2]. Seeds blown away from the fruits are believed by some communities to act as a charm to placate ancestors [2]. The fruits are used in floral decorations [4].

KNOWN HAZARDS AND SAFETY

All parts of the plant contain a milky white latex which is poisonous to both humans and livestock if ingested [8]. It is advisable to wash hands after handling plant material [4].

CONSERVATION STATUS

This species is assessed as Least Concern (LC) in South Africa according to IUCN Red List criteria [7], but its status has yet to be confirmed in the IUCN Red List [3].

SEED CONSERVATION

HARVESTING. Seeds should be harvested as soon as the fruits burst open and the seeds are visible.
PROCESSING AND HANDLING. Ripe seeds can be easily cleaned by removing the hairs and separating them from other plant material using aspirators or sieves.
STORAGE AND VIABILITY. Seeds of this species are reported to be desiccation tolerant [6]. X-ray analysis carried out on the seed lots stored at the MSB showed 70–100% of filled seeds.

PROPAGATION

SEEDS

Dormancy and pre-treatments: No pre-treatments are required, apart from removal of the silky hairs. However, the seed dormancy breaking and germination requirements are not well-investigated. Further studies under controlled conditions are needed to determine key factors for enhancing germination.
Germination, sowing and planting: Germination tests carried out LNBG showed that seeds sown in soil temperatures of 20–25°C germinated to c. 60%. Seeds were sown in a mixture of equal parts by volume of washed river sand and compost, in a container at least 10 cm deep. Seeds were sown 2 cm apart and covered with 5 mm of substrate. Seeds sown in summer (i.e. September to January) and seedlings were watered twice a week (only once a week for those sown in winter, i.e. May to August). Seeds sown in summer started to germinate after 10–15 days. Seedlings (Figure 4) should be pricked out when large enough to handle. Germination tests carried out on seeds stored at the MSB highlighted a high germination percentage (c. 90%) for seeds incubated at 25°C with 8 hours of light per day.

VEGETATIVE PROPAGATION

Can be propagated by stem and leaf cuttings.

Figure 4 – *G. physocarpus* seedlings. (Photo: W. Froneman, LNBG.)

TRADE

Plant material is available in muthi markets and in traditional Indian pharmacies in Mbombela (formerly Nelspruit).

Authors

Willem Froneman, Efisio Mattana, Lucy Shai and Jabulani Mahlangu.

References

[1] Goyder, D. J. & Nicholas, A. (2001). A revision of *Gomphocarpus* R.Br. (Apocynaceae: Asclepiadeae). *Kew Bulletin* 56 (4): 769–836.

[2] Hutchings, A., Scott, A. H., Lewis, G. & Cunningham, A. B. (1996). *Zulu Medicinal Plants: An Inventory*. University of Natal Press, Pietermaritzburg, South Africa.

[3] IUCN (2017). *The IUCN Red List of Threatened Species.* Version 2017-3. http://www.iucnredlist.org/

[4] Notten, A. (2010). *Gomphocarpus physocarpus* E.Mey. South African National Biodiversity Institute (SANBI), South Africa. http://pza.sanbi.org/

[5] Pooley, E. (1998). *A Field Guide to Wild Flowers: KwaZulu-Natal and the Eastern Region*. Natal Flora Publications Trust, Durban, South Africa.

[6] Royal Botanic Gardens, Kew (2017). *Seed Information Database (SID).* Version 7.1. http://data.kew.org/sid/

[7] von Staden, L. (2012). *Gomphocarpus physocarpus* E.Mey. In: *National Assessment: Red List of South African Plants*. Version 2017.1. http://redlist.sanbi.org/

[8] Watt, J. M. & Breyer-Brandwijk, M. G. (1962). *The Medicinal and Poisonous Plants of Southern and Eastern Africa*, 2nd edition. E. & S. Livingstone, Edinburgh & London.

Gunnera perpensa L.

TAXONOMY AND NOMENCLATURE

FAMILY. Gunneraceae

SYNONYMS. *Gunnera calthifolia* Salisb.

VERNACULAR / COMMON NAMES. River pumpkin, wild rhubarb (English); rivierpampoen (Afrikaans); ugobho, uqobho (Swati).

PLANT DESCRIPTION

LIFE FORM. Perennial herb.

PLANT. Robust, erect, rhizomatous, <1 m tall; rhizome usually creeping, <3 cm thick, yellow fleshed. **LEAVES.** Robust, tufted near apex of rhizome just above soil level, bluish green, kidney-shaped, palmately veined, hairy on both surfaces, 4–25 x 6–38 cm, margins irregularly dentate, teeth glandular. **FLOWERS.** On long slender, sparsely hairy spike, 20–90 cm tall, usually exceeding the leaves; individual flowers small, pinkish to reddish brown; female flowers at base of spike, bisexual flowers in middle, male flowers at top (Figure 1) [4].

Figure 1 – Adult plant with flowers in habitat. (Photo: W. Froneman, LNBG.)

FRUITS AND SEEDS

The fruits (drupelets) are small and fleshy, subglobose, slightly laterally compressed and thus two-crested, crowned by the calyx lobes, sessile or on pedicels of <1 mm (Figure 2) [4].

FLOWERING AND FRUITING. Flowers from September to February; fruits ripen from October to February [7].

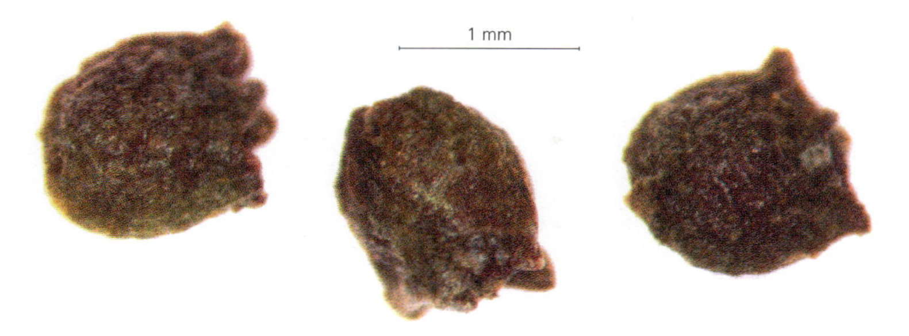

Figure 2 – External views of the dispersal units (fruits). (Photo: P. Gómez-Barreiro, RBG Kew.)

DISTRIBUTION

South Africa (Cape Provinces, Free State, Gauteng, KwaZulu-Natal, Limpopo, Mpumalanga and North West), Swaziland and Lesotho, north to Democratic Republic of Congo, Sudan, and tropical East Africa; also occurs in Madagascar [4,12,13].

HABITAT

Cold or cool, continually moist areas, mainly along upland stream banks [4], from the coast to 2,400 m a.s.l. [13].

USES

Traditional healers use a decoction of the roots (Figure 3) to remove excess fluid from the body, to expel the placenta after birth (in humans and cattle) and to relieve menstrual pain [11]. *Gunnera perpensa* is also used for treating cystitis, bladder infections and to relieve pain from rheumatism [10]. The rhizomes are applied externally for dressing wounds and for treating psoriasis [11]. Preliminary tests of crude decoctions of the rhizome showed uterotonic activity [9].

Figure 3 – *G. perpensa* rhizome and roots. (Photo: W. Froneman, LNBG.)

KNOWN HAZARDS AND SAFETY

Research suggests that *G. perpensa* is potentially toxic if used for long periods of time. Phytol (present in the stems and leaves) is a mild skin irritant [10]. Further research is needed on toxicity [5].

CONSERVATION STATUS

This widespread species is assessed as Least Concern (LC) in South Africa according to IUCN Red List criteria, but populations are declining and some have been destroyed completely by over-harvesting for the traditional medicinal plant trade [13]. Destruction of wetland habitat is also a threat [2]. Its status has yet to be confirmed in the IUCN Red List [3].

SEED CONSERVATION

HARVESTING. Fruits can be harvested from October to February.

PROCESSING AND HANDLING. Ripe fruits can be cleaned by separating them from other plant material using aspirators or sieves.

STORAGE AND VIABILITY. No information available on seed desiccation tolerance of Gunneraceae [8].

PROPAGATION

SEEDS

Dormancy and pre-treatments: No pre-treatments are required. However, further studies on seed dormancy breaking and germination requirements under controlled conditions are needed to determine key factors enhancing seed germination.

Figure 4 – *G. perpensa* seedlings. (Photo: SANBI.)

Germination, sowing and planting: Germination tests carried out at LNBG showed that the optimum soil temperatures for germination are in the 20–25°C range, with a final germination of 80%. Seeds should be sown in a container filled with a mixture of equal parts by volume of loamy soil and compost/kraal manure. Seeds should be sown evenly and covered with 5 mm of substrate and watered thoroughly. Seeds germinate after 10–15 days from sowing. Containers should have sufficient drainage holes to avoid waterlogging. Seeds and seedlings should be watered twice a week; however, during the winter months (i.e. May to August) seedlings should be watered only once a week. Seedlings should be transplanted three months after germination (Figure 4).

VEGETATIVE PROPAGATION

Large plants can be divided. Rhizomes can also be cut into sections, the optimum size being 5 cm long, although sections as small as 1 cm will sprout [1].

TRADE

Large volumes of this species are traded in muthi markets in South Africa [13].

Authors

Willem Froneman, Efisio Mattana, Vusi Lukhele, Lucy Shai, Jabulani Mahlangu, Steve Davis and Tiziana Ulian.

References

[1] Chigor, C. B. (2014). *Development of Conservation Measures for* Gunnera perpensa *L.: An Overexploited Medicinal Plant in Eastern Cape, South Africa*. DPhil. thesis, University of Fort Hare, South Africa.

[2] Glen, R. (2005). *Gunnera perpensa* L. South African National Biodiversity Institute (SANBI), South Africa. http://pza.sanbi.org/

[3] IUCN (2017). *The IUCN Red List of Threatened Species*. Version 2017-3. http://www.iucnredlist.org/

[4] Launert, E. (ed.) (1978). Haloragaceae. In: *Flora Zambesiaca*, Vol. 4. Flora Zambesiaca Managing Committee, Royal Botanic Gardens, Kew.

[5] Mwale, M. & Masika, P. J. (2011). Toxicity evaluation of the aqueous leaf extract of *Gunnera perpensa* L. *African Journal of Biotechnology* 10 (13): 2503–2513.

[6] Ngwenya, M. A., Koopman, A. & Williams, R. (2003). *Zulu Botanical Knowledge: An Introduction*. National Botanical Institute, Durban, South Africa.

[7] Pooley, E. (1998). *A Field Guide to Wild Flowers: KwaZulu-Natal and the Eastern Region*. Natal Flora Publications Trust, Durban, South Africa.

[8] Royal Botanic Gardens, Kew (2017). *Seed Information Database (SID)*. Version 7.1. http://data.kew.org/sid/

[9] Van Wyk, B.-E. & Gericke, N. (2000). *People's Plants: A Guide to Useful Plants of Southern Africa*. Briza Publications, Pretoria, South Africa.

[10] Van Wyk, B.-E., Van Oudtshoorn, B. & Gericke, N. (2009). *Medicinal Plants of South Africa*, 2nd edition. Briza Publications, Pretoria, South Africa.

[11] Watt, J. M. & Breyer-Brandwijk, M. G. (1962). *The Medicinal and Poisonous Plants of Southern and Eastern Africa*, 2nd edition. E. & S. Livingstone, Edinburgh & London.

[12] WCSP (2017). *World Checklist of Selected Plant Families*. Royal Botanic Gardens, Kew. http://wcsp.science.kew.org/

[13] Williams, V. L., Raimondo, D., Crouch, N. R., Cunningham, A. B., Scott-Shaw, C. R., Lötter, M., Ngwenya, A. M. & Dold, A. P. (2016). *Gunnera perpensa* L. In: *National Assessment: Red List of South African Plants*. Version 2017.1. http://redlist.sanbi.org/

Hypoxis hemerocallidea Fisch., C.A.Mey. & Avé-Lall.

TAXONOMY AND NOMENCLATURE

FAMILY. Hypoxidaceae

SYNONYMS. *Hypoxis obconica* Nel, *H. patula* Nel, *H. rooperi* T.Moore, *H. rooperi* var. *forbesii* Baker

VERNACULAR / COMMON NAMES. African potato, star flower (English); Afrika aartappel (Afrikaans); iNkomfe (Swati).

PLANT DESCRIPTION

LIFE FORM. Perennial herb.

PLANT. Robust, <40 cm tall; tuberous rootstock (corm) stout, vertical, 3–7 cm, turbinate to subglobose, or up to 14 x 4 cm, cylindric, hard, slimy, crowned with a ring of bristles, white or yellow orange within. **LEAVES.** Ternate, up to 40 x 1–5 cm, sickle-shaped, keeled and arched outwards, finely many-ribbed, ribs prominent tapering to the tips, dense white hairs on lower surface, margins and keel. **FLOWERS.** Inflorescences of up to 16 yellow, star-like flowers per erect stem of <30 cm, usually appearing with or just after the leaves, opening at first light and closing at midday [4,8,9,10] (Figure 1).

Figure 1 – *H. hemerocallidea* habit and flowers. (Photos: W. Froneman, LNBG.)

FRUITS AND SEEDS

The fruit is a capsule that splits across its diameter to expose small, black, glossy seeds, with a smooth testa or covered in dome-shape undulations [9] (Figure 2).

FLOWERING AND FRUITING. Flowers from August to April; fruits from November to May.

Figure 2 – External views and longitudinal section of the seeds showing the linear embryo. (Photos: P. Gómez-Barreiro, RBG Kew.)

DISTRIBUTION

South Africa (Eastern Cape, Free State, KwaZulu-Natal, Mpumalanga, Gauteng and Limpopo), Lesotho, Swaziland, Botswana, Mozambique and Zimbabwe [4,9,15].

HABITAT

Grasslands, mixed woodlands, rocky hillsides and sometimes in cultivated land [9,10], at 350–1,850 m a.s.l. [9]. Drought and fire resistant [16].

USES

The tuberous rootstock is traditionally used for treating a wide range of ailments. Weak infusions and decoctions of the corm are used as a tonic, and in the treatment of tuberculosis, testicular cancer, prostatic hypertrophy, and urinary tract infections, and as a laxative to expel intestinal worms [4,5,13]. Preparations of hypoxoside derived from *Hypoxis hemerocallidea* are used in primary health care in South Africa to boost immunity in HIV/AIDS patients [8]. *H. hemerocallidea* and other *Hypoxis* spp. contain sterols and sterolins, which are purported to be the relevant constituents involved in providing the therapeutic benefits of commercially available immune-enhancing drugs. Further research is needed to validate efficacy [7,17]. The plant is grown as an ornamental [4].

KNOWN HAZARDS AND SAFETY

Raw products can be toxic, and plant decoctions are purgative [14].

CONSERVATION STATUS

Although assessed as Least Concern (LC) in South Africa according to IUCN Red List criteria, this widespread and abundant species is declining in many areas as a result of over-collecting for medicinal uses and habitat loss as a result of agriculture and urban development [16]. It has been assessed as Data Deficient (DD) in Lesotho [12] and Swaziland [1], but its status has yet to be confirmed in the IUCN Red List [6].

SEED CONSERVATION

HARVESTING. Fruit capsules burst open between August and April. Seeds can be manually harvested from the open capsules.

PROCESSING AND HANDLING. Seeds can be cleaned by separating them from the remains of the fruit using aspirators or sieves.

STORAGE AND VIABILITY. Seeds of this species are likely to be orthodox [11]. X-ray analysis carried out on the seed lots stored at the MSB showed an average of c. 35% of empty seeds.

PROPAGATION

SEEDS

Dormancy and pre-treatments: No pre-treatments are required before sowing in the nursery, although seeds remain dormant for about a year. Seed dormancy has been reported to be due to both a coat-imposed dormancy and an embryo dormancy, with the former being largely effected by mechanical restriction. Although removing the seed coat improved germination, it is clear that other factors contribute to seed dormancy. Chemical treatments, and particularly gibberellins and cytochinins, may play a role in the release of embryo dormancy [3]. These findings suggest the presence of physiological dormancy. However, considering the seed biology of this species (see Figure 2), a morphological component may be responsible, leading to morphological or morphophysiological dormancy.

Germination, sowing and planting: Germination tests carried out at LNBG show that soil temperatures of 20–28°C lead to final germination percentages of c. 5–12%. Seeds were sown in a container filled with a mixture of equal parts by volume of washed river sand, compost and good soil. Seeds were spaced evenly across the container and covered with about 2 mm of substrate. The container was then kept in a shady place. Seeds were watered twice a week in summer (i.e. October to March) and once every three weeks in winter. The small corms are dormant in winter and must be kept dry. Seedlings were transplanted when about 10 cm high. Very different germination results have been achieved among seed lots of this species stored at the MSB, with final germination ranging from 0% to >90% for seeds incubated at 15–20°C, without any pre-treatments [11]. These differences may be explained by Hammerton *et al.* [2] who found that seed germination is related to seed density in the infructescence, and varied with harvesting sites and dates.

VEGETATIVE PROPAGATION

Can be propagated by dividing the corms [4].

TRADE

The plant and its derivatives are widely available in muthi markets and in commercial products. Plant material is available in muthi markets in Mpumalanga and in traditional pharmacies in Mbombela (formerly Nelspruit) (Figure 3).

Figure 3 – *H. hemerocallidea* corms being sold in the market at Umtamvuna. (Photo: K. Van der Walt, SANBI.)

Authors

Willem Froneman, Efisio Mattana, Vusi Lukhele, Steve Davis and Tiziana Ulian.

References

[1] Dlamini, T. S. & Dlamini, G. M. (2002). Swaziland. In: *Southern African Plant Red Data Lists*. Edited by J. S. Golding. Southern African Botanical Diversity Network Report No. 14, SABONET, Pretoria, South Africa. pp. 121–134.

[2] Hammerton, R. D., Smith, M. T. & Van Staden, J. (1989). Factors influencing seed variability and germination in *Hypoxis hemerocallidea* Fisch. & Meyer. *Seed Science and Technology* 17: 613–624.

[3] Hammerton, R. D. & Van Staden, J. (1988). Seed germination of *Hypoxis hemerocallidea*. *South African Journal of Botany* 54 (3): 277–280.

[4] Hölscher, B. (2009). *Hypoxis hemerocallidea* Fisch., C.A.Mey. & Avé-Lall. South African National Biodiversity Institute (SANBI), South Africa. http://pza.sanbi.org/

[5] Hutchings, A., Scott, A. H., Lewis, G. & Cunningham, A. B. (1996). *Zulu Medicinal Plants: An Inventory*. University of Natal Press, Pietermaritzburg, South Africa.

[6] IUCN (2017). *The IUCN Red List of Threatened Species.* Version 2017-3. http://www.iucnredlist.org/

[7] Nair, V. D. P. & Kanfer, I. (2008). Sterols and sterolins in *Hypoxis hemerocallidea* (African potato). *South African Journal of Science* 104 (7–8): 322–324.

[8] Pooley, E. (1998). *A Field Guide to Wild Flowers: KwaZulu-Natal and the Eastern Region*. Natal Flora Publications Trust, Durban, South Africa.

[9] Pope, G. V. (ed.) (2001). Hypoxidaceae. In: *Flora Zambesiaca*, Vol. 12, part 3. Flora Zambesiaca Managing Committee, Royal Botanic Gardens, Kew.

[10] Retief, E. & Herman, P. P. J. (1997). *Plants of the Northern Provinces of South Africa: Keys and Diagnostic Characters. Strelitzia* 6, National Botanical Institute, Pretoria, South Africa.

[11] Royal Botanic Gardens, Kew (2017). *Seed Information Database (SID).* Version 7.1. http://data.kew.org/sid/

[12] Talukdar, S. (2002). Lesotho. In: *Southern African Plant Red Data Lists*. Edited by J. S. Golding. Southern African Botanical Diversity Network Report No. 14, SABONET, Pretoria, South Africa. pp. 21–30.

[13] Van Wyk, B.-E., Van Oudtshoorn, B. & Gericke, N. (2009). *Medicinal Plants of South Africa*, 2nd edition. Briza Publications, Pretoria, South Africa.

[14] Watt, J. M. & Breyer-Brandwijk, M. G. (1962). *The Medicinal and Poisonous Plants of Southern and Eastern Africa*, 2nd edition. E. & S. Livingstone, Edinburgh & London.

[15] WCSP (2017). *World Checklist of Selected Plant Families.* Royal Botanic Gardens, Kew. http://wcsp.science.kew.org/

[16] Williams, V. L., Raimondo, D., Crouch, N. R., Cunningham, A. B., Scott-Shaw, C. R., Lötter, M. & Ngwenya, A. M. (2016). *Hypoxis hemerocallidea* Fisch., C.A.Mey. & Avé-Lall. In: *National Assessment: Red List of South African Plants*. Version 2017.1. http://redlist.sanbi.org/

[17] Wilt, T. J., Ishani, A., Rutks, I. & MacDonald, R. (2000). Phytotherapy for benign prostatic hyperplasia. *Public Health Nutrition* 3 (4a): 459–472.

Kigelia africana (Lam.) Benth.

TAXONOMY AND NOMENCLATURE

FAMILY. Bignoniaceae

SYNONYMS. *Kigelia aethiopica* Decne., *K. pinnata* (Jacq.) DC.

VERNACULAR / COMMON NAMES. Sausage tree (English); worsboom (Afrikaans); umVongotsi, mvongotsi (Swati).

PLANT DESCRIPTION

LIFE FORM. Tree.

PLANT. Semi-deciduous, <25(–35) m high, trunk branching low down, bark grey, branches grey brown, crown spreading (<20 m wide); variable in habit and leaf morphology, especially depending on habitat. **LEAVES.** Opposite or whorled, imparipinnate, <60 cm long. Leaflets 5–13, subsessile except the terminal leaflet, ovate, obovate or rounded, sometimes cuneate at base. **FLOWERS.** Inflorescence pendulous, very lax, 1–1.5 m long; flowers, bell-shaped, dark maroon inside, reddish brown with yellow to green or darker streaks on outside [7] (Figure 1).

Figure 1 – *K. africana* flowers (left) and fruits (above).
(Photos: L.-N. Le Roux, LNBG and T. Ulian, RBG Kew.)

FRUITS AND SEEDS

Fruits are sausage-shaped, greyish brown, pendulous berries, up to 1 m x 18 cm, weighing up to 12 kg, with lenticels when immature, massive, wood-walled, indehiscent; peduncle <1 m (Figure 1). Seeds obovoid, 10 x 7 mm, many, embedded in a fibrous pulp, with a coriaceous testa and folded cotyledons (Figure 2) [7].

FLOWERING AND FRUITING. Flowers from July to October; fruits from November to August [13].

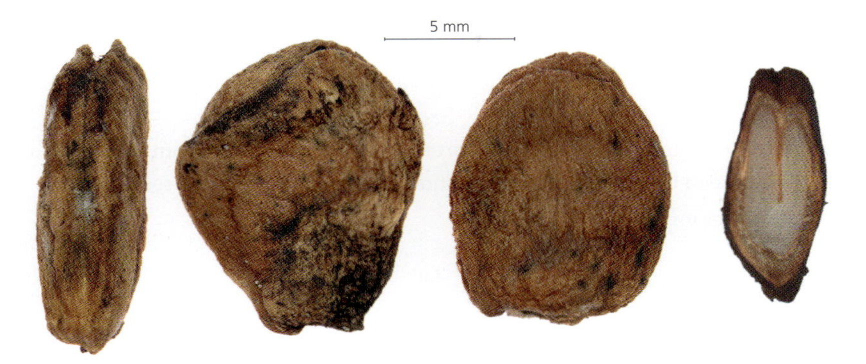

Figure 2 – External view and cross-section of the seed. (Photos: P. Gómez-Barreiro, RBG Kew.)

DISTRIBUTION

Widespread in tropical Africa, especially in drier regions [7]. Also native to South Africa (Gauteng, KwaZulu-Natal, Limpopo and Mpumalanga) and Swaziland [10]. Introduced to Madagascar, parts of the Middle East, Asia, Australia, and South and Central America [1].

HABITAT

Rainforest, open woodland, along watercourses, in riverine woodland, shrubland and high-rainfall savanna [7,13], on sandy argillaceous soils [7], rocky hillsides, loamy red clay soils and sometimes on peaty soils, from sea level to 3,000 m a.s.l. [1].

USES

Sausage tree is sacred to many African communities and has a wide variety of uses in traditional and Western medicine, including commercially available skin lotions and cosmetics [1,11]. Every part of the tree is used in herbal medicine. Powders and infusions of the bark, leaves, stems, twigs or fruits are used to clean and dress wounds, sores and ulcers [1,14]. Powdered dried fruits are used as dressings for ulcers, syphilis and rheumatism. Fruit and bark decoctions are used as enemas for stomach ailments [3]. Extracts of the bark, wood, roots and fruits have antibacterial and antifungal activity. *Kigelia africana* is renowned for anti-cancer properties, and laboratory screening has confirmed *in vitro* anti-cancer activity [2]. It is used in both traditional and Western medicine to treat malignant neoplasms, such as skin melanoma, tumours and breast cancer [1,14]. The seeds are roasted and eaten in times of famine; fruits and bark are used in traditional beer making [15]. The roots give a bright yellow dye. Naphthoquinones have been isolated from the roots and bark; it is likely that these compounds are responsible for the colouring effects [14], and play a role in antimicrobial activity [5].

KNOWN HAZARDS AND SAFETY

Ripe and unripe fruits are toxic to humans if eaten [15]. The fruit pulp can cause blisters on the mouth and skin [3]. When the large woody fruits fall from the tree, they can injure humans and livestock and can cause considerable damage to vehicles.

CONSERVATION STATUS

A widespread species assessed as Least Concern (LC) according to IUCN Red List criteria [4,10].

SEED CONSERVATION

HARVESTING. Ripe fruits are available from November to August.

PROCESSING AND HANDLING. Fruits should be opened with a sharp knife to remove the seeds from the pulp.

STORAGE AND VIABILITY. Seeds of this species are reported to be desiccation tolerant [9,12]. X-ray analysis carried out on the seed lots stored at the MSB showed that 85–95% of the seeds were filled and the remaining ones infested.

PROPAGATION

SEEDS

Dormancy and pre-treatments: No pre-treatments are required.

Germination, sowing and planting: Germination tests carried out at LNBG showed that the optimum soil temperatures for germination are in the 20–25°C range, with final germination of 50–60%. Seeds should be sown in a mixture of equal parts by volume of washed river sand, compost and loamy soil, in a container at least 10 cm deep. Seeds should be sown 4 cm apart and covered with 1 cm of substrate. Seeds and seedlings should be watered twice a week in summer,

Figure 3 – *K. africana* plants in the nursery. (Photo: W. Froneman, LNBG.)

and seedlings once a week in winter (i.e. from May to August). Seeds start to germinate after 30–40 days if planted during the summer months (i.e. from September to January). Seedlings should be transplanted when they are 10 cm tall (Figure 3). Germination experiments carried out under controlled conditions in the laboratory on seed lots stored at the MSB showed an increase in germination rate with increasing temperatures (from 11°C to 31°C), with a final germination percentage of c. 90% at 26°C, with a photoperiod of 12 hours of light per day [12].

VEGETATIVE PROPAGATION

Can be propagated from truncheons [6] and wildings [8].

TRADE

Internationally traded; *K. africana* has become a popular ingredient of cosmetics and skin lotions [14]. Plant material is also available in muthi markets and in traditional Indian pharmacies in Mbombela (formerly Nelspruit).

Authors

Willem Froneman, Efisio Mattana, Vusi Lukhele, Steve Davis and Tiziana Ulian.

References

[1] Grace, O. M. & Davis, S. D. (2002). *Kigelia africana* (Lam.) Benth. In: *Plant Resources of Tropical Africa. Precursor*. Edited by L. P. A. Oyen & R. H. M. J. Lemmens. PROTA Programme, Wageningen, The Netherlands. pp. 98–102. http://www.prota4u.org/

[2] Houghton, P. J. (2002). The sausage tree (*Kigelia pinnata*): ethnobotany and recent scientific work. *South African Journal of Botany* 68 (1): 14–20.

[3] Hutchings, A., Scott, A. H., Lewis, G. & Cunningham, A. B. (1996). *Zulu Medicinal Plants: An Inventory*. University of Natal Press, Pietermaritzburg, South Africa.

[4] IUCN (2017). *The IUCN Red List of Threatened Species*. Version 2017-3. http://www.iucnredlist.org/

[5] Jackson, S. J., Houghton, P. J., Photiou, A. & Retsas, S. (1995). Effects of fruit extracts and naphthoquinones isolated from *Kigelia pinnata* on melanoma cell lines. *British Journal of Cancer* 71 (Supplement) XXIV: 62.

[6] Joffe, P. (2003). *Kigelia africana* (Lam.) Benth. South African National Biodiversity Institute (SANBI), South Africa. http://pza.sanbi.org/

[7] Launert, E. (ed.) (1988). Bignoniaceae. In: *Flora Zambesiaca*, Vol. 8, part 3. Flora Zambesiaca Managing Committee, Royal Botanic Gardens, Kew.

[8] Orwa, C., Mutua, A., Kindt, R., Jamnadass, R. & Simons, A. (2009). *Agroforestree Database: A Tree Reference and Selection Guide*. Version 4.0. World Agroforestry Centre, Kenya. http://www.worldagroforestry.org/output/agroforestree-database

[9] Pritchard, H. W., Daws, M. I., Fletcher, B. J., Gaméné, C. S., Msanga, H. P. & Omondi, W. (2004). Ecological correlates of seed desiccation tolerance in tropical African drylands trees. *American Journal of Botany* 91 (6): 863–870.

[10] Raimondo, D., von Staden, L., Foden, W., Victor, J. E., Helme, N. A., Turner, R. C., Kamundi, D. A. & Manyama, P. A. (2009). *Red List of South African Plants. Strelitzia* 25, South African National Biodiversity Institute, Pretoria, South Africa.

[11] Royal Botanic Gardens, Kew (1999–2015). *Survey of Economic Plants for Arid and Semi-Arid Lands (SEPASAL) Database*. http://apps.kew.org/sepasalweb/sepaweb

[12] Royal Botanic Gardens, Kew (2017). *Seed Information Database (SID)*. Version 7.1. http://data.kew.org/sid/

[13] Schmidt, E., Lötter, M. & McCleland, W. (2002). *Trees and Shrubs of Mpumalanga and Kruger National Park*. Jacana Media, Johannesburg, South Africa.

[14] Van Wyk, B.-E. & Gericke, N. (2000). *People's Plants: A Guide to Useful Plants of Southern Africa*. Briza Publications, Pretoria, South Africa.

[15] Watt, J. M. & Breyer-Brandwijk, M. G. (1962). *The Medicinal and Poisonous Plants of Southern and Eastern Africa*, 2nd edition. E. & S. Livingstone, Edinburgh & London.

Pachypodium saundersii N.E.Br.

TAXONOMY AND NOMENCLATURE

FAMILY. Apocynaceae

SYNONYMS. *Pachypodium lealii* subsp. *saundersii* (N.E.Br.) G.D.Rowley

VERNACULAR / COMMON NAMES. Kudu lily (English); koedoelelie (Afrikaans); liGubaguba, phova (Swati).

PLANT DESCRIPTION

LIFE FORM. Shrub.

PLANT. Caudiciform succulent, 0.5–2 m tall; stems swollen, little-branched, glabrous, smooth, pale grey, becoming deeply longitudinally wrinkled when dry, arising from large tuber. **LEAVES.** Spirally arranged, in tight clusters at base of spines (1–4 cm), narrowly obovate, with obtuse and mucronate apex, base tapering, leathery, hairless on both sides. **FLOWERS.** Inflorescence terminal, many-flowered, very condensed; flowers pink to purple on outside of corolla, white above and greenish within, corolla tube 3–4 cm long, corolla lobes 1/2–2/3 as long as tube, asymmetrical [4,8] (Figure 1).

Figure 1 – *P. saundersii* flowers and follicles. (Photos: W. Froneman, LNBG.)

FRUITS AND SEEDS

The fruits are two horn-like follicles <15 cm (Figure 1), swollen at the base and tapering to a point. Seeds are c. 7 mm long, compressed ovoid (Figure 2), with an apical coma of pale golden hairs <5 cm long [4,8].

FLOWERING AND FRUITING. Flowers from January to June; fruits ripen in October.

Figure 2 – External view and longitudinal section of the seeds showing the embryo. The apical hairy coma has been removed during the cleaning process. (Photos: P. Gómez-Barreiro, RBG Kew.)

DISTRIBUTION

South Africa (KwaZulu-Natal, Limpopo and Mpumalanga), Swaziland, Zimbabwe and Mozambique [8,10].

HABITAT

Among rocks in dry, low-lying Bushveld or on wooded rocky ridges [8].

USES

Pachypodium saundersii is planted for good luck. It is reported to be used medicinally in Mpumalanga and is grown by succulent plant enthusiasts and as an ornamental plant [1].

KNOWN HAZARDS AND SAFETY

The closely related species *P. lealii* Welw. (of which *P. saundersii* was formerly considered a subspecies) is used as an arrow poison. It contains pachypodin, a poisonous glucoside that has similar toxicity effects to digitalis [9].

CONSERVATION STATUS

This species is assessed as Least Concern (LC) in South Africa according to IUCN Red List criteria [6], but Vulnerable (VU) in Zimbabwe due to severely fragmented habitat and collecting [5]. Its status has yet to be confirmed in the IUCN Red List [3].

SEED CONSERVATION

HARVESTING. Seeds can be collected manually as soon as the follicles burst open.

PROCESSING AND HANDLING. Seeds can be cleaned by removing the silky hairs attached to the seeds and then separating them from other plant material using aspirators or sieves.

STORAGE AND VIABILITY. No information is available on seed desiccation tolerance. However, seeds of *P. geayi* Costantin & Bois are reported to be desiccation tolerant [7]. X-ray analysis carried out on the seed lots stored at the MSB showed c. 70% filled seeds, the remainder being infested.

PROPAGATION

SEEDS

Dormancy and pre-treatments: No pre-treatments are required.

Germination, sowing and planting: Germination tests carried out at LNBG showed that the optimum soil temperatures for germination are in the 20–25°C range, with a final germination of 90%. Seeds should be sown in a mixture of equal parts by volume of washed river sand and compost, 4 cm apart and covered by 5 mm of substrate, in containers of at least 10 cm depth. Seeds and seedlings should be watered once a week. Seeds start to germinate after 10–15 days if planted in summer (i.e. September to January). Seedlings (Figure 3) should be transplanted when

Figure 3 – *P. saundersii* seedlings. (Photo: W. Froneman, LNBG.)

they are 10 cm tall. Experiments carried out under controlled conditions in the laboratory at the MSB showed a final germination percentage of 100% at 20–25°C, with a photoperiod of 8 hours of light per day [7].

VEGETATIVE PROPAGATION

Can be grafted onto *P. lamerei* Drake rootstocks [1].

TRADE

Plant material is available in muthi markets and in Indian pharmacies in Mbombela (formerly Nelspruit). International trade is regulated by CITES (the species is listed on Appendix II) [2].

Authors

Willem Froneman, Efisio Mattana, Vusi Lukhele and Tiziana Ulian.

References

[1] Bester, S. P. (2007). *Pachypodium* Lindl. South African National Biodiversity Institute (SANBI), South Africa. http://pza.sanbi.org/

[2] CITES (2017). *Convention on International Trade in Endangered Species of Wild Fauna and Flora*. https://cites.org/eng/app/index.php

[3] IUCN (2017). *The IUCN Red List of Threatened Species.* Version 2017-3. http://www.iucnredlist.org/

[4] Launert, E. (ed.) (1985). Apocynaceae. In: *Flora Zambesiaca*, Vol. 7, part 2. Flora Zambesiaca Managing Committee, Royal Botanic Gardens, Kew.

[5] Mapaura, A. & Timberlake, J. R. (2002). Zimbabwe. In: *Southern African Plant Red Data Lists*. Edited by J. S. Golding. Southern African Botanical Diversity Network Report No. 14, SABONET, Pretoria, South Africa. pp. 157–182.

[6] Raimondo, D., von Staden, L., Foden, W., Victor, J. E., Helme, N. A., Turner, R. C., Kamundi, D. A. & Manyama, P. A. (2009). *Red List of South African Plants. Strelitzia* 25, South African National Biodiversity Institute, Pretoria, South Africa.

[7] Royal Botanic Gardens, Kew (2017). *Seed Information Database (SID).* Version 7.1. http://data.kew.org/sid/

[8] Schmidt, E., Lötter, M. & McCleland, W. (2002). *Trees and Shrubs of Mpumalanga and Kruger National Park*. Jacana Media, Johannesburg, South Africa.

[9] Watt, J. M. & Breyer-Brandwijk, M. G. (1962). *The Medicinal and Poisonous Plants of Southern and Eastern Africa*, 2nd edition. E. & S. Livingstone, Edinburgh & London.

[10] WCSP (2017). *World Checklist of Selected Plant Families.* Royal Botanic Gardens, Kew. http://wcsp.science.kew.org/

Prunus africana (Hook.f.) Kalkman

TAXONOMY AND NOMENCLATURE

FAMILY. Rosaceae

SYNONYMS. *Laurocerasus africana* (Hook.f.) Browicz, *Pygeum africanum* Hook.f.

VERNACULAR / COMMON NAMES. African almond, African cherry, red stinkwood (English); rooistinkhout (Afrikaans); umDumizulu, shinhlomana (Swati).

PLANT DESCRIPTION

LIFE FORM. Tree.

PLANT. Evergreen, c. 10(–30 m) tall [21], rarely a shrub 3–5 m tall, much-branched; bark very rough, dark brown, broken into characteristic blocks (Figure 1); buttress roots often present.

LEAVES. Alternate, dark green, glossy, drooping, elliptic, finely serrate, lamina 4–15 x 2–5·5 cm (Figure 1), apex tapering, petiole 1–2 cm long, pink; crushed leaves smell of almonds.

FLOWERS. Racemes usually solitary, flowers 7–15, creamy white, in axils of scales at base of lateral shoots which may also produce leaves in their upper part, petals <2 mm long [9,18].

Figure 1 – *P. africana* stems with flowers and bark. (Photos: W. Froneman, LNBG.)

FRUITS AND SEEDS

The fruits are 5–8 x 8–12 mm, transversely ellipsoid, broader than long, slightly didymous (thus appearing as if bilocular), dry, usually glabrous, red to purplish brown [9] (Figure 2); 1–2 pyrenes (hereafter seeds) per fruit [8].

FLOWERING AND FRUITING. Flowers from October to May; fruits from September to January [18].

Figure 2 – External views of the fruits and pyrenes. (Photos: P. Gómez-Barreiro, RBG Kew.)

DISTRIBUTION

South Africa (confined to montane forests in Eastern Cape, Mpumalanga and KwaZulu-Natal) [12] and Swaziland, northwards to tropical West, Central and East Africa, from Ghana to Ethiopia [9,18]. Also native to Madagascar [9].

HABITAT

Afromontane regions, occurring in the mist belt of South Africa [21], in upland rainforest, riverine forest, or rarely solitary in plateau grasslands, usually at 1,000–2,100 m a.s.l. [9], but occasionally as high as 3,150 m [1].

USES

A multipurpose tree of local and international economic and medicinal value. In traditional medicine, the bark is used to treat chest pain [6,21]. The wood is used in carpentry and construction work [2]. The bark is used in patented medicines for the treatment of prostatic hypertrophy. The reported beneficial effects of bark extracts against prostatic adenoma is possibly due to the presence of β-sitosterol, triterpenoids and ferulic esters [21]. The tree is planted in gardens as an ornamental and for shade [12].

KNOWN HAZARDS AND SAFETY

Contains amygdalin (a poisonous cyanogenic glycoside) and hydrocyanic (prussic) acid. The bark, leaves and fruits are reported to be poisonous [6,22].

CONSERVATION STATUS

Assessed as Vulnerable (VU) in South Africa [14], where it is afforded protection under forestry legislation [16], and threatened in several other Southern Africa countries [11]. Harvesting of the bark, especially for medicinal markets in Europe, America and Asia, is the main threat. Although resilient to some bark removal, the practice of girdling the bark kills the trees [5]. Tree density in South African forests is low, and populations are declining because of overexploitation. In KwaZulu-Natal, the species is locally extinct [14]. Although assessed as Vulnerable (VU) at the global level [7], *Prunus africana* occurs in montane forests throughout sub-Saharan Africa. As a result, it has been suggested that the category Lower Risk-Near Threatened (LR-nt) applies [3]. However, the criteria and definitions of IUCN Red List categories have changed since this suggested revision, so the current global assessment needs updating [7].

SEED CONSERVATION

HARVESTING. Ripe fruits can be collected from September to January from both the plant and the ground after shedding, although the collection of fallen fruits is not recommended if they have been on the ground for more than one day [8].

PROCESSING AND HANDLING. Seeds can be easily cleaned by removing the thin fleshy pulp by rubbing through sieves with tepid water. Cleaned seeds should be spread out and left to dry at room temperature. In general, seeds should be handled in strict adherence to protocols for optimum seed storage and conservation [8].

STORAGE AND VIABILITY. The seed lot stored at the MSB showed a high percentage of empty seeds (40%). Seeds of this species are probably orthodox as they can tolerate desiccation to c. 5% moisture content and can be stored for at least three months at –20°C, as long as they are extracted from ripe, purple fruits [13,15,23].

PROPAGATION

SEEDS

Dormancy and pre-treatments: No pre-treatments are required; however, a period of after-ripening may be necessary [8].

Germination, sowing and planting: Germination tests carried out at LNBG showed that the optimum soil temperatures for germination are in the 15–20°C range, with final germination of 20–26%. Seeds were sown on a mixture of river sand, compost/kraal manure and loamy soil in a ratio of 1:1:3 by volume. Seeds were sown 4 cm apart and covered by 1.5 cm of substrate in a container that is 10 cm deep. Seeds and seedlings were watered twice a week, and seedlings were watered once a week in winter (i.e. May to August). Seeds sown in summer (i.e. September to January) started to germinate after 15–20 days. Seedlings were transplanted when 10 cm high into a container using a mixture of compost and loamy soil in a ratio of 2:3 [10]. However, higher germination percentages (70–80%) have been reported for freshly collected seeds [8]. Under controlled conditions, using fresh seeds sown on Petri dishes on 1% agar water and incubated at temperatures of 25–30°C, final germination percentages are >80% [23]. The temperature range for germination extends as low as 1°C, equivalent to temperatures in the montane habitat of this species [13,17].

VEGETATIVE PROPAGATION

Wildings are commonly used for large-scale planting [9]. *Prunus africana* can also be propagated by cuttings. The best results are obtained by using leafy cuttings (leaf area 20–25 cm^2) and rooting hormone [19]. In Cameroon, c. 10% of cuttings without hormone treatment rooted after three months [5].

TRADE

Trade in products derived from this species is on a larger scale than that derived from many other wild-collected African trees [5]. The bark is harvested in many parts of Africa for medicine. The bark (and derivative products) are exported to Europe, America, China and Japan; for example, for the manufacture of medicines to treat prostatic hypertrophy [20]. International trade in this species is regulated by CITES (the species is listed on Appendix II) [4]. Plant material is also available in muthi markets and traditional Indian pharmacies in Mbombela (formerly Nelspruit).

Authors

Willem Froneman, Efisio Mattana, Steve Davis and Tiziana Ulian.

References

[1] Beentje, H. J. (1994). *Kenya Trees, Shrubs and Lianas*. National Museums of Kenya, Nairobi, Kenya.

[2] Burkill, H. M. (1997). *The Useful Plants of West Tropical Africa. Vol. 4: Families M–R*. 2nd edition. Royal Botanic Gardens, Kew.

[3] Cable, S. & Cheek, M. R. (1998). *The Plants of Mount Cameroon. A Conservation Checklist.* Royal Botanic Gardens, Kew.

[4] CITES (2017). *Convention on International Trade in Endangered Species of Wild Fauna and Flora*. https://cites.org/eng/app/index.php

[5] Cunningham, A. B. & Mbenkum, F. T. (1993). *Sustainability of Harvesting* Prunus africana *bark in Cameroon: A Medicinal Plant in International Trade*. People and Plants Working Paper No. 2. UNESCO, Paris.

[6] Hutchings, A., Scott, A. H., Lewis, G. & Cunningham, A. B. (1996). *Zulu Medicinal Plants: An Inventory*. University of Natal Press, Pietermaritzburg, South Africa.

[7] IUCN (2017). *The IUCN Red List of Threatened Species.* Version 2017-3. http://www.iucnredlist.org/

[8] Jøker, D. (2003). *Prunus africana* (Hook.f.) Kalkman. *Seed Leaflet* No. 74. Danida Forest Seed Centre, Humlebaek, Denmark.

[9] Launert, E. (ed.) (1978). Rosaceae. In: *Flora Zambesiaca*, Vol. 4. Flora Zambesiaca Managing Committee, Royal Botanic Gardens, Kew.

[10] Mbuya, L. P., Msanga, H. P., Ruffo, C. K., Birnie, A. & Tengnäs, B. (1994). *Useful Trees and Shrubs for Tanzania: Identification, Propagation and Management for Agricultural and Pastoral Communities.* Regional Soil Conservation Unit (RCSU) and Swedish International Development Authority (SIDA), Nairobi, Kenya.

[11] National Red List (2017). *Prunus africana*. National Red List Network. http://www.nationalredlist.org/

[12] Nonjinge, S. (2006). *Prunus africana* (Hook.f.) Kalkman. South African National Biodiversity Institute (SANBI), South Africa. http://pza.sanbi.org/

[13] Pritchard, H. W. & Linington, S. H. (2002). Tree seeds and the Millennium Seed Bank Project. In: *Forest Genetic Resources No. 30*. Edited by C. Palmberg-Lerche, P. A. Iversen & P. Sigaud. Food and Agriculture Organization of the United Nations (FAO), Rome, Italy. pp. 27–30.

[14] Raimondo, D., von Staden, L., Foden, W., Victor, J. E., Helme, N. A., Turner, R. C., Kamundi, D. A. & Manyama, P. A. (2009). *Red List of South African Plants. Strelitzia* 25, South African National Biodiversity Institute, Pretoria, South Africa.

[15] Royal Botanic Gardens, Kew (2017). *Seed Information Database (SID).* Version 7.1. http://data.kew.org/sid/

[16] SA Forestry Online (2015). *Protected Trees in South Africa (2015). List of Protected Trees Under The National Forest Act, 1998 (Act No. 84 of 1998).* http://saforestryonline.co.za/indigenous/protected-trees-in-south-africa/

[17] Sacandé, M., Pritchard, H. W. & Dudley, A. E. (2004). Germination and storage characteristics of *Prunus africana* seeds. *New Forests* 27 (3): 239–250.

[18] Schmidt, E., Lötter, M. & McCleland, W. (2002). *Trees and Shrubs of Mpumalanga and Kruger National Park*. Jacana Media, Johannesburg, South Africa.

[19] Tchoundjeu, Z., Avana, M. L., Leakey, R. R. B., Simons, A. J., Assah, E., Duguma, B. & Bell, J. M. (2002). Vegetative propagation of *Prunus africana*: effects of rooting medium, auxin concentrations and leaf area. *Agroforestry Systems* 54 (3): 183–192.

[20] Van Wyk, B.-E. & Gericke, N. (2000). *People's Plants: A Guide to Useful Plants of Southern Africa*. Briza Publications, Pretoria, South Africa.

[21] Van Wyk, B.-E., Van Oudtshoorn, B. & Gericke, N. (2009). *Medicinal Plants of South Africa*, 2nd edition. Briza Publications, Pretoria, South Africa.

[22] Watt, J. M. & Breyer-Brandwijk, M. G. (1962). *The Medicinal and Poisonous Plants of Southern and Eastern Africa*, 2nd edition. E. & S. Livingstone, Edinburgh & London.

[23] Were, J., Munjuga, N., Daws, M. I., Motete, N., Erdey, D., Baxter, D., Berjak, P., Pritchard, H. W., Harris, C. & Howard, C. A. (2004). Desiccation and storage of *Prunus africana* seeds (Hook f.) Kalkm. In: *Comparative Storage Biology of Tropical Tree Seeds.* Edited by M. Sacandé, D. Jøker, M. E. Dulloo & K. A. Thomsen, International Plant Genetic Resources Institute, Rome, Italy. pp. 67–74.

Scadoxus puniceus (L.) Friis & Nordal

TAXONOMY AND NOMENCLATURE

FAMILY. Amaryllidaceae

SYNONYMS. *Gyaxis puniceus* (L.) Salisb., *Haemanthus fax-imperii* Cufod., *H. goetzei* Harms, *H. insignis* Hook., *H. magnificus* (Herb.) Herb., *H. natalensis* Hook., *H. orchidifolius* Salisb., *H. puniceus* L., *H. redouteanus* M.Roem., *H. rouperi* auct., *H. superbus* Baker

VERNACULAR / COMMON NAMES. Blood lily, red paintbrush (English); bloedlelie, rooikwas (Afrikaans); umphompo (Swati).

PLANT DESCRIPTION

LIFE FORM. Perennial herb.

PLANT. Bulbous geophyte, <75 cm tall [3]; bulb <10 cm in diameter, covered by a fleshy tunic; aerial part is dormant in winter (i.e. July–August). **LEAVES.** Shiny, with wavy margins, <40 cm long, produced after, or at the same time as, flowers; at the leaf base is a false stem and two reduced leaves speckled red brown or purple (Figure 1). **FLOWERS.** Inflorescence a many-flowered umbel, 15 cm in diameter; flowers showy, orange red, within reddish brown bracts, anthers bright yellow [6,10] (Figure 1).

Figure 1 – *S. puniceus* in flower and with fruiting head. (Photos: L.-N. Le Roux, LNBG.)

FRUITS AND SEEDS

The fruits are red berries (Figure 1) [6]. The seeds (1–3 per berry) are small (c. 4 x 3 mm), and brown (Figure 2).

FLOWERING AND FRUITING. Flowers from July to February; fruits from November to March.

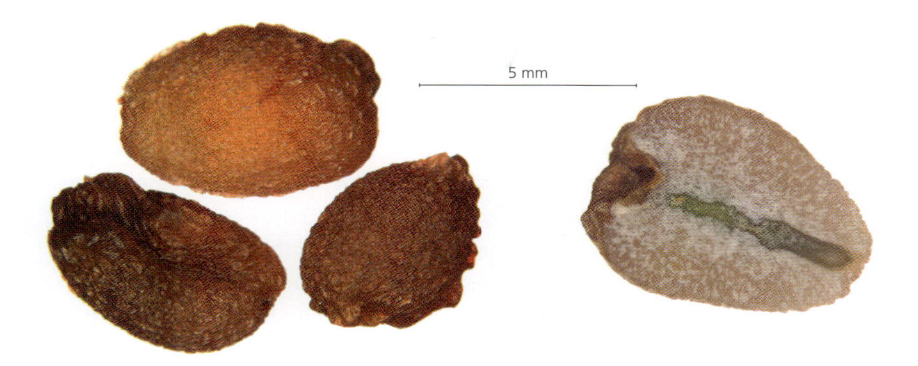

Figure 2 – External and internal views of the seeds, showing the embryo. (Photos: P. Gómez-Barreiro, RBG Kew.)

DISTRIBUTION

This species has a disjunct distribution in Southern Africa north to Ethiopia and Sudan [14]. In South Africa, it occurs in the Cape Provinces, KwaZulu-Natal, Mpumalanga, Gauteng, Free State, Limpopo and North West; it is also found in Swaziland [1,13].

HABITAT

Montane forests, coastal bushland, stream valleys, swamps and in various vegetation types at lower altitudes [14]; occurs at 15–2,100 m a.s.l. [1].

USES

Widely used in traditional medicine to treat coughs and gastrointestinal problems [12]. Root decoctions are used to treat venereal diseases. Rhizomes are an ingredient of infusions taken during pregnancy to ensure a safe childbirth [4]. A spectacular ornamental plant when in flower; the nectar and berries are eaten by birds [6].

KNOWN HAZARDS AND SAFETY

Species of *Scadoxus* and the closely related genus *Haemanthus* (to which *S. puniceus* formerly belonged) contain isoquinoline alkaloids, which are very toxic. Haemanthamine and haemanthidine have been isolated from *S. puniceus*. Indiscriminate use is potentially lethal; ingestion of the bulbs has caused fatalities [10,11].

CONSERVATION STATUS

A widespread species assessed as Least Concern (LC) in South Africa according to IUCN Red List criteria [7], but its status has yet to be confirmed in the IUCN Red List [5].

SEED CONSERVATION

HARVESTING. Ripe berries, distinguished by their red colour, can be collected directly from the plant (Figure 1).

PROCESSING AND HANDLING. Seeds can be easily cleaned by removing the fleshy pulp by rubbing through sieves with tepid water. Cleaned seeds should be spread out and left to dry at room temperature.

PROCESSING AND HANDLING. This species is reported to be desiccation sensitive; therefore long-term conservation under standard seed bank conditions is not feasible [8]. Alternative techniques, such as storage of encapsulated embryonic axes, are being investigated [9].

PROPAGATION

SEEDS

Dormancy and pre-treatments: No pre-treatments are required before sowing seeds in the nursery.
Germination, sowing and planting: Germination tests carried out at LNBG showed that the optimum soil temperatures for germination are in the 20–25°C range, with a final germination of c. 98%. Seeds should be sown in a mixture of equal parts by volume of washed river sand, compost and loamy soil, in containers that are at least 10 cm deep. Seeds should be sown 4 cm apart and 2 cm deep. Seeds and seedlings should be watered twice a week in summer and once a week in winter (i.e. May to August). Seeds start germinating after 10–15 days when planted in summer months (i.e. September to January). Seedlings should be transplanted when they are 10 cm high (Figure 3).

Figure 3 – *S. puniceus* seedlings. (Photo: W. Froneman, LNBG.)

VEGETATIVE PROPAGATION

Can be propagated by offsets removed from the parent plant in early spring when new vegetative growth begins. Offsets should have their own roots. They can be treated with a fungicide and planted in a medium of equal parts by volume of coarse river sand and finely milled bark, and kept moist. After one year they can be planted out in the ground or into pots [2].

TRADE

Plant material is available in muthi markets and in traditional Indian pharmacies. *Scadoxus puniceus* is grown internationally for the commercial horticulture trade.

Authors

Willem Froneman, Efisio Mattana, Vusi Lukhele, Karin van der Walt, Steve Davis and Tiziana Ulian.

References

[1] Conservatoire et Jardin botaniques & South African National Biodiversity Institute (SANBI) (2016). *African Plant Database*. Version 3.4.0. http://www.ville-ge.ch/musinfo/bd/cjb/africa/

[2] Duncan, G. (2001). Spectacular, rewarding *Scadoxus*. *Veld & Flora* (June 2001): 60–63.

[3] eMonocot (2017). *eMonocot: An Online Resource for Monocot Plants*. Version 1.0.5. Royal Botanic Gardens, Kew, University of Oxford, Natural History Museum, London & Natural Environment Research Council, UK. www.e-monocot.org/

[4] Hutchings, A., Scott, A. H., Lewis, G. & Cunningham, A. B. (1996). *Zulu Medicinal Plants: An Inventory*. University of Natal Press, Pietermaritzburg, South Africa.

[5] IUCN (2017). *The IUCN Red List of Threatened Species*. Version 2017-3. http://www.iucnredlist.org/

[6] Pooley, E. (1998). *A Field Guide to Wild Flowers: KwaZulu-Natal and the Eastern Region*. Natal Flora Publications Trust, Durban, South Africa.

[7] Raimondo, D., von Staden, L., Foden, W., Victor, J. E., Helme, N. A., Turner, R. C., Kamundi, D. A. & Manyama, P. A. (2009). *Red List of South African Plants. Strelitzia* 25, South African National Biodiversity Institute, Pretoria, South Africa.

[8] Sershen, N., Berjak, P. & Pammenter, N. W. (2008). Desiccation sensitivity of excised embryonic axes of selected amaryllid species. *Seed Science Research* 18 (1): 1–11.

[9] Sershen, N., Pammenter, N. W. & Berjak P. (2008). Post-harvest behaviour and short- to medium-term storage of recalcitrant seeds and encapsulated embryonic axes of selected amaryllid species. *Seed Science and Technology* 36 (1): 133–147.

[10] Turner, S. (2001). *Scadoxus puniceus* (L.) Friis & Nordal. South African National Biodiversity Institute (SANBI), South Africa. http://pza.sanbi.org/

[11] Van Wyk, B.-E. & Gericke, N. (2000). *People's Plants: A Guide to Useful Plants of Southern Africa*. Briza Publications, Pretoria, South Africa.

[12] Van Wyk, B.-E., Van Oudtshoorn, B. & Gericke, N. (2009). *Medicinal Plants of South Africa*, 2nd edition. Briza Publications, Pretoria, South Africa.

[13] WCSP (2017). *World Checklist of Selected Plant Families*. Royal Botanic Gardens, Kew. http://wcsp.science.kew.org/

[14] Zimudzi, C., Archer, R. H., Kwembeya, E. G. & Nordal, I. (2006). Synopsis of Amaryllidaceae from the *Flora Zambesiaca* area. *Kirkia* 18 (2): 151–168.

Trichilia emetica Vahl

TAXONOMY AND NOMENCLATURE

FAMILY. Meliaceae

SYNONYMS. *Trichilia roka* (Forssk.) Chiov.

VERNACULAR / COMMON NAMES. Natal mahogany (English); rooiessenhout (Afrikaans); nkuhlu, tinkuhlu (Shangaan); umKuhlu (Swati).

PLANT DESCRIPTION

LIFE FORM. Tree (occasionally shrub).

PLANT. Evergreen (sometimes deciduous), dioecious, <8–20(25) m tall, with a wide crown; bark dark grey or dark brown, rough or smooth. **LEAVES.** Opposite or alternate, <28 cm long, imparipinnate, with 3–4(5) pairs of oblong leaflets, each leaflet <15.5 x 5 cm, with apex rounded or emarginate and base rounded or cuneate; dark glossy green above, paler below (Figure 1). **FLOWERS.** Sweet-scented, pale yellow to green, in dense flower heads at ends of branches or in leaf axils, male and female flowers of similar general appearance [4,16].

Figure 1 – *T. emetica* leaves, fruits and seeds. (Photo: W. Froneman, LNBG.)

FRUITS AND SEEDS

The fruits are obovoid-globose, dehiscent capsules, opening in 2–3 valves. Seeds are black, almost completely covered by a scarlet aril [4] (Figure 2).

FLOWERING AND FRUITING. Flowers from August to November; fruits from December to April [16].

Figure 2 – External views of the seeds (left), showing the black testa and scarlet aril, and a skinned seed (right) without the black testa, showing the cotyledons and embryonic axis. (Photos: P. Gómez-Barreiro, RBG Kew.)

DISTRIBUTION

Widespread from Senegal to Ethiopia and Yemen, and throughout most of Southern Africa [3,4]. In South Africa it is found in KwaZulu-Natal, Mpumalanga and Limpopo [14,16].

HABITAT

Riparian forests, woodlands and lakeshore thickets; more rarely in savanna woodlands with deep soils, or as an emergent of thicket away from water [4], at 10–1,850 m a.s.l. [3].

USES

A multipurpose tree with a wide range of uses [10,12,15]. In traditional medicine, decoctions made from powdered bark are used to treat stomach and intestinal complaints [6,19]. Infusions of the bark or leaf are used to treat lumbago, rectal ulceration in children, as well as dysentery, kidney problems and indigestion [6,19]. Soaked bark is used as an emetic and purgative [19]. The bark is used as a fish poison and for inducing abortions [10]. Leaf or fruit poultices are used to treat skin diseases [11]. The seeds yield a good quality oil, which is rubbed into the body to treat rheumatism [19]. The wood is one of the most important timbers for woodcarving in Southern Africa, and is also used for furniture and many other items [10]. Limonoids, which are present in the seeds and bark, have antimicrobial activity and are used as insect antifeedants [10]. The seed oil is used in the manufacture of soap, cosmetics and candles [16]. *Trichilia emetica* is a fast-growing shade tree [10,12].

KNOWN HAZARDS AND SAFETY

The bark, root bark, fruits and seeds are purgative and emetic; strong doses can lead to fatal poisoning [2,19]. Cattle fed with seed residue (after oil extraction) have also been poisoned [5].

CONSERVATION STATUS

A widespread species assessed as Least Concern (LC) in South Africa according to IUCN Red List criteria [14], but its status has yet to be confirmed in the IUCN Red List [7].

SEED CONSERVATION

HARVESTING. Fruit capsules burst open between August and April. Seeds can be harvested from the open capsules.

PROCESSING AND HANDLING. Seeds can be cleaned by separating them from fruit remains using aspirators or sieves.

STORAGE AND VIABILITY. The seeds are recalcitrant [13] and can only be stored at temperatures above 0°C for a short time (Figure 3).

Figure 3 – Jar of *T. emetica* seeds stored at the MSB for short-term conservation. (Photo: P. Gómez-Barreiro, RBG Kew.)

PROPAGATION

SEEDS

Dormancy and pre-treatments: It is not clear if this species exhibits seed dormancy. The seed coat has been reported to inhibit germination [9]; however, high germination percentages (c. 95%) have been achieved with intact seeds [13]. Both percentage and speed of germination have been reported to increase significantly when the aril is removed [8]. Seeds of this species are reported as being both non-dormant (ND) and physiologically dormant (PD) [1].

Germination, sowing and planting: Germination tests carried out at LNBG showed that the optimum soil temperatures for germination are in the 20–28°C range, with a final germination of 99%. Seeds should be sown in a mixture of washed river sand, compost and soil in a ratio 2:1:1 by volume. Seeds should be sown 4 cm apart and c. 2 cm deep, the containers then placed in the shade. Seeds should be watered twice a week in summer (i.e. October to March) and once every three weeks in winter. Germination takes place 10–20 days after sowing. Seedlings should be transplanted when they are c. 10 cm high. Fresh seeds (moisture content of 40% on fresh mass basis) incubated in the laboratory under controlled conditions of 26°C, reached c. 95% germination. A positive correlation between germination rate and increased incubation temperatures was also detected when seeds were germinated in the range 11–36°C [13].

VEGETATIVE PROPAGATION

Easily propagated by cuttings taken from one-year-old shoots from young trees [17,18]. Coppice shoots and suckers can also be used for propagation [10].

TRADE

The seed oil is of industrial value for soap and candle making and is exported from East Africa [10]. Plant material is available on muthi markets in Mpumalanga and in traditional Indian pharmacies in Mbombela (formerly Nelspruit).

Authors

Willem Froneman, Efisio Mattana, Steve Davis and Tiziana Ulian.

References

[1] Baskin, C. C. & Baskin, J. M. (2014). *Seeds: Ecology, Biogeography, and Evolution of Dormancy and Germination*, 2nd edition. Academic Press, San Diego, USA.

[2] Burkill, H. M. (1997). *The Useful Plants of West Tropical Africa. Vol. 4: Families M–R.* 2nd edition. Royal Botanic Gardens, Kew.

[3] Conservatoire et Jardin botaniques & South African National Biodiversity Institute (SANBI) (2016). *African Plant Database*. Version 3.4.0. http://www.ville-ge.ch/musinfo/bd/cjb/africa/

[4] Exell, A. W., Fernandes, A. & Wild, H. (eds) (1963). Meliaceae. In: *Flora Zambesiaca*, Vol. 2, part 1. Flora Zambesiaca Managing Committee, Royal Botanic Gardens, Kew.

[5] Fox, F. W. & Norwood Young, M. E. (1982). *Food from the Veld. Edible Wild Plants of Southern Africa*. Delta, Johannesburg, South Africa.

[6] Hutchings, A., Scott, A. H., Lewis, G. & Cunningham, A. B. (1996). *Zulu Medicinal Plants: An Inventory*. University of Natal Press, Pietermaritzburg, South Africa.

[7] IUCN (2017). *The IUCN Red List of Threatened Species.* Version 2017-3. http://www.iucnredlist.org/

[8] Jøker, D. (2003). *Trichilia emetica* Vahl. *Seed Leaflet* No. 68. Danida Forest Seed Centre, Humlebaek, Denmark.

[9] Mahgembe, J. A. & Msanga, H. P. (1988). Effect of physical scarification and gibberellic acid treatments in germination of *Trichilia emetica* seed. *International Tree Crops Journal* 5 (3): 163–177.

[10] Mashungwa, G. N. & Mmolotsi, R. M. (2007). *Trichilia emetica* Vahl. In: *PROTA (Plant Resources of Tropical Africa/Ressources Végétales de l'Afrique Tropicale).* Edited by H. A. M. van der Vossen & G. S. Mkamilo. Wageningen, The Netherlands. http://www.prota4u.org/

[11] Mothogoane, M. S. (2014). *Trichilia emetica* Vahl. South African National Biodiversity Institute (SANBI), South Africa. http://pza.sanbi.org/

[12] Orwa, C., Mutua, A., Kindt, R., Jamnadass, R. & Simons, A. (2009). *Agroforestree Database: A Tree Reference and Selection Guide*. Version 4.0. World Agroforestry Centre, Kenya. http://www.worldagroforestry.org/output/agroforestree-database

[13] Pritchard, H. W., Daws, M. I., Fletcher, B. J., Gaméné, C. S., Msanga, H. P. & Omondi, W. (2004). Ecological correlates of seed desiccation tolerance in tropical African dryland trees. *American Journal of Botany* 91 (6): 863–870.

[14] Raimondo, D., von Staden, L., Foden, W., Victor, J. E., Helme, N. A., Turner, R. C., Kamundi, D. A. & Manyama, P. A. (2009). *Red List of South African Plants. Strelitzia* 25, South African National Biodiversity Institute, Pretoria, South Africa.

[15] Royal Botanic Gardens, Kew (1999–2015). *Survey of Economic Plants for Arid and Semi-Arid Lands (SEPASAL) Database*. http://apps.kew.org/sepasalweb/sepaweb

[16] Schmidt, E., Lötter, M. & McCleland, W. (2002). *Trees and Shrubs of Mpumalanga and Kruger National Park*. Jacana Media, Johannesburg, South Africa.

[17] Van Wyk, P. (1972). *Trees of the Kruger National Park*, Vol. 1. Purnell, Cape Town, South Africa.

[18] Venter, F. & Venter, J.-A. (1996). *Making The Most of Indigenous Trees*. Briza Publications, Pretoria, South Africa.

[19] Watt, J. M. & Breyer-Brandwijk, M. G. (1962). *The Medicinal and Poisonous Plants of Southern and Eastern Africa*, 2nd edition. E. & S. Livingstone, Edinburgh & London.

Warburgia salutaris (G.Bertol.) Chiov.

TAXONOMY AND NOMENCLATURE

FAMILY. Canellaceae

SYNONYMS. *Chibaca salutaris* G.Bertol., *Warburgia breyeri* R.Pott

VERNACULAR / COMMON NAMES. Pepper-bark tree, pepper leaf (English); peperbasboom (Afrikaans); isiBhaha, xibaha (Swati).

PLANT DESCRIPTION

LIFE FORM. Tree or shrub.

PLANT. Evergreen, 2.5–10(18) m tall [1,7], with rough mottled grey bark, red on inner side; branches striate and with lenticels. **LEAVES.** Alternate, oblong to oblong-lanceolate or elliptic, 4.5–11 x 1.5–2.5 cm, apex acute, base cuneate, sometimes wavy, glossy green above, paler below, aroma peppery when crushed. **FLOWERS.** <7 mm in diameter, solitary or in 3-flowered cyme, green, in leaf axils [7,14] (Figure 1).

FRUITS AND SEEDS

The fruits are round berries turning yellowish black when ripe and with a coriaceous pericarp (Figure 2). Seeds are brown, flattened, several per berry (Figure 2) [7].

FLOWERING AND FRUITING. Flowers from April to May; fruits from September to November [14].

Figure 1 – *W. salutaris*, young tree. (Photo: W. Froneman, LNBG.)

Figure 2 – *W. salutaris* fruits and freshly collected seeds. (Photos: W. Froneman, LNBG.)

DISTRIBUTION

South Africa (KwaZulu-Natal, Limpopo and Mpumalanga), Swaziland (where extremely rare) [2], Mozambique and Zimbabwe [1,17].

HABITAT

Lowland evergreen, coastal and Afromontane forests, woodlands and savanna, and rocky hillsides [1,7], at 5–1,750 m a.s.l. [1].

USES

The roots and bark are widely used in traditional medicine to treat coughs, common cold and malaria. Fresh or dried leaves are used in various dishes to give a peppery taste [9]. In the pharmaceutical trade, products derived from the bark of *Warburgia* spp. (including *W. salutaris*) are available to treat fungal, bacterial and protozoan infections [3,9,16]. The tree is grown as an ornamental [10].

KNOWN HAZARDS AND SAFETY

No known hazards reported when used in the traditional manner, but some compounds present can cause skin irritation and contact dermatitis [9]. Some products containing *Warburgia* can cause gastrointestinal upset and diarrhoea, and should not be taken during pregnancy and lactation.

CONSERVATION STATUS

This rare species has been assessed as Endangered (EN) in South Africa [11] and at the global level according to IUCN Red List criteria [4]. It is Critically Endangered (CR) in Swaziland [2] and Zimbabwe [8], and Vulnerable (VU) in Mozambique [5]. In South Africa, the tree is afforded protection under forestry legislation [13]. The main threat is unsustainable harvesting of the bark for traditional medicine [4,11] (Figure 3).

Figure 3 – Tree damaged by unsustainable harvesting of the bark. (Photo: W. Froneman, LNBG.)

SEED CONSERVATION

HARVESTING. Ripe fruits can be harvested directly from the plant.

PROCESSING AND HANDLING. Seeds can be easily cleaned by opening the fleshy pulp and removing the seeds. They should then be washed under running water to remove any remaining fruit pulp.

STORAGE AND VIABILITY. Seeds of this species are reported to be desiccation sensitive [6,15]. However, considering fruit structure, seed size and natural habitat, seeds of this species may not be recalcitrant [12].

PROPAGATION

SEEDS

Dormancy and pre-treatments: No pre-treatments are required.

Germination, sowing and planting: Germination tests carried out at LNBG showed that the optimum soil temperatures for germination are in the 20–25°C range, with final germination of c. 60–70%. Seeds should be sown 2 cm apart in a mixture of equal parts by volume of washed river sand, compost/kraal manure and loamy soil, in containers that are at least 10 cm deep, with sufficient drainage holes. Seeds should be covered with 5 mm of substrate and watered thoroughly. Germination should begin after 15–20 days. Seedlings should be watered once a week in summer, and once a fortnight in winter. Seedlings should be transplanted when they are 10 cm high (Figure 4).

VEGETATIVE PROPAGATION

Can be propagated by cuttings. LNBG achieved a success rate of 20–30% using semi-hardwood or softwood cuttings of 10 cm length, dipped in 1:3 liquid rooting hormone-water solution, and planted in washed river sand with an average soil temperature of 20°C. Automatic overhead mist

Figure 4 – *W. salutaris* seedlings in the nursery.
(Photos: W. Froneman, LNBG.)

Figure 5 – A rooted cutting.
(Photo: W. Froneman, LNBG.)

sprayers watered the cuttings for one minute every five minutes. Sufficient rooting took place after 3–4 months (Figure 5). However, the growth rate of plants produced from cuttings is much slower than that of plants raised from seed.

TRADE

Plant material is available in muthi markets in Mpumalanga and in traditional pharmacies in Mbombela (formerly Nelspruit). The demand in South Africa is such that bark is imported from Swaziland and Mozambique [11].

Authors

Willem Froneman, Efisio Mattana, Steve Davis and Tiziana Ulian.

References

[1] Conservatoire et Jardin botaniques & South African National Biodiversity Institute (SANBI) (2016). *African Plant Database*. Version 3.4.0. http://www.ville-ge.ch/musinfo/bd/cjb/africa/

[2] Dlamini, T. S. & Dlamini, G. M. (2002). Swaziland. In: *Southern African Plant Red Data Lists*. Edited by J. S. Golding. Southern African Botanical Diversity Network Report No. 14, SABONET, Pretoria, South Africa. pp. 121–134.

[3] Hutchings, A., Scott, A. H., Lewis, G. & Cunningham, A. B. (1996). *Zulu Medicinal Plants: An Inventory*. University of Natal Press, Pietermaritzburg, South Africa.

[4] IUCN (2017). *The IUCN Red List of Threatened Species*. Version 2017-3. http://www.iucnredlist.org/

[5] Izidine, S. & Bandeira, S. O. (2002). Mozambique. In: *Southern African Plant Red Data Lists*. Edited by J. S. Golding. Southern African Botanical Diversity Network Report No. 14, SABONET, Pretoria, South Africa. pp. 43–60.

[6] Kioko, J., Baxter, D. & Berjak, P. (2004). Tolerance to desiccation and storability of *Warburgia salutaris* (*ugandensis*) seeds from Kenya. In: *Comparative Storage Biology of Tropical Tree Seeds*. Edited by M. Sacandé, D. Jøker, M. E. Dulloo & K. A. Thomsen. International Plant Genetic Resources Institute, Rome, Italy. pp. 131–139.

[7] Launert, E. & Pope, G. V. (eds) (1990). Canellaceae. In: *Flora Zambesiaca*, Vol. 7, part 4. Flora Zambesiaca Managing Committee, Royal Botanic Gardens, Kew.

[8] Mapaura, A. & Timberlake, J. R. (2002). Zimbabwe. In: *Southern African Plant Red Data Lists*. Edited by J. S. Golding. Southern African Botanical Diversity Network Report No. 14, SABONET, Pretoria, South Africa. pp. 157–182.

[9] Maroyi, A. (2013). The genus *Warburgia*: a review of its traditional uses and pharmacology. *Pharmaceutical Biology* 52 (3): 378–391.

[10] Mbambezeli, G. (2004). *Warburgia salutaris* (Bertol.f.) Chiov. South African National Biodiversity Institute (SANBI), South Africa. http://pza.sanbi.org/

[11] Raimondo, D., von Staden, L., Foden, W., Victor, J. E., Helme, N. A., Turner, R. C., Kamundi, D. A. & Manyama, P. A. (2009). *Red List of South African Plants*. *Strelitzia* 25, South African National Biodiversity Institute, Pretoria, South Africa.

[12] Royal Botanic Gardens, Kew (2017). *Seed Information Database (SID)*. Version 7.1. http://data.kew.org/sid/

[13] SA Forestry Online (2015). *Protected Trees in South Africa (2015). List of Protected Trees Under The National Forest Act, 1998 (Act No. 84 of 1998)*. http://saforestryonline.co.za/indigenous/protected-trees-in-south-africa/

[14] Schmidt, E., Lötter, M. & McCleland, W. (2002). *Trees and Shrubs of Mpumalanga and Kruger National Park*. Jacana Media, Johannesburg, South Africa.

[15] Uronu, L. O. N. & Msanga, H. P. (2004). Seed desiccation tolerance of *Strychnos cocculoides*, *Ximenia americana* and *Warburgia salutaris*. In: *Comparative Storage Biology of Tropical Tree Seeds*. Edited by M. Sacandé, D. Jøker, M. E. Dulloo & K. A. Thomsen. International Plant Genetic Resources Institute, Rome, Italy. pp. 140–152.

[16] Van Wyk, B.-E., Van Oudtshoorn, B. & Gericke, N. (2009). *Medicinal Plants of South Africa*, 2nd edition. Briza Publications, Pretoria, South Africa.

[17] WCSP (2017). *World Checklist of Selected Plant Families*. Royal Botanic Gardens, Kew. http://wcsp.science.kew.org/

Seed collecting team during an expedition in the Tehuacán-Cuicatlán valley, Mexico. (Photo: FESI-UNAM.)

MEXICO

Agave kerchovei Lem.

TAXONOMY AND NOMENCLATURE

FAMILY. Asparagaceae

SYNONYMS. *Agave beaucarnei* Lem., *A. expatriata* Rose, *A. inopinabilis* Trel., *A. noli-tangere* A.Berger

VERNACULAR / COMMON NAMES. Cacayas (indigenous name); ixtle (Nahuatl); rabo de león (Spanish).

PLANT DESCRIPTION

LIFE FORM. Shrub.

PLANT. Monocarpic, solitary or tufted rosettes of 80–100 or more leaves, sometimes stemless or with a short stem (Figure 1). **LEAVES.** Lanceolate, straight or slightly curved,

Figure 1 – Habit of *A. kerchovei*. (Photo: O. Téllez-Valdés, FESI-UNAM.)

40–100 x 5–12 cm, thick at base, flat to concave on adaxial surface, convex on abaxial surface, rigid, yellowish green to green, rarely pruinose, margin usually with large teeth, 8–15 mm long, 2–5 cm apart, curved to flat, grey; apical spine 3–6 cm long, brown to grey, adaxial surface deeply grooved, abaxial surface rounded. **FLOWERS.** Spikes very dense, 2.5–5 m tall; flowers greenish to purplish, c. 3.5–4.5 cm long; ovary c. 2 cm long, fusiform, neck constricted; tepals 15–20 x 6–8 mm wide, green to reddish, linear, acute; filaments 4–5 cm, inserted on rim of tube; anthers c. 2 cm long, yellow to reddish [6].

FRUITS AND SEEDS

Fruits are oblong capsules, tightly closed, 2.5–3.5 cm long, c. 1 cm wide; apex acute to slightly rounded. Seeds (Figure 2) are lacrymiform to lunate, black, shiny, thick, surface smooth, slightly winged, 3–4 mm long, 2–3 mm wide [5,6].

FLOWERING AND FRUITING. Flowers from January to May; fruits from April to October [3,5,6].

Figure 2 – External views and longitudinal section of the seeds, showing a linear embryo.
(Photos: P. Gómez-Barreiro, RBG Kew.)

DISTRIBUTION

Endemic to Mexico. Occurs in the central region of Hidalgo to southern Oaxaca [5,6].

HABITAT

Arid and semi-arid regions, alluvial valleys and steep rocky slopes, at 1,400–1,875 m a.s.l. [6].

USES

Young flower buds ('cacayas' or 'cacallas') are eaten (roasted, fried, boiled and in preserves; Figure 3), and are used as animal fodder [1,3]. In the mid-twentieth century, local people exploited the species to obtain fibre, locally known as 'ixtle', to make bags, sacks, blankets [1], cords, baskets, 'ayates' (traditional capes used for collecting crops) and 'estropajos' (scrubbers). In the Tehuacán-Cuicatlán valley, 'quiote' (tall stems bearing the flowers) are used for roofs, fences and enclosures [3], as well as to make broom handles, musical instruments and animal sheds. The plant is grown as an ornamental [1,3].

Figure 3 – (Left) Dish prepared with buds ('cacayas') of *A. kerchovei*. (Right) Local inhabitant cutting a piece of the apex of the 'quiote' to separate the 'cacayas'. (Photos: P. Brena-Bustamante.)

KNOWN HAZARDS AND SAFETY

The sap of some *Agave* spp. can cause contact dermatitis and conjunctivitis [7]. Care should be taken when handling the plant as the teeth on the leaves are sharp.

CONSERVATION STATUS

Not considered to be either rare or threatened [4], but its status has yet to be confirmed in the IUCN Red List [8]. However, the plant is a valuable food resource for people and grazing animals. Collecting 'cacayas' reduces the reproductive capacity of individual plants as the species is monocarpic [2,3].

SEED CONSERVATION

HARVESTING. Ripe fruits should be collected at the beginning of the fruiting season [2,3]. It is not recommended to mix the fruits with other plant material to avoid increasing moisture and temperature.

PROCESSING AND HANDLING. Collected fruits should be stored in paper or cotton bags, and transported to a dark and dry place where the seeds can be separated from fruit remains by hand, sieves or an air blower.

STORAGE AND VIABILITY. X-ray analysis carried out on the seed lots stored at the MSB showed a high percentage of filled seeds (92–100%). Seeds of this species are reported to be desiccation tolerant [10].

PROPAGATION

SEEDS

Dormancy and pre-treatments: No pre-treatments are required.

Germination, sowing and planting: Seeds germinated under laboratory controlled conditions at the MSB showed high germination percentages when incubated with 8 hours of light per day at constant temperatures (15–25°C) with no pre-treatments [10]. A comparative study of two different populations was carried out at the FESI-UNAM Seed Bank. Seeds were incubated at 30°C in the light. Lower germination (37.3%) was achieved for seeds collected in San Rafael, Coxcatlán (Puebla) compared to those collected in San Gabriel Casa Blanca (Oaxaca), which reached final germination of almost 100% [3].

VEGETATIVE PROPAGATION

This species sometimes forms tight clumps of rosettes [6], in which case (as with other *Agave* spp.) division and propagation by offsets is possible.

TRADE

Flower buds are collected and sold at the Tehuacán and Teotitlán markets [8].

Authors

Fernando Peralta-Romero, Efisio Mattana, Steve Davis and Paulina Hechenleitner.

References

[1] Arias Toledo, A. A., Valverde Valdés, M. T. & Reyes Santiago, J. (2000). *Las Plantas de la Región de Zapotitlán Salinas, Puebla*. Instituto Nacional de Ecología, UNAM, Mexico.

[2] Brena-Bustamante, P. (2012). *El Aprovechamiento y la Estructura Poblacional de* Agave kerchovei *Lem., en Tehuacán Cuicatlán, Mexico*. Master's thesis, Campus Montecillo, Colegio de Postgraduados, Mexico.

[3] Brena-Bustamante, P., Lira-Saade, R., García-Moya, E., Romero-Manzanares, A., Cervantes-Maya, H., López-Carrera, M. & Chávez-Herrera, S. (2013). Aprovechamiento del escapo y los botones florales de *Agave kerchovei* en el valle de Tehuacán-Cuicatlán, Mexico. *Botanical Sciences* 91 (2): 181–186.

[4] Diario Oficial de la Federación (2010). *Norma Oficial Mexicana NOM-059-SEMARNAT-2010, Protección Ambiental — Especies Nativas de México de Flora y Fauna Silvestres — Categorías de Riesgo y Especificaciones para su Inclusión, Exclusión o Cambio — Lista de Especies en Riesgo*. Secretaría de Medio Ambiente y Recursos Naturales, Mexico, 30 December 2010.

[5] García-Mendoza, A. J. (2011). *Flora del Valle de Tehuacán-Cuicatlán, Fascículo 88, Agavaceae Dumort.* Instituto de Biología, UNAM, Mexico.

[6] Gentry, H. S. (1982). *Agaves of Continental North America*. University of Arizona Press, Tucson, Arizona, USA.

[7] Hackman, D. A. *et al*. (2006). Agave (*Agave americana*): an evidence-based systematic review by the Natural Standard Research Collaboration. *Journal of Herbal Pharmacotherapy* 6 (2): 101–122.

[8] IUCN (2017). *The IUCN Red List of Threatened Species.* Version 2017-3. http://www.iucnredlist.org/

[9] López Carrera, M. (2014). Personal communication.

[10] Royal Botanic Gardens, Kew (2017). *Seed Information Database (SID).* Version 7.1. http://data.kew.org/sid/

Amphipterygium adstringens (Schltdl.) Standl.

TAXONOMY AND NOMENCLATURE

FAMILY. Anacardiaceae

SYNONYMS. *Hypopterygium adstringens* Schltdl., *Juliania adstringens* (Schltdl.) Schltdl.

VERNACULAR / COMMON NAMES. Cuachalala, cuachalalate (Nahuatl).

Figure 1 – Leaves and unripe fruits of *A. adstringens*. (Photo: O. Téllez-Valdés, FESI-UNAM.)

PLANT DESCRIPTION

LIFE FORM. Tree.

PLANT. Deciduous, dioecious, 4–15 m tall; bark smooth, reddish grey, with corky protuberances; branches terete, apically tomentose with whitish pubescence, glabrous when mature, lenticels orbicular, elevated, yellowish; leaf scars numerous at apex; latex milky. **LEAVES.** Simple or compound, imparipinnate, with 3–7 leaflets, spathulate, oblong-lanceolate, obovate, basal and central leaflets 2–6 x 1–3 cm, terminal leaflet spathulate (a key identification feature), dentate only in distal portion, 3–6 x 1.5 x 7 cm, petioles 2–7 mm long, ribbed, densely pilose (Figure 1). **FLOWERS.** Male inflorescences 7–12 cm long, in panicles at apex of branches; female inflorescences in dense involucres, 1–1.5 cm long, 2–4-flowered [1].

FRUITS AND SEEDS

Fruits <2.5–5 cm long, 1.2–1.8 cm wide, samaroid, nodding when mature; pedicels 3–5 mm long, densely villose; wing thin and semi-transparent, pilose, glabrous with age (Figures 1&2). Seeds c. 8 x 1 mm wide, 6-ribbed, semi-woody [9], 4 seeds per fruit (usually only 2 seeds filled) (Figure 2).
FLOWERING AND FRUITING. Flowers from May to August; fruits from June to February [5].

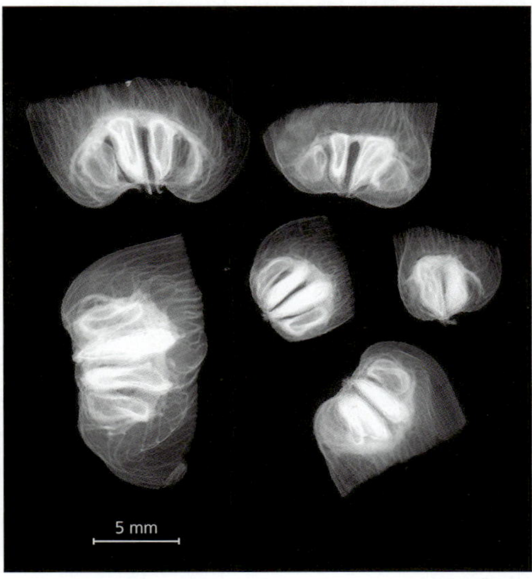

Figure 2 – External views of the fruits and an X-ray image showing their internal structure and the filled seeds in white. (Photos: P. Gómez-Barreiro, RBG Kew.)

DISTRIBUTION

Mexico (Guerrero, Jalisco, Estado de México, Michoacán, Morelos, Puebla and Oaxaca), Costa Rica, Honduras and Guatemala [1].

HABITAT

Deciduous dry forests and xerophilous scrub at altitudes of 5–1,500 m a.s.l. [1,5].

USES

The bark is astringent and has a high tannin content. It is used for the production of a red dye. The boiled bark is used in traditional medicine for treating wounds, weak gums and malaria and to prevent alopecia. It is also used to treat kidney diseases, gastritis, ulcers and other types of inflammation. The wood is used for fuel [4,5,6,7,9].

KNOWN HAZARDS AND SAFETY

No known hazards.

CONSERVATION STATUS

Not considered to be either rare or threatened in Mexico [2], but assessed according to IUCN Red List criteria as Endangered (EN) in Guatemala, where the main threat is the loss of dry forest habitat [8]. The status of this species has yet to be confirmed in the IUCN Red List [3].

SEED CONSERVATION

HARVESTING. Fruits should be collected when physiologically mature, as indicated by their brown colour.

PROCESSING AND HANDLING. Fruits should be separated from other plant material by hand and sieves, then stored in paper or cotton bags in a dark, dry place. It is best not to remove the seeds from the fruits because this can damage the seeds. The fruits are therefore the conservation unit (Figure 2).

STORAGE AND VIABILITY. X-ray analysis carried out on seed lots stored at the MSB showed a low percentage (c. 30%) of filled seeds (Figure 2). No data on seed storage behaviour are available for this or congeneric species. However, c. 95% of Anacardiaceae spp. stored at the MSB are reported to be, or are likely to be, desiccation tolerant [10]. In view of this, and the type of habitat where the species grows, it is likely that seeds of *Amphipterygium adstringens* are desiccation tolerant, although this needs to be confirmed.

PROPAGATION

SEEDS

Dormancy and pre-treatments: Fruits of *A. adstringens* underwent a pre-germination treatment at the San Rafael, Coxcatlán nursery, by soaking them in water for two days. For germination under controlled conditions, fruits were sterilised (immersed in 10% bleach solution for five minutes), re-moisturised in high humidity over water for one week at 20°C, and then scarified by removing the covering structure and chipping the coat using a scalpel [10].

Germination, sowing and planting: 70% of the fruits sown in germination trays at the San Rafael, Coxcatlán nursery germinated. Germination started at the end of the first week after sowing. The substrate used was a 3:1 mixture by volume of sterilised sieved leaf litter and agrolite. Seedlings were transplanted into black plastic bags 3–4 weeks after germination, using a mixture of equal parts by volume of sieved leaf litter and agrolite. They were watered every five days and weeded frequently by hand. Of the 800 fruits that germinated, 119 seedlings established. Germination tests under laboratory controlled conditions at the FESI-UNAM Seed Bank revealed 55% final germination for untreated fruits incubated at 30°C, with a photoperiod of 12 hours of darkness and 12 hours of light, for 20 days. 60% germination was achieved at the MSB for fruits pre-treated as above when incubated in an alternating temperature regime of 10/25°C with a photoperiod of 8/16 hours [10].

VEGETATIVE PROPAGATION

Experiments using cuttings were carried out at the San Rafael, Coxcatlán nursery. However, no vegetative propagation protocols are currently available.

TRADE

The bark is commonly traded in markets, either fresh or dry depending on the proximity of the market to collection areas. This resource has become a source of income that attracts inexperienced collectors and people with little or no knowledge of traditional uses and sustainable harvesting [4].

Authors

Fernando Peralta-Romero, Myrna Mendoza Cruz, Martin López Carrera, Efisio Mattana, Steve Davis and Paulina Hechenleitner.

References

[1] Cuevas Figueroa, X. M. (2006). A revision of the genus *Amphipterygium* (Julianiaceae). *Ibugana* 13 (1): 27–47.

[2] Diario Oficial de la Federación (2010). *Norma Oficial Mexicana NOM-059-SEMARNAT-2010, Protección Ambiental — Especies Nativas de México de Flora y Fauna Silvestres — Categorías de Riesgo y Especificaciones para su Inclusión, Exclusión o Cambio — Lista de Especies en Riesgo*. Secretaría de Medio Ambiente y Recursos Naturales, Mexico, 30 December 2010.

[3] IUCN (2017). *The IUCN Red List of Threatened Species.* Version 2017-3. http://www.iucnredlist.org/

[4] Jiménez-Merino, F. A. (2012). *Herbolaria Mexicana, Biblioteca Básica de Agricultura*, 2nd edition. Colegio de Postgraduados, Universidad Autónoma de Chapingo, Mexico.

[5] Martínez-Flores, J., Godínez, H., Eguiarte, L. & Lira, R. (2013). Impacto del descortezamiento de *Amphipterygium adstringens* (Schiede ex Schlecht.) Standl., sobre su floración y fructificación en San Rafael, Coxcatlán, Puebla. *Memorias Congreso Mexicano de Botánica*.

[6] Martínez-Moreno, D., Alvarado-Flores, R., Mendoza-Cruz, M. & Basurto-Peña, F. (2006). Plantas medicinales de cuatro mercados del estado de Puebla, México. *Bol. Soc. Bot. Méx*. 79: 79–87.

[7] Monroy-Ortiz, C. & Monroy, R. (2006). *Las Plantas, Compañeras de Siempre: la Experiencia en Morelos*. Universidad Autónoma del Estado de Morelos, Cuernavaca, Morelos, Mexico.

[8] National Red List (2017). *Amphipterygium adstringens*. National Red List Network. http://www.nationalredlist.org/

[9] Rosalinda Medina, L. (2000). *Flora del Valle de Tehuacán-Cuicatlán, Fascículo 30, Julianiaceae Hemsl.* Instituto de Biología, UNAM, Mexico.

[10] Royal Botanic Gardens, Kew (2017). *Seed Information Database (SID).* Version 7.1. http://data.kew.org/sid/

Bursera morelensis Ramírez

TAXONOMY AND NOMENCLATURE

FAMILY. Burseraceae

SYNONYMS. *Elaphrium morelense* (Ramírez) Rose

VERNACULAR / COMMON NAMES. Red cuajiote (English); coabinillo, copalillo, cuajiote rojo (Spanish).

PLANT DESCRIPTION

LIFE FORM. Tree.

PLANT. Deciduous, dioecious, <13 m tall, trunks <40 cm in diameter, outer bark reddish, exfoliating (Figure 1); branches glabrous, with an aromatic oily resin; cataphylls inconspicuous, soon deciduous. **LEAVES.** Imparipinnate, 5–11 x 1.5–4.5 cm, oblong-elliptic in general outline; petioles grooved, 1–2 cm long; leaflets 15–51, rachis inconspicuously winged. **FLOWERS.** Inflorescences in racemes or paniculate, <5 cm long, bracteoles filiform to subulate, yellowish, greenish or whitish; male flowers 3–5, petals 3–6 mm long; female flowers 3(–5), usually solitary, seldom in pairs or short panicles, similar in shape and size to male flowers [7].

Figure 1 – Exfoliating bark. (Photo: F. Peralta-Romero, FESI-UNAM.)

Figure 2 – Unripe fruits. (Photo: O. Téllez-Valdés, FESI-UNAM.)

FRUITS AND SEEDS

Fruits more than 8, peduncles 1.3–2.2 cm long, strongly thickened, recurved, glabrous, 3-valvate, 5–10 x 4–6 mm, obliquely ovoid, slightly apiculate, glabrous (Figure 2). Seeds are covered by a pale yellow pseudaril (Figure 3) [7].

FLOWERING AND FRUITING. Flowers from May to beginning of June; fruits from July to October [7].

Figure 3 – External views and longitudinal section of the seed showing a folded embryo. (Photo: P. Gómez-Barreiro, RBG Kew.)

DISTRIBUTION

Endemic to Mexico (Guanajuato, Guerrero, Hidalgo, Morelos, Puebla, Querétaro and San Luis Potosí) [7].

HABITAT

Deciduous dry forests on various substrates, at 500–1,500 m a.s.l. [7].

USES

Used in the Tehuacán-Cuicatlán Valley of Mexico for animal fodder, timber for buildings, extraction of resin and latex, dye production, live fences and soil erosion control [6]. The bark is used in traditional medicine to prepare tea as a diuretic and to disinfect wounds [5,6]. The species has been reported to have antimicrobial and anti-thrush properties [2,10]. Stems and roots are used as firewood in San Rafael, Coxcatlán [11].

KNOWN HAZARDS AND SAFETY

No known hazards.

CONSERVATION STATUS

Not considered to be either rare or threatened [3], but the status of this species has yet to be confirmed in the IUCN Red List [4].

SEED CONSERVATION

HARVESTING. Fruits of this species can be parthenocarpic [9]. In parthenocarpic fruits, the first stages of development include an unusual development of internal walls in the ovary, which invade the locule and prevent ovule development. Seeded and parthenocarpic fruits can be distinguished at maturity by their differing dehiscence: seeded fruits are dehiscent, whereas parthenocarpic fruits are only partially so [9]. Care should therefore be taken to avoid the collection of parthenocarpic fruits in the field.

PROCESSING AND HANDLING. Collected fruits should be placed in paper bags. The seeds can then be extracted from the fruits and the pseudarils removed by hand. A vertical blower can be used to remove any small remaining particles.

STORAGE AND VIABILITY. X-ray analysis carried out on seed lots stored at the MSB showed c. 45% filled seeds and a high percentage of non-viable (empty and infested) seeds. These data should be treated with caution due to the presence of parthenocarpic fruits [9], which cannot easily be detected by X-ray analysis. No data on seed storage behaviour are available for this species. However, the related *Bursera simaruba* (L.) Sarg. is probably seed desiccation tolerant [12]. Considering this, and the type of habitat where the species grows, it is likely that the seeds of *B. morelensis* are desiccation tolerant, although this needs to be confirmed.

PROPAGATION

SEEDS

Dormancy and pre-treatments: *Bursera* seeds are endozoochorous, so the seeds are protected by hard layers against physical and chemical damage [8]. The woody endocarp could result in physical (PY) or physiological (PD) seed dormancy, depending on its permeability to water, and thus on whether it allows or impedes seed imbibition [8]. The seeds of many Burseraceae spp. are reported to be either PD or to be non-dormant [1]. However, little information is available on the germination of *Bursera* spp., probably as a result of a high percentage of parthenocarpic fruits [8] and empty seeds. Further studies are required to determine the type of dormancy.

Germination, sowing and planting: Preliminary experiments have been carried out at the MSB to test the effects of different temperatures on germination. However, very few seeds germinated, confirming the need for further studies.

VEGETATIVE PROPAGATION

Experiments using cuttings were carried out at the San Rafael, Coxcatlán nursery. Cuttings from young branches were treated with a rooting hormone powder and placed in black plastic pots using a mixture of equal parts by volume of sieved and sterilised leaf litter, river sand and agrolite. The cuttings were watered every five days and weeded frequently. Out of 74 cuttings, only seven plants were established. No vegetative propagation protocol is currently available.

TRADE

Plants are not commonly traded or commercialised.

Authors

Fernando Peralta-Romero, Myrna Mendoza Cruz, Martín López Carrera, Efisio Mattana and Paulina Hechenleitner.

References

[1] Baskin, C. C. & Baskin, J. M. (2014). *Seeds: Ecology, Biogeography, and Evolution of Dormancy and Germination*, 2nd edition. Academic Press, San Diego, USA.

[2] Canales-Martínez, M. M., Rivera-Yáñez, C. R., Salas-Oropeza, J., López -Hernández, R., Jiménez-Estrada, M., Duran-Díaz, A., Flores-Ortiz, C. M., Hernández, L. B., Rosas-López, R. & Rodríguez-Monroy, M. A. (2016). Antimicrobial activity of *Bursera morelensis* Ramírez essential oil. *African Journal of Traditional, Complementary and Alternative Medicines* 14 (3): 74-82.

[3] Diario Oficial de la Federación (2010). *Norma Oficial Mexicana NOM-059-SEMARNAT-2010, Protección Ambiental — Especies Nativas de México de Flora y Fauna Silvestres — Categorías de Riesgo y Especificaciones para su Inclusión, Exclusión o Cambio — Lista de Especies en Riesgo*. Secretaría de Medio Ambiente y Recursos Naturales, Mexico, 30 December 2010.

[4] IUCN (2017). *The IUCN Red List of Threatened Species.* Version 2017-3. http://www.iucnredlist.org/

[5] Jiménez-Merino, F. A. (2012). *Herbolaria Mexicana, Biblioteca Básica de Agricultura*, 2nd edition. Colegio de Postgraduados, Universidad Autónoma de Chapingo, Mexico.

[6] Lira, R., Casas, A., Rosas-López, R., Paredes-Flores, M., Pérez-Negrón, E., Rangel-Landa, S., Solís, L., Torres, I. & Dávila, P. (2009). Traditional knowledge and useful plant richness in the Tehuacán-Cuicatlán Valley, Mexico. *Economic Botany* 63 (3): 271–287.

[7] Medina-Lemos, R. (2008). *Flora del Valle de Tehuacán-Cuicatlán, Fascículo 66, Burseraceae Kunth.* Instituto de Biología, UNAM, Mexico.

[8] Ramos-Ordoñez, M. F., del Coro Arizmendi, M. & Márquez-Guzmán, J. (2012). The fruit of *Bursera*: structure, maturation and parthenocarpy. *AoB Plants* 2012: pls027. doi: 10.1093/aobpla/pls027.

[9] Ramos-Ordoñez, M. F., Márquez-Guzman, J. & del Coro Arizmendi, M. (2008). Parthenocarpy and seed predation by insects in *Bursera morelensis*. *Annals of Botany* 102 (5): 713–722.

[10] Rivera-Yañez, C. R., Terrasas, L. I., Jimenez-Estrada, M., Campos, J. E., Flores-Ortiz, C. M., Hernández, L. B., Cruz-Sánchez, T., Garrido-Fariña, G. I., Rodríguez Monroy, M. A. & Canales-Martínez, M. M. (2017). Anti-candida activity of *Bursera morelensis* Ramirez essential oil and two compounds, α-pinene and γ-terpinene — an *in vitro* study. *Molecules*. 22: 1-13.

[11] Rosas-López, R. (2003). *Estudio Etnobotánico de San Rafael-Coxcatlán*. Bachelor's thesis, FESI-UNAM, Mexico.

[12] Royal Botanic Gardens, Kew (2017). *Seed Information Database (SID).* Version 7.1. http://data.kew.org/sid/

Castela tortuosa Liebm.

TAXONOMY AND NOMENCLATURE

FAMILY. Simaroubaceae

SYNONYMS. *Castela erecta* subsp. *texana* (Torr. & A.Gray) Cronquist

VERNACULAR / COMMON NAMES. Chaparro amargoso, venenillo (Spanish).

PLANT DESCRIPTION

LIFE FORM. Shrub.

PLANT. Dioecious, spiny, 1–2 m tall, bark bitter, branchlets light coloured, ending in stout spines bearing lateral spines. **LEAVES.** Alternate, 0.8–1.5 cm long, with a pointed or rounded tip; margins smooth, down-turned. **FLOWERS.** Axillary or terminal inflorescences with small red to orange or purplish flowers (2.5 mm long) [9], bearing a cup-like nectar disk.

Figure 1 – Branchlet with fruits. (Photo: O. Téllez-Valdés, FESI-UNAM.)

FRUITS AND SEEDS

Fruits are fleshy, shining red, roughly spherical, and slightly flattened, c. 1 cm long (Figure 1), with one seed per fruit. Seeds are enclosed in a woody endocarp (Figure 2).

FLOWERING AND FRUITING. Flowers and fruits throughout the year. Two main fruiting periods: February to June, and from September to November. Fruits may remain on the plants for several years [7].

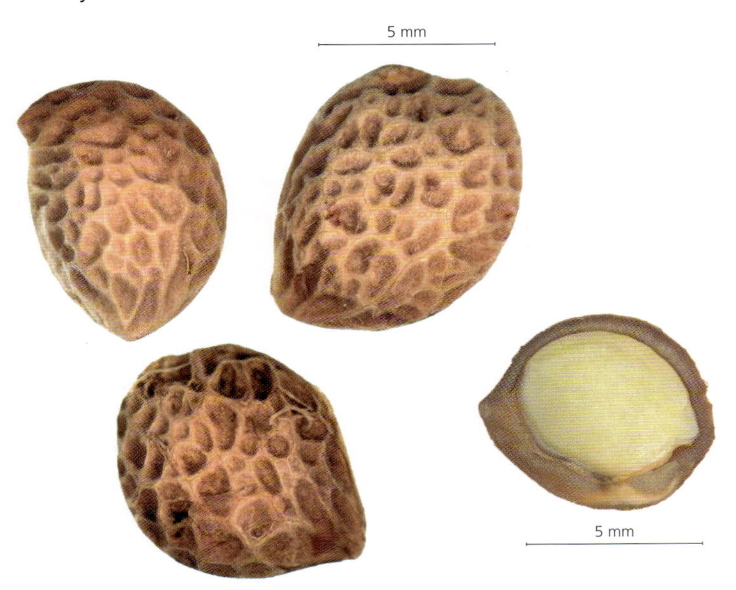

Figure 2 – External views (left) and longitudinal section (right) of the seeds.
(Photos: P. Gómez-Barreiro, RBG Kew.)

DISTRIBUTION

From Mexico (Chihuahua, Durango, Nuevo León, Oaxaca, Puebla, San Luis Potosí and Tamaulipas) and Texas to Colombia and Venezuela.

HABITAT

Xerophilous scrub and deciduous tropical woodland, mainly on limestone rocky substrates, at 700–2,100 m a.s.l. In San Rafael, it is found in gorges (barrancas) and hills (cerros).

USES

In San Rafael, aerial parts of the plant are drunk as a cold tea to treat diabetes, anger and blood pressure problems [1,2]. The plant is also used traditionally to treat amoebic dysentery, diarrhoea, trichomoniasis and diabetes [5].

KNOWN HAZARDS AND SAFETY

No known hazards.

CONSERVATION STATUS

Not considered to be either rare or threatened in Mexico [3], but the status of this species has yet to be confirmed in the IUCN Red List [4].

SEED CONSERVATION

HARVESTING. Fruits are available year-round, so seed harvesting can be carried out throughout the year.

PROCESSING AND HANDLING. Seeds can be separated from the fleshy fruits by rubbing through sieves with tepid water. Cleaned seeds should then be spread out and left to dry at room temperature.

STORAGE AND VIABILITY. X-ray analysis carried out on seed lots stored at the MSB highlighted 90–100% filled seeds. Seeds of this species are desiccation tolerant [8].

PROPAGATION

SEEDS

Dormancy and pre-treatments: Owing to their seed coat hardness, seeds should be cleaned with commercial sodium hypochlorite for 20 minutes, and then immersed in concentrated sulphuric acid for 20 minutes (stirring every five minutes). The seeds must then be washed thoroughly with water, and left in water overnight for imbibition before sowing [6].

Germination, sowing and planting: Once imbibed, the seeds can be placed on wet paper or sown in a mixture of equal parts by volume of soil and expanded perlite [6,10]. Germination rates of 40–50% have been achieved for cleaned seeds pre-treated as indicated above and incubated at 30°C, with the best results obtained with seeds stored for five years [6]. Similar germination rates were obtained without any pre-treatments (60% at 20°C), whereas 100% germination was achieved by adding 250 mg/l of gibberellic acid (GA_3) to the germination substrate [8]. Physical dormancy can be discounted because of the relatively high germination percentages achieved without scarification. Further studies are needed to confirm the presence of a physiological component of seed dormancy. Seeds are best sown individually in pots in May or June. Young plants do not require daily watering because their growth rate is slow. They do best if transplanted into a position that receives direct sunlight. In general, they can tolerate frost, although young plants may be affected by cold and wind. In windy areas, plants should be staked. The addition of liquid or solid fertiliser is beneficial for growth and development.

VEGETATIVE PROPAGATION

No protocols available.

TRADE

No data available.

Authors

Amanda Moreno Rodríguez, Efisio Mattana, César Flores, Guadalupe Zavala, Tiziana Ulian and Paulina Hechenleitner.

References

[1] Canales, M. (2005). *Base Fitoquímica del uso Tradicional de Plantas para el Tratamiento de Enfermedades de Posible Origen*. Bachelor's thesis, FESI-UNAM, Mexico.

[2] Canales Martínez, M., Hernández Delgado, T., Caballero Nieto, J., Romo de Vivar, A., Durán Díaz, A. & Lira Saade, R. (2006). Análisis cuantitativo del conocimiento tradicional de las plantas medicinales en San Rafael, Coxcatlán, Valle de Tehuacán-Cuicatlán, Puebla, México. *Acta Botanica Mexicana* 75: 21–43.

[3] Diario Oficial de la Federación (2010). *Norma Oficial Mexicana NOM-059-SEMARNAT-2010, Protección Ambiental — Especies Nativas de México de Flora y Fauna Silvestres — Categorías de Riesgo y Especificaciones para su Inclusión, Exclusión o Cambio — Lista de Especies en Riesgo*. Secretaría de Medio Ambiente y Recursos Naturales, Mexico, 30 December 2010.

[4] IUCN (2017). *The IUCN Red List of Threatened Species.* Version 2017-3. http://www.iucnredlist.org/

[5] Mendoza Orozco, M. & Godínez Álvarez, H. (2007). El chaparro amargoso, ¿atrapado sin salida? *Ciencias* 86: 34–36.

[6] Moreno Rodríguez, A. (2009). *Perfiles de Acumulación de Metabolitos Secundarios en 3 Especies Medicinales de San Rafael, Coxcatalán, Puebla, Bajo Diferentes Practicas de Cultivo*. Master's thesis, FESI-UNAM, Mexico.

[7] Pavón, N. P. & Briones, O. (2001). Phenological patterns of nine perennial plants in an intertropical semi-arid Mexican scrub. *Journal of Arid Environments* 49 (2): 265–277.

[8] Royal Botanic Gardens, Kew (2017). *Seed Information Database (SID).* Version 7.1. http://data.kew.org/sid/

[9] Standley, P. C. (1979). Trees and shrubs of Mexico. *Contributions from the United States National Herbarium* 23: 1–1721.

[10] Zavala-Hernández, M. G. (2013). *Perfiles de acumulación de metabolitos secundarios en* Castela tortuosa *Liebm. en diferentes prácticas de cultivo*. Biology thesis, FESI-UNAM, Mexico.

Ceiba aesculifolia subsp. *parvifolia*
(Rose) P.E.Gibbs & Semir

TAXONOMY AND NOMENCLATURE

FAMILY. Malvaceae

SYNONYMS. *Ceiba parvifolia* Rose

VERNACULAR / COMMON NAMES. Pochote (Spanish modified indigenous name).

PLANT DESCRIPTION

LIFE FORM. Tree.

PLANT. Deciduous, 8–10 m tall; trunk aculeate, bark covered with corky protuberances (Figure 1). **LEAVES.** Compound, palmate, leaflets 5–7, ovate, rounded, 2–4 x 1.3–1.8 cm, petioles 2.5–3.5 cm long. **FLOWERS.** Succulent, 13 cm long, yellowish white, petals coiled, stamens long [7].

Figure 1 – Bark covered by corky protuberances. (Photo: O. Téllez-Valdés, FESI-UNAM.)

FRUITS AND SEEDS

Fruits are capsules 8 cm long, ellipsoid, with a white, fibrose mass developed from the endocarp (Figure 2). Seeds are 5–10 mm long, reniform or pyriform with a blackish brown and smooth testa (Figure 3) [4].

FLOWERING AND FRUITING. Flowers from December to January [4]; fruits until March.

Figure 2 – Fruits of *C. aesculifolia* subsp. *parvifolia*. (Photo: O. Téllez-Valdés, FESI-UNAM.)

Figure 3 – External views (above) and longitudinal section (below) of the seeds, showing a folded embryo. (Photos: P. Gómez-Barreiro, RBG Kew.)

DISTRIBUTION

Mexico (Guerrero, Morelos, Puebla, Tabasco, Yucatán, Chiapas and Oaxaca) and Guatemala [7].

HABITAT

Dry deciduous forests and dry valleys in arid and semi-arid regions [4].

USES

In San Rafael, Coxcatlán (Puebla), the bark is used to prepare tea, to treat diabetes, kidney pain, tumours and gastritis and to heal wounds [7]. Phenolic compounds that are present in the bark have antibacterial, antifungal and antioxidant activities that are probably responsible for the medicinal properties effective against kidney maladies and tumours, and for the wound-healing capacity [6]. The bark is carved to make small figurines. In pre-Hispanic times, the fruit fibre was used to make cloth for princesses and kings, and as filling for cushions, pillows, mattresses and soft furnishings. Cushions made from the fibre of this species are popular with those suffering from asthma or allergies to wool and feathers [7]. Flowers and immature seeds are edible, and provide oil that can be used to make soap [1].

KNOWN HAZARDS AND SAFETY

The trunks and branches are spiny.

CONSERVATION STATUS

Not considered to be either rare or threatened in Mexico [3], but the status of this species has yet to be confirmed in the IUCN Red List [5].

SEED CONSERVATION

HARVESTING. Fruits can be collected using long poles.
PROCESSING AND HANDLING. Fruits should be stored in paper or cotton bags and taken to a dark, dry place where the seeds can be extracted from the fruits by hand.
STORAGE AND VIABILITY. X-ray analysis carried out on seed lots stored at the MSB highlighted c. 85% filled seeds. Seeds of this species are reported to be desiccation tolerant [8].

PROPAGATION

SEEDS

Dormancy and pre-treatments: High germination percentages have been achieved without any pre-treatments [8]. However, further studies are needed to determine the type of seed dormancy. Before sowing at the San Rafael, Coxcatlán nursery, seeds were soaked in water for 24 hours.
Germination, sowing and planting: Of seeds sown in germination trays at the San Rafael, Coxcatlán nursery, 50% germinated within one week. A 3:1 mixture by volume of sieved, sterilised leaf litter and agrolite was used. Seedlings were transplanted into black plastic bags 3–4 weeks after germination, using equal parts by volume of sieved leaf litter and agrolite. Seedlings were watered by hand every five days. Of 400 germinated seeds, 77 seedlings were established. Seed lots stored at the MSB tested under laboratory controlled conditions germinated to high percentages (>95%) when incubated at constant temperatures (15, 20 and 25°C) with 8 hours of light per day [8].

VEGETATIVE PROPAGATION

No protocols are available. Experiments using cuttings were carried out at the San Rafael, Coxcatlán nursery. Of 83 cuttings placed in pots containing equal parts of agrolite and leaf litter, only five became established.

TRADE

Fruits and seeds are sold and traded in local markets [1,2].

Authors

Lirio Jazmín Sánchez Hernández, Myrna Mendoza Cruz, Martin López Carrera, Efisio Mattana, Steve Davis and Paulina Hechenleitner.

References

[1] Avendaño-Goméz, A., Lira-Saade, R., Dávila-Aranda, P., Casas-Fernández, A. & De La Torre-Almaraz, R. (2007). Caracterización de un hongo asociado a la planta macho del pochote (*Ceiba aesculifolia* (H. B. & K.) Britten & Baker f. subsp. *parvifolia* (Rose) P.E.Gibbs & Semir) en Tehuacán-Cuicatlán, México. *Agrociencia* 42 (2): 205–215.

[2] Cervantes, H. (2014). Personal communication.

[3] Diario Oficial de la Federación (2010). *Norma Oficial Mexicana NOM-059-SEMARNAT-2010, Protección Ambiental — Especies Nativas de México de Flora y Fauna Silvestres — Categorías de Riesgo y Especificaciones para su Inclusión, Exclusión o Cambio — Lista de Especies en Riesgo*. Secretaría de Medio Ambiente y Recursos Naturales, Mexico, 30 December 2010.

[4] Gibbs, P. & Semir, J. (2004). A taxonomic revision of the genus *Ceiba* Mill. (Bombacaceae). *Anales del Jardín Botánico de Madrid* 60 (2): 259–300.

[5] IUCN (2017). *The IUCN Red List of Threatened Species.* Version 2017-3. http://www.iucnredlist.org/

[6] Orozco, J., Rodriguez-Monroy, M. A., Martínez, K. E., Flores, C. M., Jiménez-Estrada, M., Durán, A., Rosas López, R., Hernández, L. B. & Canales, M. (2013). Evaluation of some medicinal properties of *Ceiba aesculifolia* subsp. *parvifolia*. *Journal of Medicinal Plants Research* 7 (7): 309–314.

[7] Orozco, M. J. (2010). *Estudio Comparativo de Algunas Actividades Biológicas de las Cortezas de* Ceiba aesculifolia *subsp.* parvifolia, Juliana adstringens *y* Cyrtocarpa procera *de San Rafael, Coxcatlán, Puebla*. Master's thesis, FESI-UNAM, Mexico.

[8] Royal Botanic Gardens, Kew (2017). *Seed Information Database (SID).* Version 7.1. http://data.kew.org/sid/

Cyrtocarpa procera Kunth

TAXONOMY AND NOMENCLATURE

FAMILY. Anacardiaceae

SYNONYMS. *Dasycarya grisea* Liebm.

VERNACULAR / COMMON NAMES. Chupandílla (indigenous name); chupandia (Spanish).

PLANT DESCRIPTION

LIFE FORM. Tree.

PLANT. Deciduous, 4–12 m tall, with a wide canopy; trunk <50 cm in diameter, branching 1 m above ground; outer bark greyish, inner bark reddish. **LEAVES.** Compound, with 9–13 paired leaflets, rounded, apex acute, tomentose (Figure 1). **FLOWERS.** White, grouped in panicles c. 6 cm long; petals 3 mm long [6,7].

Figure 1 – *C. procera* showing fruits and compound leaves. (Photo: O. Téllez-Valdés, FESI-UNAM.)

FRUITS AND SEEDS

Fruits 1.5–2.5 cm long, oblong, pubescent, juicy, yellow green (Figure 1) [6]. Carpels 5, with up to 3 filled seeds enclosed in a woody endocarp (Figure 2).

FLOWERING AND FRUITING. Flowers from March to May [3]; fruits from May to June [6].

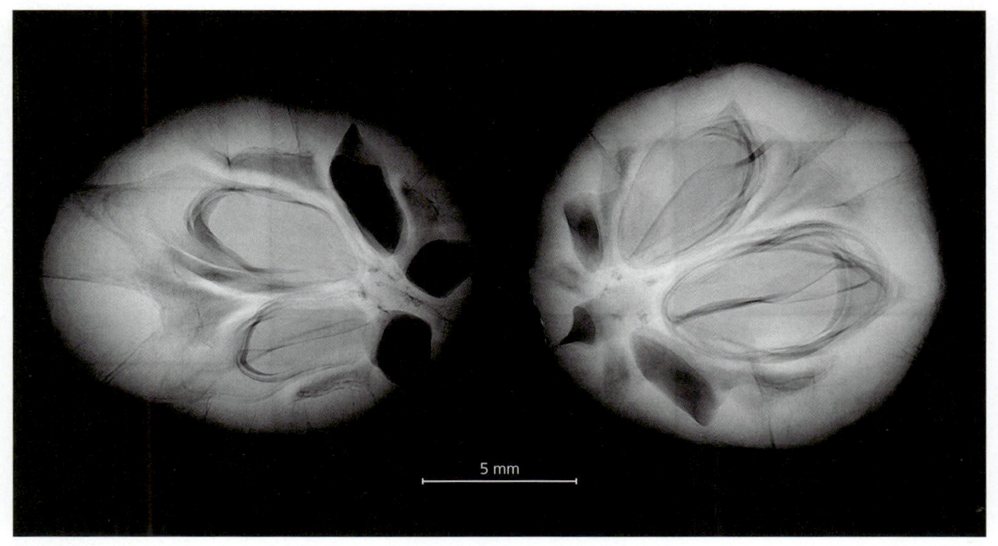

Figure 2 – Above: External views (left) and sections showing the seeds (right) of the woody endocarp. Below: X-ray image showing filled seeds and empty carpels. (Photos: P. Gómez-Barreiro, RBG Kew.)

DISTRIBUTION

Endemic to Mexico (Colimo, Durango, Guerrero, Jalisco, Estado de México, Michoacán, Morelos, Nayarit, Oaxaca and Puebla) [3,6,7,8].

HABITAT

Tropical deciduous dry forests, spiny forests, mesophilous mountain forests, oak and pine forests, from warm to temperate and semi-dry climates [7], at 500–1,350 m a.s.l. [6].

USES

The fruits are edible, and the leaves, seeds and fruits are used as animal food [7]. The bark and fruit are used in traditional medicine to treat diarrhoea and dysentery, and as a laxative [6,9]. Laboratory analysis of the fruit has indicated antibacterial, antioxidant and anti-inflammatory activities [5]. The wood is used in construction, to make fences and handicrafts, and as firewood. The bark is used to produce soap [7].

KNOWN HAZARDS AND SAFETY

No known hazards.

CONSERVATION STATUS

Not considered to be either rare or threatened in Mexico [2], but the status of this species has yet to be confirmed in the IUCN Red List [4].

SEED CONSERVATION

HARVESTING. Fruits can be collected from May to June [6].
PROCESSING AND HANDLING. Fruits should be stored in paper or cotton bags and kept in a dark and dry place until they are delivered to the seed bank, where the stony endocarp should be removed from the fruit pulp.
STORAGE AND VIABILITY. X-ray analysis carried out on seed lots stored at the MSB highlighted a high percentage of filled seeds (>90%). No information is available on the seed desiccation tolerance of this and related species.

PROPAGATION

SEEDS

Dormancy and pre-treatments: No information is available on seed dormancy breaking and germination requirements.
Germination, sowing and planting: 100% of seeds germinated in 20 days at the FESI-UNAM Seed Bank when incubated at 30°C, with a photoperiod of 12 hours of darkness and 12 hours of light.

VEGETATIVE PROPAGATION

No protocols available.

TRADE

Fruits of *C. procera* are commercially traded [1].

Authors

Lirio Jazmín Sánchez Hernández, Efisio Mattana and Paulina Hechenleitner.

References

[1] Cervantes, H. (2014). Personal communication.

[2] Diario Oficial de la Federación (2010). *Norma Oficial Mexicana NOM-059-SEMARNAT-2010, Protección Ambiental — Especies Nativas de México de Flora y Fauna Silvestres — Categorías de Riesgo y Especificaciones para su Inclusión, Exclusión o Cambio — Lista de Especies en Riesgo*. Secretaría de Medio Ambiente y Recursos Naturales, Mexico, 30 December 2010.

[3] Fonseca, R. M. (2005). Una nueva especie de *Cyrtocarpa* (Anacardiaceae) de México. *Acta Botánica Mexicana* 71: 45–52.

[4] IUCN (2017). *The IUCN Red List of Threatened Species.* Version 2017-3. http://www.iucnredlist.org/

[5] Martinez-Elizalde, K. S., Jiminez-Estrada, M., Mateo Flores, C., Barbo Hernandez, L., Rosas-Lopez, R., Duran-Diaz, A., Nieto-Yañez, O. J., Barbosa, E., Rodriguez-Monroy, M. A. & Canales-Martinez, M. (2015). Evaluation of the medicinal properties of *Cyrtocarpa procera* Kunth fruit extracts. *BMC Complementary and Alternative Medicine* 15: 74.

[6] Medina-Lemos, R. & Fonseca, R. M. (2009). *Flora del Valle Tehuacán-Cuicatlán, Fascículo 71, Anacardiaceae Lindl.* Instituto de Biología, UNAM, Mexico.

[7] Standley, P. C. (1979). Trees and shrubs of Mexico. *Contributions from the United States National Herbarium* 23: 1–1721.

[8] Tropicos.org (2017). *Cyrtocarpa procera* Kunth. In: *Tropicos*. Missouri Botanical Garden, Saint Louis, Missouri, USA. http://www.tropicos.org/Name/50145943

[9] UNAM (2009). *Biblioteca Digital de la Medicina Tradicional Mexicana*. UNAM, Mexico. http://www.medicinatradicionalmexicana.unam.mx/index.php

Escontria chiotilla (F.A.C.Weber ex K.Schum.) Rose

TAXONOMY AND NOMENCLATURE

FAMILY. Cactaceae

SYNONYMS. *Cereus chiotilla* F.A.C.Weber ex K.Schum., *Myrtillocactus chiotilla* (F.A.C.Weber) P.V.Heath

VERNACULAR / COMMON NAMES. Chiotilla, geotilla, jiotilla, quiotilla (Spanish modified indigenous names).

PLANT DESCRIPTION

LIFE FORM. Cactus.

PLANT. Arborescent, <7 m tall; main trunk short, branches <2 m long and 8–20 cm wide (Figure 1), bright green, with 7–8 ribs with acute margins. **LEAVES.** Reduced to spines; central spines 1(–5), 1–7 cm long, unequal, one longer, straight, subulate greyish brown; radial spines 10–20, 0.5–1 cm long, subulate, straight, grey (Figure 2). **FLOWERS.** Infundibuliform, yellow, 3–4.5 cm long, c. 2 cm wide, at the apex of plant, bracts chartaceous (Figure 2) [1,3].

Figure 1 – Habit of *E. chiotilla*.
(Photo: D. Franco-Estrada, FESI-UNAM.)

FRUITS AND SEEDS

Fruits purple, globose, <5 cm in diameter [1], scaly (Figure 2), with c. 400 seeds per fruit [12]. Seeds black, 1.3–1.7 mm long, 0.8–1.3 mm wide, shiny, wrinkled, with a keel on the dorsal region (Figure 3) [9].

FLOWERING AND FRUITING. Flowers in two different seasons: between March and May, and between July and August. Fruits from April to May, and from September to November [3].

Figure 2 – Scaly fruits. (Photo: H. Cervantes-Maya, FESI-UNAM.)

Figure 3 – External views (top) and longitudinal section (bottom) of the seeds. (Photos: P. Gómez-Barreiro, RBG Kew.)

DISTRIBUTION

Endemic to Mexico (Guerrero, Michoacán, Oaxaca and Puebla) [3].

HABITAT

Deciduous dry forests and xerophilous shrub, often in large associations known as 'quiotillales', at 600–2,000 m a.s.l. [3].

USES

The fruits are harvested from both wild and silviculturally managed populations [2]. The fruits are used to produce drinks and alcoholic beverages, preserves, and eaten fresh when in season. They are also used to produce jam and fruit concentrates, and as sweetener for ice cream and sherbet. The seeds, stems and flowers are also edible [4]. *Escontria chiotilla* is also used for forage and firewood, and is planted as a live fence [6].

KNOWN HAZARDS AND SAFETY

Care should be taken when handling the plant as the spines are sharp.

CONSERVATION STATUS

Assessed as Least Concern (LC) according to IUCN Red List criteria [10]. *Escontria chiotilla* is not included in the national list of protected species in Mexico [8].

SEED CONSERVATION

HARVESTING. Fruits are harvested using a long pole with scissors. Harvesting takes place from April to May, and from September to November.

PROCESSING AND HANDLING. Harvested ripe fruits should be kept in paper or cotton bags until they are delivered to the seed bank, where they can be cut open with a knife and the seeds extracted from the pulp. Seeds should then be washed with tap water to remove the pulp, and dried at room temperature. The seeds can then be passed through a sieve or cleaned with pressurised air to remove any remaining dry pulp.

STORAGE AND VIABILITY. X-ray analysis carried out on seed lots stored at the MSB revealed high percentages of filled seeds (>80%). No data are available on the seed desiccation tolerance of this or related species.

PROPAGATION

SEEDS

Dormancy and pre-treatments: Seeds of this species are reported to reach high germination percentages at constant temperatures (20–30°C), without any pre-treatments [13]. However, at the nursery in San Rafael, Coxcatlán, seeds were surface sterilised, using commercial sodium hypochlorite at 2% in water for 7 minutes, then washed and left to dry for 12–24 hours prior to sowing. Further studies are needed to fully understand the seed dormancy and germination requirements.

Germination, sowing and planting: Under laboratory controlled conditions, 100% of the seeds germinated in 20 days at the FESI-UNAM Seed Bank when incubated at 30°C, with a photoperiod of 12 hours of light per day. At the San Rafael, Coxcatlán nursery, seeds were sown in germination trays using a 3:1 mixture by volume of agrolite and sieved leaf litter. The leaf litter was sterilised by exposing it to sunlight for 1–2 days. The seeds were scattered evenly on the watered substrate and then covered by a fine layer of substrate. The seedbed was then covered with a sealed dome to reduce water evaporation. The seedlings were transplanted into plastic bags or pots after 60–90 days, when they were 1–2 cm tall, using a mixture of equal parts by volume of sterilised sieved leaf litter and fine gravel (e.g. 'tepojal', 'tepezil' or 'tezontle'). A layer of coarse gravel was placed in the bottom of containers to ensure good drainage. Two transplants were made to plastic pots at three months and six months after germination. Seedlings were watered every five days, then once every 15 days after three months. Hand weeding was carried out frequently. The plants were ready to be planted in the field after 6–12 months. Out of 1,000 seeds sown, 300 plants were established.

VEGETATIVE PROPAGATION

Stem cuttings are difficult to root [5,11]. No protocols are available.

TRADE

Fruits are sold in small markets in Zapotitlán Salinas, Puebla [3] and in Oaxaca [1,2]. International trade of plant material is regulated by CITES (listed on Appendix II) [7].

Authors

Daniel Franco-Estrada, Myrna Mendoza Cruz, Martin López Carrera, Efisio Mattana, Steve Davis and Paulina Hechenleitner.

References

[1] Anderson, E. F. (2001). *The Cactus Family*. Timber Press, Portland, Oregon, USA.

[2] Arellano, E. & Casas, A. (2003). Morphological variation and domestication of *Escontria chiotilla* (Cactaceae) under silvicultural management in the Tehuacán Valley, central Mexico. *Genetic Resources and Crop Evolution* 50: 439–543.

[3] Arias, S., Gama-López, S., Guzmán-Cruz, L. U. & Vázquez-Benítez, B. (2012). *Flora del Valle de Tehuacán-Cuicatlán, Fascículo 95, Cactaceae Juss.*, 2nd edition. Instituto de Biología, UNAM & CONABIO, Mexico.

[4] Arias Toledo, A. A., Valverde Valdés, M. T. & Reyes Santiago, J. (2000). *Las Plantas de la Región de Zapotitlán Salinas, Puebla*. Instituto Nacional de Ecología, UNAM, Mexico.

[5] Borg, J. (1959). *Cacti: A Gardener's Handbook for Their Identification and Cultivation*, 3rd edition. Blandford Press, London.

[6] Casas, A. (2002). Uso y manejo de cactáceas columnares mesoamericanas. *CONABIO Biodiversitas* 40: 18–23.

[7] CITES (2017). *Convention on International Trade in Endangered Species of Wild Fauna and Flora*. https://cites.org/eng/app/index.php

[8] Diario Oficial de la Federación (2010). *Norma Oficial Mexicana NOM-059-SEMARNAT-2010, Protección Ambiental — Especies Nativas de México de Flora y Fauna Silvestres — Categorías de Riesgo y Especificaciones para su Inclusión, Exclusión o Cambio — Lista de Especies en Riesgo*. Secretaría de Medio Ambiente y Recursos Naturales, Mexico, 30 December 2010.

[9] Franco, E. D. (2013). *Catálogo Ilustrado de Cactáceas Columnares del Valle de Tehuacán-Cuicatlán*. Bachelor's thesis, FESI-UNAM, Mexico.

[10] IUCN (2017). *The IUCN Red List of Threatened Species*. Version 2017-3. http://www.iucnredlist.org/

[11] López-Gómez, R., Díaz-Pérez, J. C. & Flores-Martínez, G. (2000). Vegetative propagation of three species of cacti: pitaya (*Stenocereus griseus*), tunillo (*Stenocereus stellatus*) and jiotilla (*Escontria chiotilla*). *Agrociencia* 34 (3): 363–367.

[12] Loza-Cornejo, S., López-Mata, L. & Terrazas, T. (2008). Morphological seed traits and germination of six species of Pachycereeae. *Journal of the Professional Association for Cactus Development* 10: 71–84.

[13] Seal, C. E., Flores, J., Ceroni Stuva, A., Dávila Aranda, P., León-Lobos, P., Ortega-Baes, P., Galíndez, G., Aparicio-González, M. A., Castro Cepero, V., Daws, M. I., Eason, M., Flores Ortiz, C. M., del Fueyo, P. A., Olwell, P., Ordoñez, C., Peñalosa Castro, I., Quintanar Zúñiga, R., Ramírez Bullón, N., Rojas-Aréchiga, M., Rosas, M., Sandoval, A., Stuppy, W., Ulian, T., Vázquez Medrano, J., Walter, H., Way, M. & Pritchard, H. W. (2009). *The Cactus Seed Biology Database*, Release 1. Royal Botanic Gardens, Kew.

Gymnolaena oaxacana (Greenm.) Rydb.

TAXONOMY AND NOMENCLATURE

FAMILY. Compositae

SYNONYMS. *Dyssodia oaxacana* Greenm.

VERNACULAR / COMMON NAMES. Cempazúchil de monte, molito, zempoalxóchitl chiquito (Spanish modified indigenous names).

Figure 1 – Inflorescence of *G. oaxacana*. (Photo: O. Téllez-Valdés, FESI-UNAM.)

PLANT DESCRIPTION

LIFE FORM. Shrub.

PLANT. Up to 1.5 m tall, stems terete, striated, puberulent. **LEAVES.** Mainly opposite, 3–10 cm long, 0.6–2 mm wide, ovate to lanceolate, glabrous to sparsely puberulent, with numerous oil glands; leaf base rounded, apex apiculate, margins serrate to dentate. **FLOWERS.** Solitary capitula, 3–8 cm long, peduncles bracteate; bracts 1–7, lanceolate, larger near the head; ligulate flowers in general 5–8, orange, 0.6–1 cm long, 3–5 mm wide, linear to oblong; disc flowers 15–40, yellow, ligule 0.9–1.2 cm long (Figure 1) [8].

FRUITS AND SEEDS

Cypselas 4–6 mm long, pappus sericeous with c. 20 squamulae, divided in unequal 5–9 bristles (Figure 2) [8].

FLOWERING AND FRUITING. Flowers from June to December; fruits from July to January [8].

Figure 2 – External views of the cypselas and longitudinal section. (Photos: P. Gómez-Barreiro, RBG Kew.)

DISTRIBUTION

Mexico, endemic to the Valley of Tehuacán-Cuicatlán [2,8].

HABITAT

Deciduous dry forests and xerophilous scrub, on limestone slopes, marl and shale, at altitudes of 760–1,900 m a.s.l. [8].

USES

The aerial parts of the plant are boiled and orally administered to treat diarrhoea [4]. The plant is also used as animal fodder [1,6].

KNOWN HAZARDS AND SAFETY

No known hazards.

CONSERVATION STATUS

Not considered to be either rare or threatened [3], but the status of this species has yet to be confirmed in the IUCN Red List [5].

SEED CONSERVATION

HARVESTING. Ripe cypselas can be collected from the plants in November.
PROCESSING AND HANDLING. Fruits should be stored in paper bags until they are delivered to the seed bank, where they can be separated by hand from other plant material.
STORAGE AND VIABILITY. Low percentages of filled seeds have been detected by X-ray analysis of the seed lots stored at the MSB, with percentages of empty seeds reaching 70%. Seeds of this species are reported to be desiccation tolerant [7].

PROPAGATION

SEEDS

Dormancy and pre-treatments: High germination percentages have been achieved without any pre-treatments [7]. However, further studies are needed to confirm the lack of any class of seed dormancy.
Germination, sowing and planting: The FESI-UNAM Seed Bank achieved a 40% germination rate when seeds were incubated at 30°C, with a photoperiod of 12 hours of darkness and 12 hours of light for 20 days. 100% germination was achieved at the MSB on banked seed lots at constant temperatures (15–25°C) with a photoperiod of 8 hours of light per day [7].

VEGETATIVE PROPAGATION

No protocols available.

TRADE

The plant is collected from ravines near human settlements and used when needed.

Authors

Fernando Peralta-Romero, Efisio Mattana and Paulina Hechenleitner.

References

[1] Casas, A., Valiente-Banuet, A., Viveros, J. L., Caballero, J., Cortés, L., Dávila, P., Lira, R. & Rodríguez, I. (2001). Plant resources of the Tehuacán-Cuicatlán Valley, Mexico. *Economic Botany* 55 (1): 129–166.

[2] Comisión Nacional para el Conocimiento & Uso de la Biodiversidad (CONABIO) (2011). *La Biodiversidad en Puebla: Estudio de Estado, Mexico.* CONABIO, Gobierno del Estado de Puebla, Benemérita Universidad Autónoma de Puebla, Mexico.

[3] Diario Oficial de la Federación (2010). *Norma Oficial Mexicana NOM-059-SEMARNAT-2010, Protección Ambiental — Especies Nativas de México de Flora y Fauna Silvestres — Categorías de Riesgo y Especificaciones para su Inclusión, Exclusión o Cambio — Lista de Especies en Riesgo.* Secretaría de Medio Ambiente y Recursos Naturales, Mexico, 30 December 2010.

[4] Hernández, T., Canales, M., Avila, J. G., Durán, A., Caballero, J., Romo de Vivar, A. & Lira, R. (2003). Ethnobotany and antibacterial activity of some plants used in traditional medicine of Zapotitlán de las Salinas, Puebla (México). *Journal of Ethnopharmacology* 88 (2–3): 181–188.

[5] IUCN (2017). *The IUCN Red List of Threatened Species.* Version 2017-3. http://www.iucnredlist.org/

[6] Rosas-López, R. (2003). *Estudio Etnobotánico de San Rafael-Coxcatlán.* Bachelor's thesis. FESI-UNAM, Mexico.

[7] Royal Botanic Gardens, Kew (2017). *Seed Information Database (SID).* Version 7.1. http://data.kew.org/sid/

[8] Villarreal-Quintanilla, J. A., Villaseñor-Ríos, J. L. & Medina-Lemos, R. (2008). *Flora del Valle de Tehuacán-Cuicatlán, Fascículo 62, Asteraceae Bercht. & J.Presl.* Instituto de Biología, UNAM, Mexico.

Gymnosperma glutinosum (Spreng.) Less.

TAXONOMY AND NOMENCLATURE

FAMILY. Compositae

SYNONYMS. *Baccharis fasciculosa* Klatt, *B. glutinosa* (Spreng.) Hook. & Arn., *Gymnosperma corymbosum* DC., *G. multiflorum* DC., *G. scoparium* DC., *Selloa glutinosa* Spreng., *S. multiflora* Kuntze, *S. scoparia* Kuntze, *Xanthocephalum glutinosum* (Spreng.) Shinners

VERNACULAR / COMMON NAMES. Gumhead (English); zotla (Nahuatl); popote, popotillo, tatalencho (Spanish).

PLANT DESCRIPTION

LIFE FORM. Perennial herb or subshrub.

PLANT. Erect, <1(–2) m tall, glabrous or almost glabrous, resinous and sticky. **LEAVES.** Alternate, linear-elliptic to narrowly lanceolate, <8.5 cm long, 1 cm wide, acute, entire, trinervate, densely dotted on both sides, sessile or almost so. **FLOWERS.** Inflorescence of numerous heads in dense terminal corymbiform groups, with round heads. Flowers bisexual; stamens alternate with corona lobes (Figure 1) [1,8].

Figure 1 – Inflorescences of *G. glutinosum*. (Photo: O. Téllez-Valdés, FESI-UNAM.)

FRUITS AND SEEDS

Fruits are oblong achenes, 1–1.5 mm long, with 4 to 5 ribs and tiny hairs (Figure 2).

FLOWERING AND FRUITING. Flowers mainly from June to November, but may be found in flower from April to December [5]; fruits mainly from August to November.

Figure 2 – External views of the seeds. (Photo: P. Gómez-Barreiro, RBG Kew.)

DISTRIBUTION

Southern United States to Guatemala [1]. In Mexico, it is found in Chihuahua, Coahuila, Nuevo León, Tamaulipas and San Luis Potosí [5].

HABITAT

Open and semi-desert areas in very diverse habitats: including dry, rocky or gravelly slopes, loamy flats, mountains and stream beds. Very common in agricultural areas, dry deciduous forests and xerophilous scrub (including creosote bush) [1,5], at 300–2,700 m a.s.l.

USES

In traditional medicine, a poultice made from the leaves and stems is used to treat rheumatism, sprains, fractures and inflammation [2]. The leaves are placed over the abdomen to assist in childbirth [10]. A tea is prepared from the flowers [2] and aerial parts as an analgesic and to treat diarrhoea [6,12].

KNOWN HAZARDS AND SAFETY

No known hazards.

CONSERVATION STATUS

A widespread species that is not considered to be either rare or threatened in Mexico [4], but its status has yet to be confirmed in the IUCN Red List [7].

SEED CONSERVATION

HARVESTING. Achenes can be harvested by hand.

PROCESSING AND HANDLING. The fruits can be passed through a sieve with a mesh size of 300 µm to obtain the seeds.

STORAGE AND VIABILITY. X-ray analysis carried out on seed lots stored at the MSB showed that the percentage of filled seeds varies consistently between c. 65% and 90% as a result of some seed lots being infested. Seeds of this species are reported to be desiccation tolerant [11]. High germination rates (70–85%) have been reported for both freshly collected seeds and those stored at –20°C [9].

PROPAGATION

SEEDS

Dormancy and pre-treatments: Seeds can be soaked in commercial 1% diluted chlorine for five minutes and then rinsed three times with tap water to remove the chlorine completely. The seeds are then soaked in water for a day before sowing [9].

Germination, sowing and planting: Germination percentages of c. 75% for seeds pre-treated as above and incubated at 30°C (maximum 33.1°C, minimum 27.4°C) have been reported [9]. However, higher germination percentages (86–96%) were achieved for untreated seeds stored at the MSB and incubated at lower temperatures (10, 15 and 25°C) [11], as also reported by Baskin *et al.* (1998) [3]. Seeds are non-dormant at maturity and germinate to 95–98% at simulated autumn temperatures [3]. Seedlings are transplanted from agar to the nursery to stimulate the development of the radicle. Medium-sized seedlings are planted in a 2:1 mixture by volume of soil and agrolite, enriched with Hoagland solution, and are then placed in well-lit positions with relatively high temperatures for at least a month. The timing of transplanting and aftercare are critical for successful propagation.

VEGETATIVE PROPAGATION

No protocols available.

TRADE

No data available.

Authors

Amanda Moreno Rodríguez, Efisio Mattana, Cesar Flores, Tiziana Ulian, Steve Davis and Paulina Hechenleitner.

References

[1] Alipi, A. M. & Mondragón Pichardo, J. (2009). *Gymnosperma glutinosum* (Spreng.) Less. In: *Malezas de Mexico*. Edited by H. Vibrans. CONABIO, Mexico. http://www.conabio.gob.mx/malezasdemexico/asteraceae/gymnosperma-glutinosum/fichas/ficha.htm

[2] Arias Toledo, A. A., Valverde Valdés, M. T. & Reyes Santiago, J. (2000). *Las Plantas de la Región de Zapotitlán Salinas, Puebla*. Instituto Nacional de Ecología, UNAM, Mexico.

[3] Baskin, C. C., Baskin, J. M. & Van Auken, O. W. (1998). Role of temperature in dormancy break and/or germination of autumn-maturing achenes of eight perennial Asteraceae from Texas, U.S.A. *Plant Species Biology* 13 (1): 13–20.

[4] Diario Oficial de la Federación (2010). *Norma Oficial Mexicana NOM-059-SEMARNAT-2010, Protección Ambiental — Especies Nativas de México de Flora y Fauna Silvestres — Categorías de Riesgo y Especificaciones para su Inclusión, Exclusión o Cambio — Lista de Especies en Riesgo*. Secretaría de Medio Ambiente y Recursos Naturales, Mexico, 30 December 2010.

[5] eFloras.org (2006). *Flora of North America*, Vol. 20: Asteraceae. www.efloras.org

[6] Hernández, T., Canales, M., Avila, J. G., Durán, A., Caballero, J., Romo de Vivar, A. & Lira, R. (2003). Ethnobotany and antibacterial activity of some plants used in traditional medicine of Zapotitlán de las Salinas, Puebla (México). *Journal of Ethnopharmacology* 88 (2–3): 181–188.

[7] IUCN (2017). *The IUCN Red List of Threatened Species*. Version 2017-3. http://www.iucnredlist.org/

[8] McVaugh, R. (1984). *Flora Novo-Galiciana: A Descriptive Account of the Vascular Plants of Western Mexico*, Vol. 12: Compositae. University of Michigan Press, Ann Arbor, USA.

[9] Moreno Rodríguez, A. (2009). *Perfiles de Acumulación de Metabolitos Secundarios en 3 Especies Medicinales de San Rafael, Coxcatlán, Puebla, Bajo Diferentes Prácticas de Cultivo*. Master's thesis, FESI-UNAM, Mexico.

[10] Olivas-Sánchez, M. P. (1999). *Plantas Medicinales del Estado de Chihuahua*. Centro de Estudios Biológicos, Universidad Autónoma de Juárez, Mexico.

[11] Royal Botanic Gardens, Kew (2017). *Seed Information Database (SID)*. Version 7.1. http://data.kew.org/sid/

[12] Serrano, R. (2004). *Estudio Comparativo de la Actividad Antibacteriana de* Gymnosperma glutinosum *(Spreng.) Less. de dos Localidades: San Rafael Coxcatlán, Puebla y Tepeji del Rio, Hidalgo*. Bachelor's thesis, FESI-UNAM, Mexico.

Hylocereus undatus (Haw.) Britton & Rose

TAXONOMY AND NOMENCLATURE

FAMILY. Cactaceae

SYNONYMS. *Cereus tricostatus* Rol.-Goss., *C. trigonus* var. *guatemalensis* Eichlam, *C. undatus* Pfeiff., *C. undulatus* D.Dietr., *Hylocereus guatemalensis* (Eichlam) Britton & Rose, *H. tricostatus* (Rol.-Goss.) Britton & Rose

VERNACULAR / COMMON NAMES. Dragon fruit, queen of the night (English); pitahaya, pitaya (indigenous names).

PLANT DESCRIPTION

LIFE FORM. Epiphytic cactus.

PLANT. Climbing, scrambling or shrubby; stems 3–6 cm wide, ribs 3, margins wavy, aerial roots present. **LEAVES.** Reduced to spines, 1–3(–5) per areole, spines 2–4 mm long, c. 1 mm wide at base, conical, straight to slightly curved, whitish to grey. **FLOWERS.** Spectacular, funnel-shaped, opening at night, white, 26–30(–35) cm long, 15–20 cm in diameter (Figure 1) [1,2].

Figure 1 – Flower of *H. undatus*. (Photo: O. Téllez-Valdés, FESI-UNAM.)

FRUITS AND SEEDS

Fruits are purple, with a white flesh, 10–12 x c. 7 cm (Figure 2). Seeds are black, c. 2.8 x c. 2.5 mm (Figure 3) [2].

FLOWERING AND FRUITING. Flowers between May and August; fruits from July to September [2].

Figure 2 – Longitudinal section of the fruit with seeds. (Photo: O. Téllez-Valdés, FESI-UNAM.)

Figure 3 – External views (left) and longitudinal section (right) of the seeds. (Photos: P. Gómez-Barreiro, RBG Kew.)

DISTRIBUTION

The native range of this species is not known. *Hylocereus undatus* has long been in cultivation; escapes from cultivation have become widely naturalised. It is found in USA and Mexico, through Central America to South America, and throughout the Caribbean region, from sea level to 2,750 m a.s.l. [1,2,8]. It is also naturalised in parts of tropical Asia, Australia, and on several Pacific Ocean islands, and can be invasive (e.g. in Florida and in parts of Australia and South Africa).

HABITAT

Deciduous dry forests, xerophilous scrub and cultivated in gardens [2].

USES

Hylocereus undatus is commercially grown for its edible fruits. In Mexico, the fruits are harvested from naturalised and cultivated plants. In traditional medicine, the flowers are used to prepare tea to treat cardiovascular conditions. The stems are used to treat common cold [2]. Widely grown by cactus enthusiasts and planted as an ornamental [1,2], *H. undatus* is also used as a rootstock for grafting other species of cacti.

KNOWN HAZARDS AND SAFETY

Care should be taken when handling the plant as the spines are sharp.

CONSERVATION STATUS

Widespread and assessed as Data Deficient (DD) according to IUCN Red List criteria because the native distribution is unknown [8]. *Hylocereus undatus* is not included in the national list of protected species in Mexico [6].

SEED CONSERVATION

HARVESTING. Fruits are harvested by hand using a long pole.
PROCESSING AND HANDLING. Harvested fruits should be kept in paper or cotton bags until they are delivered to the seed bank, where they can be cut open with a knife to extract the seeds from the pulp. The seeds should then be washed under tap water and dried at room temperature, after which they can be passed through a sieve or cleaned with pressurised air to separate the seeds from the dried pulp.
STORAGE AND VIABILITY. X-ray analysis carried out on seed lots stored at the MSB revealed a high percentage of filled seeds (c. 100%). Seeds of this species are reported to be desiccation tolerant [10].

PROPAGATION

SEEDS

Dormancy and pre-treatments: High germination percentages have been achieved on seeds stored at the MSB without any pre-treatments. Seeds of this species do not exhibit dormancy [4]. However, further studies are needed to confirm the lack of any class of seed dormancy. Prior to sowing at the San Rafael, Coxcatlán nursery, seeds were surface sterilised using 4% sodium hypochlorite for 30 minutes.
Germination, sowing and planting: Seed lots stored at the MSB germinated to c. 80–90% when incubated in the light (8 or 12 hours of light per day) at constant temperatures (21–25°C), whereas c. 70% germinated at higher temperatures (i.e. 30°C).

VEGETATIVE PROPAGATION

Readily establishes from stem cuttings. In studies carried out by FESI-UNAM staff, cuttings were treated with three different concentrations (10, 100 and 1,000 ppm) of indole-3-acetyl-*p*-nitrophenyl ester (IAP) and planted in plastic bags. Rooting success and the number of roots per plant were evaluated after two weeks. Rooted cuttings were transplanted to plastic bags using 'tezontle' (a porous, volcanic gravel) as the substrate. The most successful treatment was the application of 10 ppm IAP (resulting in 100% of cuttings rooting). Commercially propagated from both seeds and cuttings [7]. A protocol for *in vitro* propagation has been published [9].

TRADE

A commercial fruit crop (grown in several regions of the New World, tropical Asia and warmer parts of Europe) that is internationally traded. Locally, in Zapotitlán, Salinas (Puebla), there are organisations concerned with developing the economic potential of the species [3]. International trade of plant material is regulated by CITES (listed on Appendix II) [5].

Authors

Daniel Franco-Estrada, Efisio Mattana, Steve Davis and Paulina Hechenleitner.

References

[1] Anderson, E. F. (2001). *The Cactus Family*. Timber Press, Portland, Oregon, USA.

[2] Arias, S., Gama-López, S., Guzmán-Cruz, L. U. & Vázquez-Benítez, B. (2012). *Flora del Valle de Tehuacán-Cuicatlán, Fascículo 95, Cactaceae Juss.*, 2nd edition. Instituto de Biología, UNAM & CONABIO, Mexico.

[3] Arias Toledo, A. A., Valverde, Valdés, M. T. & Reyes Santiago, J. (2000). *Las Plantas de la Región de Zapotitlán Salinas, Puebla.* Instituto Nacional de Ecología, UNAM, Mexico.

[4] Baskin, C. C. & Baskin, J. M. (2014). *Seeds: Ecology, Biogeography, and Evolution of Dormancy and Germination*, 2nd edition. Academic Press, San Diego, USA.

[5] CITES (2017). *Convention on International Trade in Endangered Species of Wild Fauna and Flora*. https://cites.org/eng/app/index.php

[6] Diario Oficial de la Federación (2010). *Norma Oficial Mexicana NOM-059-SEMARNAT-2010, Protección Ambiental — Especies Nativas de México de Flora y Fauna Silvestres — Categorías de Riesgo y Especificaciones para su Inclusión, Exclusión o Cambio — Lista de Especies en Riesgo*. Secretaría de Medio Ambiente y Recursos Naturales, Mexico, 30 December 2010.

[7] El-Obeidy, A. A. (2006). Mass propagation of pitaya (dragon fruit). *Fruits* 61 (5): 313–319.

[8] IUCN (2017). *The IUCN Red List of Threatened Species.* Version 2017-3. http://www.iucnredlist.org/

[9] Qingzhu Hua, Pengkun Chen, Wanqing Liu, Yuewen Ma, Ruiwei Liang, Lu Wang, Zehuai Wang, Guibing Hu & Yonghua Qin (2015). A protocol for rapid *in vitro* propagation of genetically diverse pitaya. *Plant Cell, Tissue and Organ Culture* 120 (2): 741–745.

[10] Royal Botanic Gardens, Kew (2017). *Seed Information Database (SID).* Version 7.1. http://data.kew.org/sid/

Jatropha neopauciflora Pax

TAXONOMY AND NOMENCLATURE

FAMILY. Euphorbiaceae

SYNONYMS. *Jatropha harmsiana* Mattf., *Mozinna pauciflora* Rose

VERNACULAR / COMMON NAMES. Sangre de drago (Spanish).

PLANT DESCRIPTION

LIFE FORM. Shrub.

PLANT. Deciduous, <4 m tall; branches pendulous, bark dark brown, exfoliating [9].

LEAVES. Simple, verticillate, obovate, <5 cm long; the sap turns reddish or black when exposed to air. **FLOWERS.** Pink or red, bell-shaped, 10 mm long (Figure 1) [7].

Figure 1 – Flowers and fruit of *J. neopauciflora*.
(Photos: O. Téllez-Valdés, FESI-UNAM.)

FRUITS AND SEEDS

Fruits are 3 cm wide, green, bilocular and bilobed (Figure 1). Seeds are 1.5 x 1.4 mm, ovate, green brown (Figure 2) [7].

FLOWERING AND FRUITING. Flowers from April to May [6]; ripe fruits can be found in July and August.

Figure 2 – External views of the seeds, section and an excised embryo. (Photos: P. Gómez-Barreiro, RBG Kew.)

DISTRIBUTION

Endemic to Mexico (Oaxaca and Puebla) [10], mainly in the Valle de Tehuacán-Cuicatlán [9].

HABITAT

Xerophilous scrub and deciduous dry forests, in association with *Bursera arida* (Rose) Standl., *B. galeottiana* Engl., *Cephalocereus columna–trajani* (Karw. ex Pfeiff.) K.Schum., *Echinocactus platyacanthus* Link & Otto, *Manihot pauciflora* Brandegee, *Beaucarnea gracilis* Lem., *Castela tortuosa* Liebm. and species of *Hechtia*, *Agave*, *Opuntia*, *Fouquieria*, *Mammillaria* and *Coryphantha. Jatropha neopauciflora* is found in areas that have a semi-dry to subhumid climate and a long dry season, in soils derived from fine-grained sedimentary rocks, sandstones, conglomerate and limestones, at altitudes of 1,140–1,900 m a.s.l. [7].

USES

In traditional medicine, the latex is used to strengthen gums and teeth, and to treat oral thrush, throat infections and athlete's foot [1,9]. Laboratory studies have indicated that the latex has antibacterial and anti-inflammatory activity and antioxidant properties [3,4]. The sap is also used by local communities to make soap and shampoo [1,7] and is used to produce a red dye [9]. *Jatropha neopauciflora* is sometimes used as a live fence and for shading, and its branches are used for firewood [7].

Figure 3 – Workshop with the local communities on the uses of *J. neopauciflora*. (Photo: E. Mattana, RBG Kew.)

KNOWN HAZARDS AND SAFETY

No known hazards.

CONSERVATION STATUS

Not considered to be either rare or threatened in Mexico [2], but the status of this species has yet to be confirmed in the IUCN Red List [5].

SEED CONSERVATION

HARVESTING. Ripe fruits can be collected in July or August.

PROCESSING AND HANDLING. Physiologically mature fruits should be kept in paper bags and stored in a shaded and dry place until they are delivered to the seed bank, where the seeds can be extracted.

STORAGE AND VIABILITY. X-ray analysis carried out on a seed lot stored at the MSB revealed high percentages (30–90%) of empty seeds. No data are available on seed storage behaviour; however, four other *Jatropha* spp. whose seeds are stored at the MSB are reported to be desiccation tolerant [8].

PROPAGATION

SEEDS

Dormancy and pre-treatments: Preliminary tests carried out at the MSB showed that seed scarification (i.e. partially removing the seed covering structure) increased germination success, suggesting the presence of physical or physiological dormancy. Further studies are needed to determine the class of seed dormancy and the optimum temperature for germination. Before sowing at the San Rafael, Coxcatlán nursery, seeds were soaked in water for 24 hours.

Germination, sowing and planting: Preliminary tests carried out on seeds stored at the MSB highlighted a 43% germination rate for scarified seeds incubated at 25°C with 8 hours of light per day. 40% germination was achieved at the FESI-UNAM Seed Bank for seeds soaked in commercial bleach solution and incubated at 30°C with 12 hours of light per day. At the San Rafael, Coxcatlán nursery, seeds were sown in a 3:1 mixture by volume of sieved and sterilised leaf litter and agrolite. Seedlings were transplanted into black plastic bags three or four weeks after germination, using equal parts by volume of sieved leaf litter and agrolite. Watering and weeding were carried out every five days. Out of 400 germinated seeds, 123 plants have established.

VEGETATIVE PROPAGATION

Experiments using cuttings were carried out at the San Rafael, Coxcatlán nursery. Cuttings taken from young branches and treated with a hormone rooting powder were planted in black plastic pots using equal parts by volume of sieved and sterilised leaf litter, river sand and agrolite. They were watered every five days and frequently weeded. Out of 84 rooted cuttings, 41 plants were established.

TRADE

Jatropha neopauciflora is not traded or commercialised.

Authors

Lirio Jazmín Sánchez Hernández, Myrna Mendoza Cruz, Martin López Carrera, Efisio Mattana and Paulina Hechenleitner.

References

[1] Arias Toledo, A. A., Valverde Valdés, M. T. & Reyes Santiago, J. (2000). *Las Plantas de la Región de Zapotitlán Salinas, Puebla*. Instituto Nacional de Ecología, UNAM, Mexico.

[2] Diario Oficial de la Federación (2010). *Norma Oficial Mexicana NOM-059-SEMARNAT-2010, Protección Ambiental — Especies Nativas de México de Flora y Fauna Silvestres — Categorías de Riesgo y Especificaciones para su Inclusión, Exclusión o Cambio — Lista de Especies en Riesgo*. Secretaría de Medio Ambiente y Recursos Naturales, Mexico, 30 December 2010.

[3] Hernández, A. B., Alarcón-Aguilar, F. J., Jiménez-Estrada, M., Hernández-Portilla, L. B., Flores-Ortiz, C. M., Rodríguez-Monroy, M. A. & Canales-Martinez, M. (2017). Biological properties and chemical composition of *Jatropha neopauciflora* Pax. *African Journal of Traditional, Complementary and Alternative Medicines* 14 (1): 32-42.

[4] Hernandez-Hernandez, A. B., Alarcon-Aguilar, F. J., Almanza-Peréz, J. C., Nieto-Yañez, O., Olivares-Sanchez, J. M., Duran-Diaz, A., Rodriguez-Monroy, M. A. & Canales-Martinez, M. M. (2017). Antimicrobial and anti-inflammatory activities, wound-healing effectiveness and chemical characterization of the latex of *Jatropha neopauciflora* Pax. *Journal of Ethnopharmacology* 204: 1–7.

[5] IUCN (2017). *The IUCN Red List of Threatened Species*. Version 2017-3. http://www.iucnredlist.org/

[6] Reyes, S. J., Brachet, C. I., Pérez, J. C & Gutiérrez de la Rosa, A. (2004). *Cactáceas y Otras Plantas Nativas de la Cañada Cuicatlán, Oaxaca, México*. Sociedad Mexicana de Cactologia, Mexico.

[7] Rodríguez Acosta, M., Jiménez Merino, F. A. & Coombes, J. A. (2010). *Plantas de Importancia Económica en el Estado de Puebla*. Benemérita Universidad Autónoma de Puebla, Mexico.

[8] Royal Botanic Gardens, Kew (2017). *Seed Information Database (SID)*. Version 7.1. http://data.kew.org/sid/

[9] Téllez Valdés, O., Reyes, M., Dávila Aranda, P., Gutiérrez García, K., Téllez Poo, O., Álvarez Espino, R. X., González Romero, A. V, Rosas Ruiz, I., Ayala Razo, M., Murguía Romero, M. & Guzmán, U. (undated). *Guía Ecoturística. Las Plantas del Valle de Tehuacan-Cuicatlan*. Volkswagen, FESI-UNAM & Millennium Seed Bank Project, Royal Botanic Gardens, Kew.

[10] WCSP (2017). *World Checklist of Selected Plant Families*. Royal Botanic Gardens, Kew. http://wcsp.science.kew.org/

Lantana camara L.

TAXONOMY AND NOMENCLATURE

FAMILY. Verbenaceae

SYNONYMS. Several subspecies recognised, but all previously published infraspecific varieties and forms are now considered to be synonyms.

VERNACULAR / COMMON NAMES. Blacksage, lantana, yellow sage (English); cinco negritos (Spanish).

PLANT DESCRIPTION

LIFE FORM. Shrub.

PLANT. Vigorous, erect or subscandent, evergreen, <3(–5) m tall, foetid; stems slightly quadrangular, armed with recurved prickles. **LEAVES.** Usually opposite, ovate or ovate-oblong, 2–12 x 2–7 cm; apex acute-acuminate; base rounded or narrowly cuneate; adaxially scabrous-strigose; abaxially pilose-hirsute or glabrous; petioles 5–20 mm long. **FLOWERS.** Sessile; calyx 1–3 mm long, cup-shaped, internally glabrous; corolla usually orange red (but can be white to red) with yellow centre, tube 8–13 mm long; stamens 4; flowers grouped in terminal, rounded inflorescences (Figure 1) [3,11,12]. *Lantana camara* is highly variable, having been in cultivation for more than 300 years. As a result, there are hundreds of hybrids and cultivars [3].

Figure 1 – Inflorescence of *L. camara*. (Photo: O. Téllez-Valdés, FESI-UNAM.)

FRUITS AND SEEDS

Fruits are spherical drupes, 5 x 4–6 mm, fleshy, juicy, black to deep purple, shiny, glabrous (Figure 2). Seeds two per fruit, 6 x 3.5 mm, ovoid (Figure 3) [16].

FLOWERING AND FRUITING. Flowers and fruits all year round [16].

Figure 2 – Ripe fruits.
(Photo: O. Téllez-Valdés, FESI-UNAM.)

Figure 3 – External views (left) and cross-section (right) of the endocarps showing the seeds.
(Photos: P. Gómez-Barreiro, RBG Kew.)

DISTRIBUTION

Native to Mexico and tropical America, but introduced to many tropical countries [8,15]. In Mexico it is found in Baja California Sur, Campeche, Chiapas, Chihuahua, Coahuila, Colima, Distrito Federal, Durango, Guanajuato, Guerrero, Hidalgo, Jalisco, Estado de México, Michoacán, Morelos, Nayarit, Nuevo León, Oaxaca, Puebla, Querétaro, Quintana Roo, Sinaloa, Sonora, Tabasco, Tamaulipas, Veracruz and Yucatán [14,15]. *Lantana camara* is invasive and considered to be one of the world's worst weeds. It is established and expanding in many regions of the world, often as a result of clearing of forest for timber or agriculture [3].

HABITAT

Found in a wide variety of forests (including deciduous dry forests, semi-evergreen and evergreen forests, spiny forests, mesophilous mountain forests, oak, pine and mixed forests) [13], xerophilous scrub, ruderal vegetation, grassland, and secondary vegetation, abandoned fields and roadsides, at 700–2,250 m a.s.l. [16], extending to coastal areas where introduced, and forming dense thickets.

USES

Used in traditional medicine to treat a wide range of digestive, respiratory and gynaecological disorders [7]. The oil obtained from boiling the flowers, stems and leaves is used to treat toothache and earache. The plant is also used to treat liver conditions, rheumatism, bile and amoebic infections, as well as headache, kidney pain, epilepsy, chest pains, cramp, hair loss, inflammation, skin rashes, heart conditions, diabetes, tumours and ulcers. It is also used to treat insect, scorpion and snake bites, and is a diuretic [13]. *Lantana camara* is widely used as an ornamental in parks, gardens and hotel grounds [10,12].

KNOWN HAZARDS AND SAFETY

The fruits are poisonous if eaten. Ingestion of unripe fruits has led to child fatalities [17]. Although dry leaves are eaten by animals, the aerial parts (including the fruits) are toxic to grazing animals and their consumption can lead to death, resulting in major economic losses. The leaves contain toxic pentacyclic triterpenoids which cause hepatotoxicity and photosensitivity in grazing animals such as sheep, goats, cattle and horses [6]. Contact with the plant can cause skin irritation. *Lantana camara* is a serious weed in many parts of the world and has allelopathic properties which can reduce the vigour of nearby plant species. It has a tendency to crowd out other species, disrupting succession and decreasing biodiversity [3]. It is declared a noxious weed in some countries (e.g. Australia, where there are restrictions on the sale of the plant and land owners in some states are required by law to eradicate it).

CONSERVATION STATUS

Widespread and abundant; not considered to be either rare or threatened in Mexico [2], but its status has yet to be confirmed in the IUCN Red List [4].

SEED CONSERVATION

HARVESTING. Ripe fruits can be collected from August to November.

PROCESSING AND HANDLING. Fruits should be stored in paper or cotton bags and kept in a dark and dry place until they are delivered to the seed bank, where the seeds can be extracted from the fruit pulp and then dried as soon as possible.

STORAGE AND VIABILITY. X-ray analysis carried out on seed lots stored at the MSB highlighted a high percentage of empty seeds (>50%). Seeds of this species are likely to be desiccation tolerant [9].

PROPAGATION

SEEDS

Dormancy and pre-treatments: The removal by hand of the fresh pericarp is reported to improve seed germination [1]. However, low germination percentages are reported in the literature for scarified seeds (e.g. [9]). Further studies are needed to determine dormancy loss and germination requirements.

Germination, sowing and planting: The germination rate is low under both laboratory conditions and in the field (with estimates of 4–45%) [3]. Germination experiments carried out at the FESI-UNAM Seed Bank revealed 50% final germination when seeds were incubated at 30°C, with a photoperiod of 12 hours of darkness per day. Seed lots stored at the MSB achieved germination rates lower than 60% under different conditions and treatments [9], confirming the need for further studies on seed biology and ecology.

VEGETATIVE PROPAGATION

Lantana camara can be propagated by semi-hardwood and hardwood cuttings [10]. The use of a hormone rooting powder increases the success rate [5].

TRADE

Products derived from *L. camara* are not traded or commercialised. In some countries, trade in *L. camara* is restricted by law because of its invasiveness.

Authors

Lirio Jazmín Sánchez Hernández, Efisio Mattana, Steve Davis and Paulina Hechenleitner.

References

[1] Baskin, C. C. & Baskin, J. M. (2014). *Seeds: Ecology, Biogeography, and Evolution of Dormancy and Germination*, 2nd edition. Academic Press, San Diego, USA.

[2] Diario Oficial de la Federación (2010). *Norma Oficial Mexicana NOM-059-SEMARNAT-2010, Protección Ambiental — Especies Nativas de México de Flora y Fauna Silvestres — Categorías de Riesgo y Especificaciones para su Inclusión, Exclusión o Cambio — Lista de Especies en Riesgo*. Secretaría de Medio Ambiente y Recursos Naturales, Mexico, 30 December 2010.

[3] Global Invasive Species Database (2006). *Lantana camara*. http://www.iucngisd.org/gisd/speciesname/Lantana+camara

[4] IUCN (2017). *The IUCN Red List of Threatened Species.* Version 2017-3. http://www.iucnredlist.org/

[5] Koike, Y., Matsushima, K. & MItarai, Y. (2017). The influence of indole 3-butyric acid on hardwood propagation of *Lantana camara* L. In: *The IAFOR International Conference on Sustainability, Energy & the Environment — Hawaii 2017 Official Conference Proceedings*. http://papers.iafor.org/papers/iicseehawaii2017/IICSEEHawaii2017_33548.pdf

[6] Lonare, M. K., Sharma, M., Hajare, S. W. & Borekar, V. I. (2012). *Lantana camara*: overview on toxic to potent medicinal properties. *International Journal of Pharmaceutical Sciences and Research* 3 (9): 3031–3035.

[7] Márquez Alonso, C. & Lara Ochoa, F. (1999). *Plantas Medicinales de México II: Composición, Usos y Actividad Biológica*. UNAM, Mexico.

[8] Martínez, M. (1979). *Catálogo de Nombres Vulgares y Científicos de Plantas Mexicanas*. Fondo de Cultura Económica, Mexico.

[9] Royal Botanic Gardens, Kew (2017). *Seed Information Database (SID).* Version 7.1. http://data.kew.org/sid/

[10] Royal Horticultural Society (2017). *Lantana camara*. https://www.rhs.org.uk/Plants/99415/Lantana-camara/Details

[11] Rzedowski, J. & Rzedowski, G. C. (2002). Verbenaceae. In: *Flora del Bajío y de Regiones Adyacentes, Fascículo 100*. Instituto de Ecología-Centro Regional del Bajío, Consejo Nacional de Ciencia y Tecnología y Comisión Nacional para el Conocimiento y Uso de la Biodiversidad, Mexico.

[12] Téllez Valdés, O., Reyes, M., Dávila Aranda, P., Gutiérrez García, K., Téllez Poo, O., Álvarez Espino, R. X., González Romero, A. V, Rosas Ruiz, I., Ayala Razo, M., Murguía Romero, M. & Guzmán, U. (undated). *Guía Ecoturística. Las Plantas del Valle de Tehuacan-Cuicatlan*. Volkswagen, FESI-UNAM & Millennium Seed Bank Project, Royal Botanic Gardens, Kew.

[13] UNAM (2009). *Biblioteca Digital de la Medicina Tradicional Mexicana*. UNAM, Mexico. http://www.medicinatradicionalmexicana.unam.mx/index.php

[14] Villaseñor, R. J. L. & Espinosa García, F. J. (1998). *Catálogo de Malezas de México*. UNAM, Consejo Nacional Consultivo Fitosanitario & Fondo de Cultura Económica, Mexico.

[15] WCSP (2017). *World Checklist of Selected Plant Families.* Royal Botanic Gardens, Kew. http://wcsp.science.kew.org/

[16] Willmann, D., Schmidt, E.-M., Heinrich, M. & Rimpler, H. (2000). *Flora del Valle de Tehuacán-Cuicatlán, Fascículo 27, Verbenaceae J.St.-Hil.* Instituto de Biología, UNAM, Mexico.

[17] Wolfson, S. L. & Solomons, T. W. (1964). Poisoning by fruit of *Lantana camara*: an acute syndrome observed in children following ingestion of the green fruit. *American Journal of Diseases of Children* 107 (2): 173–176.

Lippia origanoides Kunth

TAXONOMY AND NOMENCLATURE

FAMILY. Verbenaceae

SYNONYMS. *Lippia berlandieri* M.Martens & Galeotti, *L. graveolens* Kunth, *L. mattogrossensis* Moldenke, *L. palmeri* S.Watson, *L. pendula* Rusby, *Lantana origanoides* M.Martens & Galeotti

VERNACULAR / COMMON NAMES. Mexican oregano (English); orégano, orégano Mexicano (Spanish).

Figure 1 – Inflorescences of *L. origanoides*. (Photo: E. Mattana, RBG Kew.)

PLANT DESCRIPTION

LIFE FORM. Shrub.

PLANT. Up to 3 m tall, stems usually densely hairy, rarely hispid, aromatic. **LEAVES.** Usually opposite, sometimes ternate, hirsute, rarely sessile, petioles 0.1–2.4 cm long. **FLOWERS.** Grouped in frondose inflorescences (Figure 1), calyx 1–2 mm long, hirsute, corolla 2–6 mm long [8].

FRUITS AND SEEDS

Fruits are dry schizocarps, with thin dry pericarp and mesocarp (Figure 2), readily splitting at maturity into two mericarps [3]. Nutlets 1–2 mm long [9].

FLOWERING AND FRUITING. Flowers and fruits from March to October.

Figure 2 – External views of the nutlets and longitudinal section showing the spatulate embryo.
(Photo: P. Gómez-Barreiro, RBG Kew.)

DISTRIBUTION

Tropical and subtropical America, from northern Argentina, Uruguay and Paraguay in the south, to northern South America (including Bolivia, Brazil, Colombia, Guyana and Venezuela), throughout Mesoamerica (from Costa Rica to Mexico) to southern USA (New Mexico and Texas). Also native to Trinidad and Tobago [13].

HABITAT

Xerophilous scrub throughout semi-arid regions in Mexico [10], on sandy clayey soils and limestone rocks.

USES

The leaves are used mainly as a culinary herb for food seasoning and flavouring beverages [1]. In traditional medicine, *Lippia origanoides* is used to treat pain, inflammation and spasms and to induce abortion. A decoction of the stems and leaves is used to treat respiratory and gastrointestinal problems, especially coughs, fever, diarrhoea and stomach pains. As 'agua de tiempo', it is used as an emmenagogue and to induce abortion, and as 'baños de asiento' (i.e. hot baths). The leaves are applied to wounds as 'pasmo' (i.e. cooling) and to treat 'granos' (skin rashes). The leaf juice is used as drops, or mixed with cooking oil, to treat earache [12].

KNOWN HAZARDS AND SAFETY

No known hazards.

CONSERVATION STATUS

Not considered to be either rare or threatened in Mexico [2], but the status of this species has yet to be confirmed in the IUCN Red List [5]. In Mexico, large quantities of the plant are collected, with 50% of harvesting being illegal, placing this resource at risk [10].

SEED CONSERVATION

HARVESTING. Seed collecting is carried out from March to October by gently rubbing the inflorescences between two rubber layers (e.g. by using rubber gloves).

PROCESSING AND HANDLING. The harvested material is sieved to separate the seeds from other plant material.

STORAGE AND VIABILITY. X-ray analysis carried out on seed lots stored at the MSB found c. 95% filled seeds. No data are available on desiccation tolerance, but many *Lippia* spp. have desiccation tolerant seeds [11].

PROPAGATION

SEEDS

Dormancy and pre-treatments: Seeds can be soaked in commercial 1% diluted chlorine for 5 minutes and then rinsed with tap water 3 times to remove the chlorine. The cleaned seeds can then be soaked in water for 24 hours before sowing [7]. Seeds of this species are suspected to be non-dormant because applied treatments (gibberellic acid [GA_3] and potassium nitrate [KNO_3]) did not improve seed germination compared to untreated seeds. However, further studies are needed to exclude any type of physiological dormancy [6]. Base temperature for germination (Tb, °C) and the thermal constant for 50% germination (S, °Cd) of untreated seeds were 2.5°C and 76.9°Cd respectively [6].

Germination, sowing and planting: Germination rates of c. 75% have been reported for seeds pre-treated as above and incubated at 30°C [7]. Cleaned and soaked seeds can be sown on moist paper, on a porous substrate such as 'tezontle' (a porous, extrusive, igneous, volcanic rock) or in a 2:1 mixture of soil and expanded perlite. The species is not demanding in terms of soil moisture and grows in dry soils. Seeds can be sown individually in 7.5 cm diameter pots. The best time for sowing is in the months of May–June to avoid the coldest period of the year. When the plants are well-developed, they can be transplanted to larger pots or planted directly in the open ground. Watering can be daily, or every 2–3 days; care should be taken not to overwater. Although tolerant of dry, nutrient-poor soils, the addition of silt or organic fertiliser can improve establishment and growth. Increased survival rate of seedlings (from c. 30 to c. 90%) was achieved by watering with Hoagland solution and early transplanting to pots [7].

VEGETATIVE PROPAGATION

Can be propagated by stem cuttings [4].

TRADE

Dry leaves are sold in markets [9]. The national production of oregano in Mexico in 2005 was c. 4,000 tonnes, obtained mainly from wild populations in the states of Chihuahua, Durango, Coahuila, Guanajuato, and Querétaro; 50% of the harvest was gathered illegally and mainly exported to USA and Europe [10].

Authors

Amanda Moreno Rodríguez, Efisio Mattana, César Flores, Tiziana Ulian and Paulina Hechenleitner.

References

[1] CONAFOR (undated). Lippia graveolens *Kunth. Ficha Técnica para la Reforestación.* http://www.conafor.gob.mx:8080/documentos/docs/13/940Lippia%20graveolens.pdf

[2] Diario Oficial de la Federación (2010). *Norma Oficial Mexicana NOM-059-SEMARNAT-2010, Protección Ambiental — Especies Nativas de México de Flora y Fauna Silvestres — Categorías de Riesgo y Especificaciones para su Inclusión, Exclusión o Cambio — Lista de Especies en Riesgo*. Secretaría de Medio Ambiente y Recursos Naturales, Mexico, 30 December 2010.

[3] França, F. & Atkins, S. (2009). Neotropical Verbenaceae. In: *Neotropikey — Interactive Key and Information Resources for Flowering Plants of the Neotropics*. Edited by W. Milliken, B. Klitgård & A. Baracat. http://www.kew.org/science/tropamerica/neotropikey/families/Verbenaceae.htm

[4] Herrera-Moreno, A. M., Carranza, C. E. & Chácon-Sánchez, M. I. (2013). Establishment of propagation methods for growing promising aromatic plant species of the *Lippia* (Verbenaceae) and *Tagetes* (Asteraceae) genera in Colombia. *Agronomía Colombiana* 31 (1): 27–37.

[5] IUCN (2017). *The IUCN Red List of Threatened Species.* Version 2017-3. http://www.iucnredlist.org/

[6] Mattana, E., Sacande, M., Sanogo, A. K., Lira, R., Gómez-Barreiro, P., Rogledi, M. & Ulian, T. (2017). Thermal requirements for seed germination of underutilised *Lippia* species. *South African Journal of Botany* 109: 223–230.

[7] Moreno Rodríguez, A. (2009). *Perfiles de Acumulación de Metabolitos Secundarios en 3 Especies Medicinales de San Rafael, Coxcatalán, Puebla, Bajo Diferentes Practicas de Cultivo*. Master's thesis, UNAM, Mexico.

[8] Ocampo-Velázquez, R. V., Malda-Barera, G. X. & Suárez-Ramos, G. (2009). Reproductive biology of Mexican oregano (*Lippia graveolens* Kunth) in three exploitation conditions. *Agrociencia* 43 (5): 475–482.

[9] O'Leary, N., Denham, S. S., Salimena, F. & Múlgura, M. E. (2012). Species delimitation in *Lippia* section *Goniostachyum* (Verbenaceae) using the phylogenetic species concept. *Botanical Journal of the Linnean Society* 170 (2): 197–219.

[10] Osorno-Sánchez, T., Flores-Jaramillo, D., Hernández-Sandoval, L. & Linding-Cisneros, R. (2009). Management and extraction of *Lippia graveolens* in the arid lands of Queretaro, Mexico. *Economic Botany* 63 (3): 314–318.

[11] Royal Botanic Gardens, Kew (2017). *Seed Information Database (SID)*. Version 7.1. http://data.kew.org/sid/

[12] UNAM (2009). *Biblioteca Digital de la Medicina Tradicional Mexicana*. UNAM, Mexico. http://www.medicinatradicionalmexicana.unam.mx/index.php

[13] WCSP (2017). *World Checklist of Selected Plant Families.* Royal Botanic Gardens, Kew. http://wcsp.science.kew.org/

Myrtillocactus geometrizans (Mart. ex Pfeiff.) Console

TAXONOMY AND NOMENCLATURE

FAMILY. Cactaceae

SYNONYMS. *Myrtillocactus geometrizans* var. *grandiareolatus* (Bravo) Backeb., *M. grandiareolatus* Bravo

VERNACULAR / COMMON NAMES. Bilberry cactus (English); garambullo (Spanish modified indigenous name).

PLANT DESCRIPTION

LIFE FORM. Cactus.

PLANT. Arborescent, candelabra-like, <7 m tall; main trunk short, branches 6–12 cm wide; 5–7 ribs, margins rounded, distinctly blue green. **LEAVES.** Reduced to spines; central spine 1 or absent, 1–7 cm long, 2–6 mm wide, flat on sides, rigid, porrect, grey; radial spines 4–9, 0.2–3 cm long, rigid, subulate, grey, red when young. **FLOWERS.** Infundibuliform, yellowish green, 2–4 cm long, 2.5–3.5 cm wide; pericarp and receptacle tube with vestigial bracts and naked areoles (Figure 1) [1,2].

Figure 1 – Flowers of *M. geometrizans*. (Photo: O. Téllez-Valdés, FESI-UNAM.)

FRUITS AND SEEDS

Fruits dark purple, 1–2 x 0.8–2 cm, naked (Figure 2). Seeds black seeds, c. 1.3–1.7 x 1–1.5 mm, wrinkled, with a dorsal keel (Figure 3) [11].

FLOWERING AND FRUITING. Flowers from February to April; fruits from July to September [2].

Figure 2 – Naked fruits.
(Photo: O. Téllez-Valdés, FESI-UNAM.)

Figure 3 – External views (left) and longitudinal section (right) of the seeds. (Photos: P. Gómez-Barreiro, RBG Kew.)

DISTRIBUTION

Endemic to Mexico (Aguascalientes, Durango, Guanajuato, Guerrero, Hidalgo, Jalisco, Estado de México, Michoacán, Nuevo León, Oaxaca, Puebla, Querétaro, San Luis Potosí, Tamaulipas, Veracruz and Zacatecas) [2].

HABITAT

Deciduous dry forests, xerophilous scrub, less frequent in grasslands and oak forests, forming communities known as 'garambullales', at 550–2,000 m a.s.l. [2].

USES

Ripe fruits are eaten when fresh and taste similar to cranberry, or may be dried like raisins [1] or preserved as jam. In the municipality of Zapotitlán, Salinas (Puebla), the fruits are used to prepare an alcoholic beverage known as 'garambullo liquor', which is popular in the region. *Myrtillocactus geometrizans* is frequently planted as a hedge and boundary marker, and is planted on cultivated slopes to prevent soil erosion [7]. Occasionally, central branches of large specimens are removed and the space used to store animal fodder and maize (*Zea mays* L.) [3,6]. Widely grown by cactus enthusiasts and often planted as an ornamental on account of its attractive bluish green stems [5]. *Myrtillocactus geometrizans* is also used as a rootstock for grafting slower-growing cacti.

KNOWN HAZARDS AND SAFETY

Care should be taken when handling the plant as the spines are sharp.

CONSERVATION STATUS

Assessed as Least Concern (LC) according to IUCN Red List criteria [12]. *Myrtillocactus geometrizans* is not included in the national list of protected species in Mexico [10].

SEED CONSERVATION

HARVESTING. Fruits can be gathered using a long pole.

PROCESSING AND HANDLING. Harvested ripe fruits should be kept in paper or cotton bags until they are delivered to the seed bank, where they can be cut open with a knife and the seeds extracted from the pulp. The seeds should then be washed with tap water to remove the pulp, and dried at room temperature, after which they can be passed through a sieve or cleaned with pressurised air to separate the seeds from any remaining dry pulp.

STORAGE AND VIABILITY. X-ray analysis carried out on seed lots stored at the MSB showed high percentages of filled seeds (70–100%). Seeds of this species are reported to be desiccation tolerant [15].

PROPAGATION

SEEDS

Dormancy and pre-treatments: High germination percentages were achieved when seeds were incubated in a wide range of temperatures without any pre-treatments [15]. Seeds of this species are reported to be non-dormant [4].

Germination, sowing and planting: Seeds germinate to high percentages (>60%) when incubated in the light at constant temperatures of >20°C, reaching 100% germination at 30°C [15]. However, according to data published by Noble [13] optimum seed germination temperatures are in the 17–25°C range. 80% germination was reached at the FESI-UNAM Seed Bank when seeds were incubated in the light at 30°C for 20 days. Germination tests were carried out at the San Rafael, Coxcatlán nursery to develop a seed propagation protocol.

VEGETATIVE PROPAGATION

The base of stem cuttings should be allowed to dry for 2–3 weeks before planting in a well-drained soil mixture [14]. A study has been carried out on vegetative propagation techniques, including the type of substrate [16], but as yet there is no definitive propagation protocol.

TRADE

The fruits are sold in local markets in Teotitlán, Tehuacán and Ajalpán [8]. International trade of plant material is regulated by CITES (listed on Appendix II) [9].

Authors

Daniel Franco-Estrada, Efisio Mattana, Steve Davis and Paulina Hechenleitner.

References

[1] Anderson, E. F. (2001). *The Cactus Family*. Timber Press, Portland, Oregon, USA.

[2] Arias, S., Gama-López, S., Guzmán-Cruz, L. U. & Vázquez-Benítez, B. (2012). *Flora del Valle de Tehuacán-Cuicatlán, Fascículo 95, Cactaceae Juss.*, 2nd edition. Instituto de Biología, UNAM & CONABIO, Mexico.

[3] Arias Toledo, A. A., Valverde Valdés, M. T. & Reyes Santiago, J. (2000). *Las Plantas de la Región de Zapotitlán Salinas, Puebla*. Instituto Nacional de Ecología, UNAM, Mexico.

[4] Baskin, C. C. & Baskin, J. M. (2014). *Seeds: Ecology, Biogeography, and Evolution of Dormancy and Germination*, 2nd edition. Academic Press, San Diego, USA.

[5] Borg, J. (1959). *Cacti: A Gardener's Handbook for Their Identification and Cultivation*, 3rd edition. Blandford Press, London.

[6] Casas, A. (2002). Uso y manejo de cactáceas columnares mesoamericanas. *CONABIO Biodiversitas* 40: 18–23.

[7] Casas, A., Pickersgill, B., Caballero, J. & Valiente-Banuet, A. (1997). Ethnobotany and domestication in xoconochtli, *Stenocereus stellatus* (Cactaceae) in the Tehuacán Valley and La Mixteca Baja, México. *Economic Botany* 51 (3): 279–292.

[8] Cervantes, H. (2014). Personal communication.

[9] CITES (2017). *Convention on International Trade in Endangered Species of Wild Fauna and Flora*. https://cites.org/eng/app/index.php

[10] Diario Oficial de la Federación (2010). *Norma Oficial Mexicana NOM-059-SEMARNAT-2010, Protección Ambiental — Especies Nativas de México de Flora y Fauna Silvestres — Categorías de Riesgo y Especificaciones para su Inclusión, Exclusión o Cambio — Lista de Especies en Riesgo*. Secretaría de Medio Ambiente y Recursos Naturales, Mexico, 30 December 2010.

[11] Franco, E. D. (2013). *Catálogo Ilustrado de Cactáceas Columnares del Valle de Tehuacán-Cuicatlán*. Bachelor's thesis, FESI-UNAM, Mexico.

[12] IUCN (2017). *The IUCN Red List of Threatened Species*. Version 2017-3. http://www.iucnredlist.org/

[13] Noble, P. S. (ed.) (2002). *Cacti: Biology and Uses*. University of California Press, Berkeley, California, USA.

[14] Royal Botanic Garden, Edinburgh (2012). *Myrtillocactus geometrizans* (C.Mart.) Console. http://www.rbge.org.uk/the-gardens/plant-of-the-month/plant-profiles/myrtillocactus-geometrizans

[15] Seal, C. E., Flores, J., Ceroni Stuva, A., Dávila Aranda, P., León-Lobos, P., Ortega-Baes, P., Galíndez, G., Aparicio-González, M. A., Castro Cepero, V., Daws, M. I., Eason, M., Flores Ortiz, C. M., del Fueyo, P. A., Olwell, P., Ordoñez, C., Peñalosa Castro, I., Quintanar Zúñiga, R., Ramírez Bullón, N., Rojas-Aréchiga, M., Rosas, M., Sandoval, A., Stuppy, W., Ulian, T., Vázquez Medrano, J., Walter, H., Way, M. & Pritchard, H. W. (2009). *The Cactus Seed Biology Database*, Release 1. Royal Botanic Gardens, Kew.

[16] Serna, L. (2012). *Germinación y Establecimiento de Seedlings de* Myrtillocactus geometrizans *(Mart.) Console (Cactaceae) en la Cuenca del Río de Zapotitlán (San Rafael, Coxcatlán)*. Master's thesis, UNAM, Mexico.

Neobuxbaumia tetetzo (F.A.C.Weber ex K.Schum.) Backeb.

TAXONOMY AND NOMENCLATURE

FAMILY. Cactaceae

SYNONYMS. *Pachycereus tetetzo* (F.A.C.Weber ex J.M.Coult.) Ochot., *Pilocereus tetetzo* (F.A.C.Weber ex J.M.Coult.) F.A.C.Weber ex K.Schum., *Carnegiea tetetzo* (F.A.C.Weber ex J.M.Coult.) P.V.Heath

VERNACULAR / COMMON NAMES. Tetecho (indigenous name).

Figure 1 – Habit of *N. tetetzo.* (Photo: D. Franco-Estrada, FESI-UNAM.)

PLANT DESCRIPTION

LIFE FORM. Cactus.

PLANT. Columnar, <15 m tall, main trunk 30–60(–70) cm wide, with 13–17(–20) ribs; single column when young, up to 16 secondary branches at maturity, each 10–20 cm wide (Figure 1).

LEAVES. Reduced to spines; central spines 0–3, 3–12 cm long, one longer, subulate, reddish brown to greyish or blackish; radial spines 2–13, 0.5–2 cm long, acicular, greyish brown to blackish.

FLOWERS. Tubular-infundibuliform, greenish white, 4.7–5.5 cm long, arranged around the apex; pericarp and receptacle tube with fleshy bracts and areoles with hairs and occasional spines (Figure 2) [1,2].

Figure 2 – Floral buds. (Photo: O. Téllez-Valdés, FESI-UNAM.)

Figure 3 – Preserves of floral buds of *N. tetetzo*. (Photo: F. Peralta-Romero, FESI-UNAM.)

FRUITS AND SEEDS

Fruits green to red, 3–4 x 2.5–3 cm, scaly. Seeds dark brown, c. 1.5–2 x 1–2 mm wide, shiny, smooth, with a keel from the dorsal region to the apical region [4].

FLOWERING AND FRUITING. Flowers between May and July; fruits ripen between June and July [2].

DISTRIBUTION

Endemic to Mexico (Oaxaca and Puebla) [2].

HABITAT

Deciduous dry forests and xerophilous scrub on calcareous soils, at 600–1,900 m a.s.l. [2].

USES

The floral buds (known as 'tetechas') are cut and boiled, then cooked with vinegar or pickled (Figure 3). The fruits ('zalehitas') are collected and eaten when dried. The seeds are crushed with chilli pepper to make a hot sauce or 'salsa'. The trunk (known locally as 'cuilotes') is cut into longitudinal sections (known as 'huacal') and dried. The resultant planks are used to build rustic houses [3,5].

KNOWN HAZARDS AND SAFETY

The fruits can be irritant to the mouth, tongue and throat if large quantities are eaten [3]. Care should be taken when handling the plant to avoid any sharp spines.

CONSERVATION STATUS

Assessed as Least Concern (LC) according to IUCN Red List criteria [8]. *Neobuxbaumia tetetzo* is not included in the national list of protected species in Mexico [7].

SEED CONSERVATION

HARVESTING. Fruits are harvested using a pole with scissors.

PROCESSING AND HANDLING. Fruits are cut open with a knife to extract the seeds from the pulp. The seeds are washed under tap water and dried at room temperature, and then passed through a sieve to separate the seeds from dried pulp remains.

STORAGE AND VIABILITY. Viability of seeds stored at the FESI-UNAM Seed Bank was 95% as assessed by cut test. This was confirmed by 90% seed germination under conditions of 30°C in the light for 120 days. No information is available on seed desiccation tolerance for this species.

PROPAGATION

SEEDS

Dormancy and pre-treatments: This species has been propagated in the San Rafael, Coxcatlán nursery. Seeds were surface sterilised in a mixture of water with 2% sodium hypochlorite for seven minutes, and then rinsed several times with tap water, and left to dry for 12–24 hours. Seeds of this species are reported to be non-dormant [4].

Germination, sowing and planting: According to data published by Noble [9], optimum seed germination temperatures are in the 15–30°C range. Seeds were germinated at the San Rafael, Coxcatlán nursery following the pre-treatment above. They were sown in germination trays in a 3:1 mixture by volume of agrolite and sieved leaf litter. The leaf litter had been sterilised by exposure to sunlight in transparent plastic bags for 1–2 days. The seeds were scattered evenly over the watered substrate and then covered by a fine layer of substrate. Transplanting into plastic bags or pots took place after 60–90 days when the seedlings were 1–2 cm tall, using a mixture of equal parts by volume of sterilised sieved leaf litter and fine gravel (e.g. 'tepojal', 'tepezil' or 'tezontle'). A layer of coarse gravel was placed in the bottom of containers to ensure good drainage. Two transplants into black plastic pots were made at three months and six months after germination. Watering took place every five days, then every 15 days after three months. Hand weeding was carried out frequently. The plants were ready to be planted in the field after 6–12 months.

VEGETATIVE PROPAGATION

The base of stem cuttings should be allowed to dry before planting in a well-drained soil mixture. As yet, there is no definitive propagation protocol.

TRADE

Flower buds of *N. tetetzo* are sold mainly in local markets (Figure 4) [5]. The international trade of this plant is regulated by CITES (listed on Appendix II) [6].

Figure 4 – Floral buds of *N. tetetzo* sold on the local market. (Photo: H. Cervantes Maya, FESI-UNAM.)

Authors

Daniel Franco-Estrada, Myrna Mendoza Cruz, Martín López Carrera, Efisio Mattana, Steve Davis and Paulina Hechenleitner.

References

[1] Anderson, E. F. (2001). *The Cactus Family*. Timber Press, Portland, Oregon, USA.

[2] Arias, S., Gama-López, S., Guzmán-Cruz, L. U. & Vázquez-Benítez, B. (2012). *Flora del Valle de Tehuacán-Cuicatlán, Fascículo 95, Cactaceae Juss.*, 2nd edition. Instituto de Biología, UNAM & CONABIO, Mexico.

[3] Arias Toledo, A. A., Valverde Valdés, M. T. & Reyes Santiago, J. (2000). *Las Plantas de la Región de Zapotitlán Salinas, Puebla.* Instituto Nacional de Ecología, UNAM, Mexico.

[4] Baskin, C. C. & Baskin, J. M. (2014). *Seeds. Ecology, Biogeography, and Evolution of Dormancy and Germination*, 2nd edition. Academic Press, San Diego, USA.

[5] Casas, A. (2002). Uso y manejo de cactáceas columnares mesoamericanas. *CONABIO Biodiversitas* 40: 18–23.

[6] CITES (2017). *Convention on International Trade in Endangered Species of Wild Fauna and Flora*. https://cites.org/eng/app/index.php

[7] Diario Oficial de la Federación (2010). *Norma Oficial Mexicana NOM-059-SEMARNAT-2010, Protección Ambiental — Especies Nativas de México de Flora y Fauna Silvestres — Categorías de Riesgo y Especificaciones para su Inclusión, Exclusión o Cambio — Lista de Especies en Riesgo*. Secretaría de Medio Ambiente y Recursos Naturales, Mexico, 30 December 2010.

[8] IUCN (2017). *The IUCN Red List of Threatened Species.* Version 2017-3. http://www.iucnredlist.org/

[9] Noble, P. S. (ed.) (2002). *Cacti: Biology and Uses.* University of California Press, Berkeley, California, USA.

Prosopis laevigata (Humb. & Bonpl. ex Willd.) M.C.Johnst.

TAXONOMY AND NOMENCLATURE

FAMILY. Leguminosae – Caesalpinioideae

SYNONYMS. *Acacia laevigata* Willd., *Mimosa laevigata* (Willd.) Poir.

VERNACULAR / COMMON NAMES. Smooth mesquite (English); mezquite (Spanish).

PLANT DESCRIPTION

LIFE FORM. Tree or shrub.

PLANT. Up to 12 m tall, trunk thorny, 30–60 cm in diameter, bark brown (coffee coloured), twigs glabrous, spines axillary. **LEAVES.** Pinnate or bipinnate, 2.5–12 cm long, with 20–30 pairs of glabrous, linear, oblong leaflets. **FLOWERS.** Racemes 4–10 cm long, petals greenish white yellow, 3–4 mm long, calyx 1 mm long (Figure 1) [5].

Figure 1 – Inflorescences, bark and fruits of *P. laevigata*. (Photos: O. Téllez-Valdés, FESI-UNAM.)

FRUITS AND SEEDS

The fruits are samaras, 9–17 cm long, 0.7–1.4 cm wide, yellow, linear, glabrous, straight or slightly curved. The segments are rounded or rectangular in cross-section, and shorter in length than width. The seeds are longitudinally positioned in the legume (Figure 2) [3].

FLOWERING AND FRUITING. Flowers from February to May [6]; fruits from May to August.

Figure 2 – External views of the seeds. (Photo: P. Gómez-Barreiro, RBG Kew.)

DISTRIBUTION

Southern USA to Central America [13]. In Mexico, the species is found in the central highlands of northern Mexico, the lowlands of southern Tamaulipas, and in parts of Oaxaca, Morelos, Puebla and Chiapas [5]. In Puebla, it is abundant in Zapotitlán Salinas.

HABITAT

Hillsides, depressions and floodplains [5].

USES

In traditional medicine, the bark is used to treat bee stings, toothache, eye infections, rheumatism and fever; young roots are used as a diuretic and for bronchitis [2]. The plant produces a resin that has comparable properties to gum arabic derived from *Senegalia senegal* (L.) Britton [syn. *Acacia senegal* (L.) Willd.], which is used in the food industry. An insect locally known as 'cocopaches' (*Thasus* sp., Coreidae), a leaf-footed bug which is a very popular dish among the Popoloca people in Zapotitlán Salinas, feeds mainly on the sap, young leaves and green seed pods of mezquite [1,2]. *Prosopis laevigata* is used by the community of San Rafael close to the Tehuacán-Cuicatlán valley as a source of firewood and construction material, as well as for food and forage [11]. *Prosopis laevigata* is a nurse plant for the establishment of the cactus *Hylocereus undatus* (Haw.) Britton & Rose [12].

KNOWN HAZARDS AND SAFETY

Care should be taken when handling the plant as the thorns are sharp.

CONSERVATION STATUS

Assessed as Lower Risk-Least Concern (LR-lc) according to previous IUCN Red List criteria [7]. *Prosopis laevigata* is not included in the national list of protected species in Mexico [4]; however, it is commonly harvested as high-quality firewood, which results in constant reduction of wild populations [2].

SEED CONSERVATION

HARVESTING. Seed collection takes place from July to September. Samaras are collected directly from the plants.

PROCESSING AND HANDLING. Collected fruits are dried for 5–8 days to complete their maturation. Fully ripe fruits are soaked in water to soften their shells, then gently crushed to extract the seeds. The seeds are separated from the soft endocarp and fruit using sieves, and then left to dry.

STORAGE AND VIABILITY. X-ray analysis carried out on seed lots stored at the MSB showed that the percentage of filled seeds varies consistently from 40% to 90% due to the presence of infested seeds. Seeds of this species are desiccation tolerant [10].

PROPAGATION

SEEDS

Dormancy and pre-treatments: Seeds may be scarified by soaking them in commercial sodium hypochlorite for 1 hour, or for 30 minutes in concentrated sulphuric acid, and then washed several times in running water and soaked in water for 24 hours. At the MSB, seeds have been scarified mechanically by removing the covering structures or chipping the seed coat [10]. This species is propagated at the San Rafael, Coxcatlán nurseries using similar pre-treatments as in the laboratory controlled conditions, i.e. chemical scarification with 100% sulphuric acid for 20 minutes, soaking in water for 24 hours, mechanical scarification by making a small cut in the seed coat, and then soaking in water for 24 hours.

Germination, sowing and planting: After being scarified, a high percentage of seeds germinate when incubated under light at 20–25°C [10]. Seeds can be sown directly in the field, or in the nursery for subsequent transplanting. Raising the plants in the nursery has the advantage of allowing the plants to develop thorns, allowing protection from herbivores when the plants are transferred to the field. Furthermore, seedlings with well-established leaves and root systems can take advantage of the rainy season for their establishment. The recommended substrate is a mixture of 70% soil, 15% sand and 15% dried manure by volume, placed in black polyethylene bags (25 cm deep and 15 cm in diameter, with a small cut in the corner for drainage). 1–2 pre-treated seeds are sown in each bag at a depth of 1.5–2 cm. Germination occurs in 2–3 days; 20 cm high seedlings can be transferred to the field after four months. Planting can start in March, before the rainy season, or can be carried out in the autumn/winter period if protection against cold is provided. Watering should be carried out weekly. The soil should be moist but not waterlogged. Seedlings should be protected from full sun by planting them in shaded areas at a spacing of 5 x 5 m (i.e. 400 plants per hectare). Plants established at this spacing should be pruned at a height of 1.8 m to encourage the development of lateral branches, which would otherwise be within the reach of browsing animals. Although not requiring fertiliser, the addition of 5 kg of animal manure to each plant will promote better growth. By the third year, it is recommended that this species is associated with other legumes, forage grasses and shrubs, or with other crops, such as *Opuntia ficus-indica* (Mill.) L. (nopal) or *Hylocereus undatus* (Haw.) Britton & Rose (pitahaya) [2]. Seeds germinated in the San Rafael, Coxcatlán nurseries were planted in plastic trays covered with a sealed plastic dome to reduce evaporation and the need to water. A 3:1 mixture by volume of agrolite and sifted leaf litter was used. The leaf litter had been sterilised by exposure to sunlight for 1–2 days. Two transplants were made to black plastic pots (at one month and six months after germination), using a 3:1 mixture of black soil and agrolite. Watering took place every five days, weeding was frequent, and root pruning took place at six months or 12 months. The plants were then ready to be transplanted to the field. Out of 1,900 seeds sown, 278 plants were established.

VEGETATIVE PROPAGATION

Can be propagated by cuttings (but only 30% typically root) or from prunings (which save up to two months in plant development) [8]. Also propagated by air layering [9].

TRADE

No data available.

Authors

Amanda Moreno Rodríguez, Efisio Mattana, Myrna Mendoza Cruz, Martín López Carrera, Cesar Flores, Tiziana Ulian and Paulina Hechenleitner.

References

[1] Acuña, A. M., Caso, L., Aliphat, M. M. & Vergara, C. H. (2011). Edible insects as part of the traditional food system of the Popoloca town of Los Reyes Metzontla, Mexico. *Journal of Ethnobiology* 31 (1): 150–169.

[2] Arias Toledo, A. A., Valverde Valdés, M. T. & Reyes Santiago, J. (2000). *Las Plantas de la Región de Zapotitlán Salinas, Puebla*. Instituto Nacional de Ecología, UNAM, Mexico.

[3] Calderón de Rzedowski, G. & Rzedowski, J. (2001). *Flora Fanerogámica del Valle de México*. Instituto de Ecología, A. C. & Comisión Nacional para el Conocimiento & Uso de la Biodiversidad, Pátzcuaro, Michoacán, Mexico.

[4] Diario Oficial de la Federación (2010). *Norma Oficial Mexicana NOM-059-SEMARNAT-2010, Protección Ambiental — Especies Nativas de México de Flora y Fauna Silvestres — Categorías de Riesgo y Especificaciones para su Inclusión, Exclusión o Cambio — Lista de Especies en Riesgo*. Secretaría de Medio Ambiente y Recursos Naturales, Mexico, 30 December 2010.

[5] Ffolliott, P. F. & Thames, J. L. (1983). *Handbook on Taxonomy of* Prosopis *in Mexico, Peru and Chile*. Food and Agriculture Organization of the United Nations (FAO), Rome, Italy.

[6] Gómez Lorence, F., Signoret Poillon, J. & Abuín Moreiras, M. C. (1970). *Mezquites y Huizaches: Algunos Aspectos de la Economía, Ecología y Taxonomía de los Géneros,* Prosopis *y* Acacia *en México*. Instituto Mexicano de Recursos Naturales Renovables, Mexico.

[7] IUCN (2017). *The IUCN Red List of Threatened Species*. Version 2017-3. http://www.iucnredlist.org/

[8] Olvera Hernández, M. T. (2006). *Evaluación de los Reguladores Auxínicos AIA, AIB y p-Nitrofenil-indol-3-acetato en la Rizogénesis de* Prosopis laevigata, Cercidium praecox *y* Mimosa luisana *de Zapotitlán Salinas, Puebla*. Thesis, FESI-UNAM, Mexico.

[9] Ramírez-Malagón, R., Delgado-Bernal, E., Borodanenko, A., Pérez-Moreno, L., Barrera-Guerra, J. L., Núñez-Palenius, H. G. & Ochoa-Alejo, N. (2014). Air layering and tiny-air layering techniques for mesquite [*Prosopis laevigata* (H. B. ex Willd.) Johnst.M.C.] tree propagation. *Arid Land Research and Management* 28 (1): 118–128.

[10] Royal Botanic Gardens, Kew (2017). *Seed Information Database (SID)*. Version 7.1. http://data.kew.org/sid/

[11] Sánchez Hernández, D. (2012). *Disponibilidad Especial y Temporal de las Plantas Útiles más Importantes de San Rafael, Municipio de Coxcatlán, Puebla*. Bachelor's thesis, FESI-UNAM, Mexico.

[12] Valiente-Banuet, A. & Ezcurra, E. (1991). Shade as a cause of the association between the cactus *Neobuxbaumia tetetzo* and the nurse plant *Mimosa luisana* in the Tehuacán Valley, Mexico. *Journal of Ecology* 79 (4): 961–971.

[13] WCSP (2017). *World Checklist of Selected Plant Families*. Royal Botanic Gardens, Kew. http://wcsp.science.kew.org/

Stenocereus pruinosus (Otto ex Pfeiff.) Buxb.

TAXONOMY AND NOMENCLATURE

FAMILY. Cactaceae

SYNONYMS. *Echinocactus pruinosus* Otto ex Pfeiff., *Lemaireocereus pruinosus* (Otto ex Pfeiff.) Britton & Rose

VERNACULAR / COMMON NAMES. Pitayo (Spanish).

PLANT DESCRIPTION

LIFE FORM. Cactus.

PLANT. Columnar, tree-like, <5 m tall, sparsely to much-branched, stems 8–12 cm wide, widening towards the base, apex markedly pruinose, whitish, becoming glaucous dark green; ribs 5–8, widely separated, with wavy margins. **LEAVES.** Reduced to spines; central spines 1–4, 1.5–4 cm long, subulate, ascendant, rigid, grey; radial spines 5–8, 1–2 cm long, subulate, radiate, rigid, yellow to grey. **FLOWERS.** Infundibuliform, white to pinkish white, 7–9.5 cm long, receptacle tube larger than perianth (Figure 1) [1,2].

Figure 1 – Habit of *S. pruinosus*. (Photo: D. Franco-Estrada, FESI-UNAM.)

FRUITS AND SEEDS

Fruits red, purple or greenish orange, 4–8 x 3.5–6 cm, spiny, with deciduous spines (Figure 2). Seeds black, c. 2 x 1–2 mm, wrinkled, with a keel from the dorsal region to the apical (Figure 3) [8].

FLOWERING AND FRUITING. Flowers from March to September; fruits from May to September [1].

Figure 2 – Spiny fruits. (Photo: O. Téllez-Valdés, FESI-UNAM.)

Figure 3 – External views of the seeds, showing the dorsal keel, and longitudinal section showing the embryo. (Photos: P. Gómez-Barreiro, RBG Kew.)

DISTRIBUTION

Endemic to Mexico (Chiapas, Guanajuato, Guerrero, Michoacán, Oaxaca, Puebla, Querétaro, Tamaulipas and Veracruz) [1,2].

HABITAT

Deciduous dry forests, xerophilous scrub and secondary vegetation, at 500–1,900 m a.s.l. [1].

USES

The fruits, locally known as 'pitayas', are eaten fresh or used to make jam. A cold drink ('aguas frescas') is made by soaking the fruits in water and sugar [2,5]. *Stenocereus pruinosus* is sometimes planted as a hedge and boundary marker [3]. It is also used for fodder and firewood [4].

KNOWN HAZARDS AND SAFETY

Care should be taken when handling the plant as the spines are sharp.

CONSERVATION STATUS

Assessed as Least Concern (LC) according to IUCN Red List criteria [9]. *Stenocereus pruinosus* is not included in the national list of protected species in Mexico [7].

SEED CONSERVATION

HARVESTING. Fruits are harvested from wild plants and from silviculturally managed wild plants, using a pole with scissors.

PROCESSING AND HANDLING. Fruits are cut open with a knife to extract the seeds from the pulp. The seeds are then washed under tap water and dried at room temperature, after which they are passed through a sieve or blown with an air blower, to separate them from any remaining dried pulp.

STORAGE AND VIABILITY. X-ray analysis carried out on seed lots stored at the MSB showed that the percentage of filled seeds is always >90%. Seeds of this species are reported to be desiccation tolerant [10].

PROPAGATION

SEEDS

Dormancy and pre-treatments: At the San Rafael, Coxcatlán nursery, seeds were surface sterilised in 2% sodium hypochlorite for seven minutes, and then rinsed several times with tap water and left to dry for 12–24 hours.

Germination, sowing and planting: Seeds were germinated at the San Rafael, Coxcatlán nursery following the pre-treatment described above. They were sown in germination trays in a 3:1 mixture by volume of agrolite and sieved leaf litter. The leaf litter had been sterilised by exposure to sunlight in transparent plastic bags for 1–2 days. The seeds were scattered evenly over the watered substrate and then covered by a fine layer of substrate. The seedbed was then covered with a sealed dome to reduce evaporation. Transplanting into plastic bags or pots took place after 60–90 days when the seedlings were 1–2 cm tall, using a mixture of equal parts by volume of sterilised sieved leaf litter and fine gravel (e.g. 'tepojal', 'tepezil' or 'tezontle'). A layer of coarse gravel was placed in the bottom of containers to ensure good drainage. Two transplants were made to black plastic pots at three months and six months after germination. Watering took place every five days, then every 15 days after three months. Hand weeding was carried out frequently. The plants were ready to be planted in the field after 6–12 months. Seed germination experiments carried out under laboratory controlled conditions on seed lots stored at the MSB highlighted high germination percentages (88–100%) for untreated seeds incubated in the light, at temperatures of 20–30°C.

VEGETATIVE PROPAGATION

The base of stem cuttings should be allowed to dry before planting in a well-drained soil mixture. As yet, there is no definitive propagation protocol.

TRADE

Fruits of *S. pruinosus* are sold in local markets, such as those in Teotitlán, Tehuacán and Ajalpán [5]. International trade in this plant is regulated by CITES (listed on Appendix II) [6].

Authors

Daniel Franco-Estrada, Myrna Mendoza Cruz, Martín López Carrera, Efisio Mattana, Steve Davis and Paulina Hechenleitner.

References

[1] Anderson, E. F. (2001). *The Cactus Family*. Timber Press, Portland, Oregon, USA.

[2] Arias, S., Gama-López, S., Guzmán-Cruz, L. U. & Vázquez-Benítez, B. (2012). *Flora del Valle de Tehuacán-Cuicatlán, Fascículo 95, Cactaceae Juss.*, 2nd edition. Instituto de Biología, UNAM & CONABIO, Mexico.

[3] Arias Toledo, A. A., Valverde Valdés, M. T. & Reyes Santiago, J. (2000). *Las Plantas de la Región de Zapotitlán Salinas, Puebla*. Instituto Nacional de Ecología, UNAM, Mexico.

[4] Casas, A. (2002). Uso y manejo de cactáceas columnares mesoamericanas. *CONABIO Biodiversitas* 40: 18–23.

[5] Cervantes, H. (2014). Personal communication.

[6] CITES (2017). *Convention on International Trade in Endangered Species of Wild Fauna and Flora*. https://cites.org/eng/app/index.php

[7] Diario Oficial de la Federación (2010). *Norma Oficial Mexicana NOM-059-SEMARNAT-2010, Protección Ambiental — Especies Nativas de México de Flora y Fauna Silvestres — Categorías de Riesgo y Especificaciones para su Inclusión, Exclusión o Cambio — Lista de Especies en Riesgo*. Secretaría de Medio Ambiente y Recursos Naturales, Mexico, 30 December 2010.

[8] Franco, E. D. (2013). *Catálogo Ilustrado de Cactáceas Columnares del Valle de Tehuacán-Cuicatlán*. Bachelor's thesis, FESI-UNAM, Mexico.

[9] IUCN (2017). *The IUCN Red List of Threatened Species*. Version 2017-3. http://www.iucnredlist.org/

[10] Royal Botanic Gardens, Kew (2017). *Seed Information Database (SID)*. Version 7.1. http://data.kew.org/sid/

Stenocereus stellatus (Pfeiff.) Riccob.

TAXONOMY AND NOMENCLATURE

FAMILY. Cactaceae

SYNONYMS. *Lemaireocereus stellatus* (Pfeiff.) Britton & Rose

VERNACULAR / COMMON NAMES. Pitayo (Spanish); xoconochtli, xoconostle (Nahuatl).

PLANT DESCRIPTION

LIFE FORM. Cactus.

PLANT. Columnar, <4 m tall, shrubby or tree-like, branching near base, stems dark to bluish green, 7–16 cm wide; ribs (7)8–12(13), margins wavy. **LEAVES.** Reduced to spines; central spines 1–4 (usually 3), of which 2 divergent upwards, 1.5–2 cm long, and 1 downward pointing, 1.5–3(5) cm long, acicular, rigid, grey; radial spines 5–13, 0.5–1.2(1.6) cm long, acicular, rigid, yellowish white, turning grey. **FLOWERS.** Tubular-infundibuliform, white to pale pink, 4.5–6 cm long, apicular, receptacle tube longer than perianth (Figure 1) [1,2].

Figure 1 – Habit of *S. stellatus*.
(Photo: D. Franco-Estrada, FESI-UNAM.)

FRUITS AND SEEDS

Fruits red, 3.5–4 cm long, 3–3.5(4) cm wide, spiny, spines deciduous (Figure 2). Seeds black, c. 1.5–2 x 1–1.5 mm wide, wrinkled, with a dorsal keel (Figure 3) [8].

FLOWERING AND FRUITING. Flowers from June to September; fruits from July to October [2].

Figure 2 – Spiny fruits and stems. (Photo: O. Téllez-Valdés, FESI-UNAM.)

Figure 3 – External views of the seeds. (Photo: P. Gómez-Barreiro, RBG Kew.)

DISTRIBUTION

Endemic to Mexico (Morelos, Oaxaca and Puebla) [2].

HABITAT

Tropical deciduous forests, xerophilous scrub and secondary vegetation, at 600–2,300 m a.s.l. [2].

USES

The fruits and seeds are edible [3,4]. The fruits are eaten fresh or used to make jam, as well as a cold drink ('aguas frescas') [3]. The plant is used as a hedge and boundary marker [3,4]. Dry stems are used as firewood in traditional pottery making. *Stenocereus stellatus* is also used for fodder and for making an alcoholic beverage ('colonche') [4,5], and is planted on cultivated slopes to prevent soil erosion [5]. Different phenotypes are distinguished, named and classified by indigenous people according to fruit characteristics, especially the size, colour and flavour of the pulp, the spininess and the thickness of the peel [5].

KNOWN HAZARDS AND SAFETY

Care should be taken when handling the plant as the spines are sharp.

CONSERVATION STATUS

Assessed as Least Concern (LC) according to IUCN Red List criteria [9]. *Stenocereus stellatus* is not included in the national list of protected species in Mexico [7].

SEED CONSERVATION

HARVESTING. Fruits are harvested from wild plants, silviculturally managed wild plants and cultivated plants, using a pole with scissors [5].

PROCESSING AND HANDLING. Fruits are cut open with a knife to extract the seeds from the pulp. The seeds are then washed under tap water and dried at room temperature, after which they are passed through a sieve or blown with an air blower to separate them from any remaining dried pulp.

STORAGE AND VIABILITY. X-ray analysis carried out on the seed lot stored at the MSB showed a high percentage of filled seeds (88%). No information is available on seed storage behaviour; however, the seeds of *S. stellatus* are likely to be desiccation tolerant like those of the closely related *S. pruinosus* (Otto ex Pfeiff.) Buxb. [11].

PROPAGATION

SEEDS

Dormancy and pre-treatments: At the San Rafael, Coxcatlán nursery, seeds were surface sterilised in 2% sodium hypochlorite for seven minutes, and then rinsed several times with tap water and left to dry for 12–24 hours.

Germination, sowing and planting: Seeds were germinated at the San Rafael, Coxcatlán nursery following the pre-treatment described above. They were sown in germination trays in a 3:1 mixture by volume of agrolite and sieved leaf litter. The leaf litter had been sterilised by exposure to sunlight in transparent plastic bags for 1–2 days. The seeds were scattered evenly over the watered substrate and then covered with a fine layer of substrate. The seedbed was then covered with a sealed dome to reduce evaporation. Transplanting into plastic bags or pots took place after 60–90 days when the seedlings were 1–2 cm tall, using a mixture of equal parts by volume of sterilised sieved leaf litter and fine gravel (e.g. 'tepojal', 'tepezil' or 'tezontle'). A layer of coarse gravel was placed in the bottom of containers to ensure good drainage. Two transplants were made to black plastic pots at three months and six months after germination. Watering took place every five days, then every 15 days after three months. Hand weeding was carried out frequently. The plants were ready to be planted in the field after 6–12 months. Under laboratory controlled conditions at the MSB, high germination percentages were achieved, reaching a maximum rate of 94% at 20°C. Higher temperatures resulted in lower germination rates (i.e. 83% and 69% at 25°C and 30°C, respectively).

VEGETATIVE PROPAGATION

Stenocereus stellatus is propagated by cuttings and grown in orchards ('huertas' and 'solares'). Branches of 1–1.5 m long are cut and left to dry in the sun for two weeks to prevent fungal infection of the cut surfaces. Planting is carried out a few weeks before the start of the rainy season, between the end of April and the beginning of May. Branches are planted in holes enriched with goat manure [4]. López-Gómez *et al.* [10] determined that the best results were obtained using apical cuttings of 0.5 m in length.

TRADE

There is considerable demand for fruits and other products derived from *S. stellatus*, which are commercially important and widely available in local markets [5]. International trade in this plant is regulated by CITES (listed on Appendix II) [6].

Authors

Daniel Franco-Estrada, Myrna Mendoza Cruz, Martín López Cabrera, Tiziana Ulian, Efisio Mattana, Steve Davis and Paulina Hechenleitner.

References

[1] Anderson, E. F. (2001). *The Cactus Family*. Timber Press, Portland, Oregon, USA.

[2] Arias, S., Gama-López, S., Guzmán-Cruz, L. U. & Vázquez-Benítez, B. (2012). *Flora del Valle de Tehuacán-Cuicatlán, Fascículo 95, Cactaceae Juss.*, 2nd edition. Instituto de Biología, UNAM & CONABIO, Mexico.

[3] Arias Toledo, A. A., Valverde Valdés, M. T. & Reyes Santiago, J. (2000). *Las Plantas de la Región de Zapotitlán Salinas, Puebla*. Instituto Nacional de Ecología, UNAM, Mexico.

[4] Casas, A. (2002). Uso y manejo de cactáceas columnares mesoamericanas. *CONABIO Biodiversitas* 40: 18–23.

[5] Casas, A., Pickersgill, B., Caballero, J. & Valiente-Banuet, A. (1997). Ethnobotany and domestication in xoconochtli, *Stenocereus stellatus* (Cactaceae) in the Tehuacán Valley and La Mixteca Baja, México. *Economic Botany* 51 (3): 279–292.

[6] CITES (2017). *Convention on International Trade in Endangered Species of Wild Fauna and Flora*. https://cites.org/eng/app/index.php

[7] Diario Oficial de la Federación (2010). *Norma Oficial Mexicana NOM-059-SEMARNAT-2010, Protección Ambiental — Especies Nativas de México de Flora y Fauna Silvestres — Categorías de Riesgo y Especificaciones para su Inclusión, Exclusión o Cambio — Lista de Especies en Riesgo*. Secretaría de Medio Ambiente y Recursos Naturales, Mexico, 30 December 2010.

[8] Franco, E. D. (2013). *Catálogo Ilustrado de Cactáceas Columnares del Valle de Tehuacán-Cuicatlán*. Bachelor's thesis, FESI-UNAM, Mexico.

[9] IUCN (2017). *The IUCN Red List of Threatened Species.* Version 2017-3. http://www.iucnredlist.org/

[10] López-Gómez, R., Díaz-Pérez, J. C. & Flores-Martínez, G. (2000). Vegetative propagation of three species of cacti: pitaya (*Stenocereus griseus*), tunillo (*Stenocereus stellatus*) and jiotilla (*Escontria chiotilla*). *Agrociencia* 34 (3): 363–367.

[11] Royal Botanic Gardens, Kew (2017). *Seed Information Database (SID).* Version 7.1. http://data.kew.org/sid/

Varronia bullata subsp. *humilis* (Jacq.) Feuillet

TAXONOMY AND NOMENCLATURE

FAMILY. Boraginaceae

SYNONYMS. *Cordia bullata* var. *globosa* (Jacq.) Govaerts, *C. globosa* (Jacq.) Kunth, *C. globosa* var. *humilis* (Jacq.) I.M.Johnst., *C. globosa* subsp. *humilis* (Jacq.) Borhidi, *Varronia globosa* subsp. *humilis* (Jacq.) Borhidi

VERNACULAR / COMMON NAMES. Chilito, rompe camisa, yerba de la sangre (Spanish).

PLANT DESCRIPTION

LIFE FORM. Shrub.

PLANT. Up to 2 m tall; stems strigose-villose or hirsute to glabrescent, trichomes moniliform, eglandular. **LEAVES.** c. 1–5 x 0.3–2.5 cm, symmetric, ovate to oblong-ovate, with acute base, apex acuminate to acute, serrate, bi-serrate to bi-crenate; adaxial surface estrigose, abaxial surface densely villose, glandular; petioles c. 0.5–2.5 cm long. **FLOWERS.** White, pale yellow to greenish, heterostylous, sessile, in terminal, globose heads of 1–2 cm in diameter, comprising c. 40 flowers, peduncles c. 0.5–3 cm long; corolla bell-shaped (Figure 1) [5].

Figure 1 – Flowers and plant.
(Photos: O. Téllez-Valdés, FESI-UNAM.)

FRUITS AND SEEDS

One-seeded drupes, c. 4–4.5 x 3 mm wide, subglobose, light yellow and greenish with brown spots, tuberculate (Figure 2) [5].

FLOWERING AND FRUITING. Flowers and fruits from June to August [5].

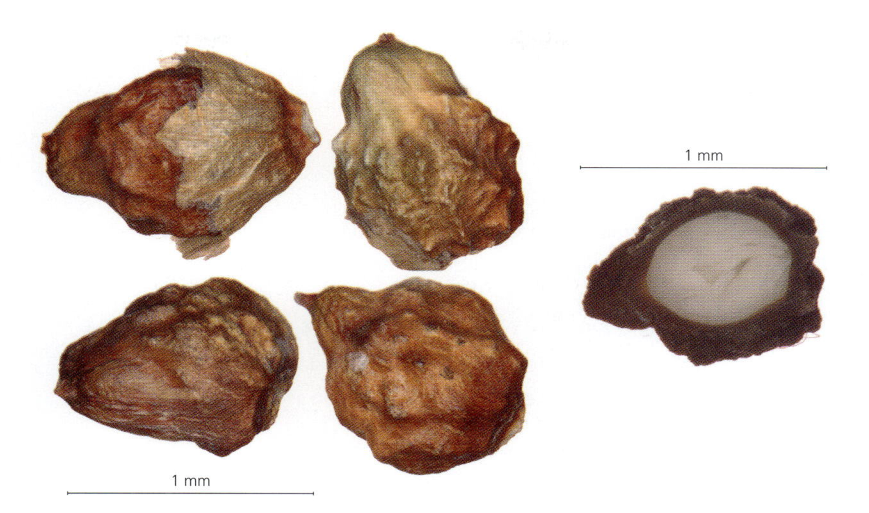

Figure 2 – External views of the fruits and longitudinal section of the seed showing a folded embryo. (Photos: P. Gómez-Barreiro, RBG Kew.)

DISTRIBUTION

From Mexico to South America, including the Caribbean region. In Mexico, it occurs in Campeche, Chiapas, Guerrero, Jalisco, Morelos, Puebla, Quintana Roo, San Luis Potosí, Sinaloa and Yucatán [5].

HABITAT

Deciduous dry forests, on reddish soils, at altitudes of 1,300–1,800 m a.s.l. [5].

USES

Used by inhabitants of San Rafael, Coxcatlán, Puebla in the treatment of infectious diseases and gastrointestinal disorders [1,3]. An infusion of the aerial parts of the plant is used for treating fungal infections of the skin, and gastrointestinal and throat conditions of possible infectious origin, as well as for its antitussive, astringent, haemostatic and tonic properties [3]. Laboratory research on the essential oil derived from aerial parts of the plant has indicated antimicrobial activity [6].

KNOWN HAZARDS AND SAFETY

No known hazards.

CONSERVATION STATUS

Not considered to be either rare or threatened in Mexico [2], but assessed according to IUCN Red List criteria as Least Concern (LC) in Cuba (as *Varronia globosa* var. *humilis*) [7]. The status of this species has yet to be confirmed in the IUCN Red List [4].

SEED CONSERVATION

HARVESTING. Fruits can be collected manually from the plants in August and September.

PROCESSING AND HANDLING. Harvested fruits should be placed in paper bags until they are delivered to the seed bank, where seeds can be manually extracted from the dry fruits. An air blower can be used to separate the seeds from other plant material.

STORAGE AND VIABILITY. X-ray analysis carried out on seed lots stored at the MSB showed a high percentage of empty seeds, with only c. 60% of seeds filled. No data are available on the seed storage behaviour of this species and its relatives.

PROPAGATION

SEEDS

Dormancy and pre-treatments: Seed dormancy breaking and germination requirements are not well-investigated. Further studies under controlled conditions are needed to determine key factors for enhancing seed germination.

Germination, sowing and planting: Few seeds (20%) germinated when incubated for 20 days at the FESI-UNAM Seed Bank at 30°C with a photoperiod of 12 hours of darkness and 12 hours of light. In addition, only low germination percentages were achieved on seeds stored at the MSB, confirming the need for further research.

VEGETATIVE PROPAGATION

Can be propagated by cuttings. Experiments on 15 cm long cuttings indicate that the best results are obtained with a mixture of equal parts by volume of soil and coconut fibre [8].

TRADE

The plant is not traded or available in local markets; it is collected when needed.

Authors

Fernando Peralta-Romero, Efisio Mattana, Steve Davis and Paulina Hechenleitner.

References

[1] Alvarado-Zavaleta, Miguel, M. I., Canales, M., Jiménez, M., Meráz, S., García, A. M., Ávila, J. G. & Hernández, T. (2011). Actividad antimicrobiana de *Cordia globosa* (Jacq.) Kunth. *Rev. Latinoamer. Quim.* 38 (Suplemento Especial).

[2] Diario Oficial de la Federación (2010). *Norma Oficial Mexicana NOM-059-SEMARNAT-2010, Protección Ambiental — Especies Nativas de México de Flora y Fauna Silvestres — Categorías de Riesgo y Especificaciones para su Inclusión, Exclusión o Cambio — Lista de Especies en Riesgo*. Secretaría de Medio Ambiente y Recursos Naturales, Mexico, 30 December 2010.

[3] Hernández, T., Canales, M., Avila, J. G., Durán, A., Caballero, J., Romo de Vivar, A. & Lira, R. (2003). Ethnobotany and antibacterial activity of some plants used in traditional medicine of Zapotitlán de las Salinas, Puebla (México). *Journal of Ethnopharmacology* 88 (2–3): 181–188.

[4] IUCN (2017). *The IUCN Red List of Threatened Species.* Version 2017-3. http://www.iucnredlist.org/

[5] Lira-Charco, E. & Ochotereno, H. (2012). *Flora del Valle de Tehuacán-Cuicatlán, Fascículo 110, Boraginaceae Juss.* Instituto de Biología, UNAM, Mexico.

[6] Melissa, M., García-Bores, A., Meraz, S., Piedra, E., Ávila, M., Serrano, R., Orozco, J., Jiménez-Estrada, M., Chavarría, J. C., Peñalosa, I., Ávila, J. G. & Hernández, T. (2016). Antimicrobial activity of essential oil of *Cordia globosa*. *African Journal of Pharmacy and Pharmacology* 10 (11): 179–184.

[7] National Red List (2017). *Varronia globosa* var. *humilis*. National Red List Network. http://www.nationalredlist.org/

[8] Paulino, R. de C., Henriques, G. P. de S. A., Sousa Neto, H. L., Coelho, M. de F. B., Dombroski, J. L. D. & Lopes, C. P. de A. (2011). Influência do substrato na propagação por e staquia de *Cordia globosa*. *Cadernos de Agroecologia* 6 (2): 1–5.

Varronia curassavica Jacq.

TAXONOMY AND NOMENCLATURE

FAMILY. Boraginaceae

SYNONYMS. *Cordia brevispicata* M.Martens & Galeotti, *C. curassavica* (Jacq.) Roem. & Schult., *C. divaricata* Kunth, *C. graveolens* Kunth, *C. peruviana* var. *mexicana* A.DC., *Lithocarpum curassavicum* (Jacq.) Kuntze, *L. divaricatum* (Kunth) Kuntze

VERNACULAR / COMMON NAMES. Black sage, wild sage (English); barredor, bola negra (Spanish).

PLANT DESCRIPTION

LIFE FORM. Shrub.

PLANT. Deciduous, <3 m tall, young branches glandular resinous, scattered or densely hispidulous, shortly hirsute or strigose. **LEAVES.** Alternate, 16 x 1–3 cm, often on very short branches, agglomerated, slightly varied in shape, thin, sometimes with tiny hairs or glands, and short and pubescent petioles (Figure 1). **FLOWERS.** Sessile, in pedunculate, terminal spikes; calyx bell-shaped, granulose, pubescent; corolla white or greenish white, tubular, 5-lobed, lobes recurved; stamens 5, filaments pubescent; style twice divided (Figure 1) [3,5].

Figure 1 – Inflorescence (left) and plant with unripe (green) and ripe (red) fruits. (Photos: O. Téllez-Valdés, FESI-UNAM.)

FRUITS AND SEEDS

Fruits are globose, 4–5 mm long, fleshy, red, ovoid, one-seeded drupes, with a persistent calyx (Figures 1&2) [5].

FLOWERING AND FRUITING. Flowers and fruits from August to November.

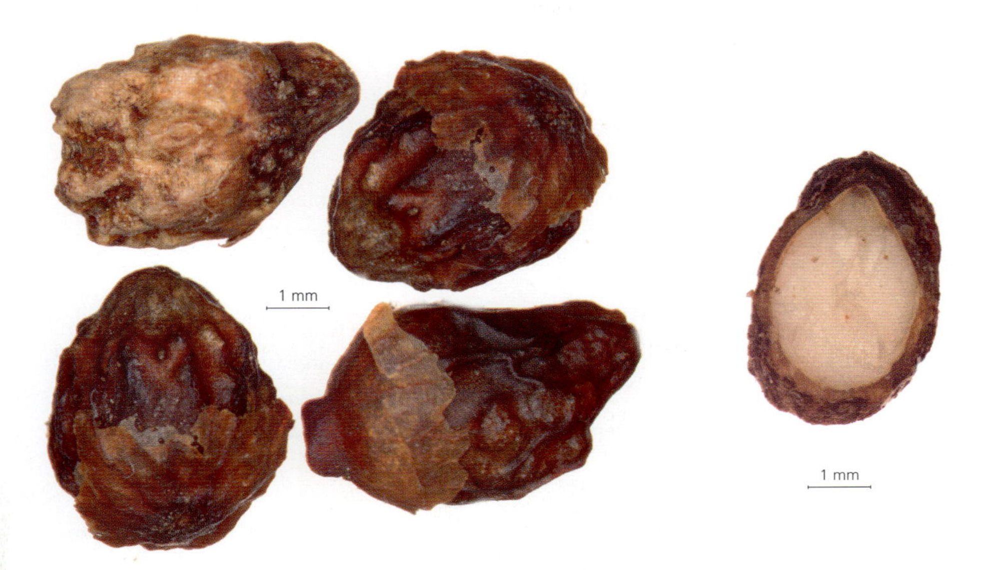

Figure 2 – External views of the fruits (left) and longitudinal section (right) of the seed showing a folded embryo. (Photos: P. Gómez-Barreiro, RBG Kew.)

DISTRIBUTION

Native to Mexico (Baja California, Baja California Sur, Sonora, Campeche, Chiapas and Quintana Roo), Central America, northern South America and the Caribbean (Greater Antilles and Lesser Antilles) [1,5]. It is naturalised in parts of South-East Asia and on some islands of the Pacific and Indian Oceans (e.g. Mauritius) and can be invasive [3,9].

HABITAT

Xerophilous scrub and deciduous dry forest, from near sea level to c. 1,000 m a.s.l. [5].

USES

The fruits are edible; tea made from the leaves is used to treat stomach infections. The plant is also used as animal fodder and to produce honey [8].

KNOWN HAZARDS AND SAFETY

No known hazards.

CONSERVATION STATUS

A widespread species that is not considered to be either rare or threatened in Mexico [2], but its status has yet to be confirmed in the IUCN Red List [4].

SEED CONSERVATION

HARVESTING. Ripe fruits can be collected in November.

PROCESSING AND HANDLING. Collected fruits should be stored in paper or cotton bags and kept in a dark and dry place until they are delivered to the seed bank, where they can be separated manually from other plant material.

STORAGE AND VIABILITY. X-ray analysis carried out on the seed lot stored at the MSB showed a high percentage of filled seeds (77%). Seeds of this species are reported to be desiccation tolerant [6].

PROPAGATION

SEEDS

Dormancy and pre-treatments: High germination percentages have been achieved for scarified seeds (i.e. by removing the covering structure) [8], suggesting the presence of physical or physiological dormancy. However, further studies should be carried out to identify the type of seed dormancy present.

Germination, sowing and planting: 100% of intact seeds germinated when incubated at the FESI-UNAM Seed Bank at 30°C, with a photoperiod of 12 hours darkness and 12 hours light for 20 days (Figure 3). 100% germination was also achieved at the MSB for scarified seeds incubated at constant 25°C and at an alternating temperature regime of 20/35°C with 8 hours of light per day [6].

VEGETATIVE PROPAGATION

Propagated successfully from 10–15 cm cuttings at Universidad Nacional del Nordeste, Corrientes, Argentina. Experiments suggest that the use of rooting hormone powder is not required [7].

TRADE

No data available.

Figure 3 – Seedlings of *V. curassavica*. (Photo: L. García, FESI-UNAM.)

Authors

Lirio Jazmín Sánchez Hernández, Efisio Mattana and Paulina Hechenleitner.

References

[1] Arellano Rodríguezi, J. A., Salvador Flores, J., Tun Garrido, J. & Cruz Bojórquez, M. M. (2003). *Nomenclatura, Forma de Vida, Uso, Manejo y Distribución de las Especies Vegetales de la Península de Yucatán, Fascículo 20*. Universidad Nacional Autónoma de Yucatán, Mexico.

[2] Diario Oficial de la Federación (2010). *Norma Oficial Mexicana NOM-059-SEMARNAT-2010, Protección Ambiental — Especies Nativas de México de Flora y Fauna Silvestres — Categorías de Riesgo y Especificaciones para su Inclusión, Exclusión o Cambio — Lista de Especies en Riesgo*. Secretaría de Medio Ambiente y Recursos Naturales, Mexico, 30 December 2010.

[3] Heike, V. (2009). *Cordia curassavica* (Jacq.) Roem. & Schult. (Boraginaceae). In: *Malezas de Mexico*. Edited by H. Vibrans. CONABIO, Mexico. http://www.conabio.gob.mx/malezasdemexico/boraginaceae/cordia-curassavica/fichas/ficha.htm

[4] IUCN (2017). *The IUCN Red List of Threatened Species.* Version 2017-3. http://www.iucnredlist.org/

[5] Nash, L. D. & Moreno, N. P. (1981). *Flora de Veracruz, Fascículo 18*: *Boraginaceae*. Instituto Nacional de Investigaciones Sobre Recursos Bióticos, Xalapa, Vera Cruz, Mexico.

[6] Royal Botanic Gardens, Kew (2017). *Seed Information Database (SID).* Version 7.1. http://data.kew.org/sid/

[7] Schroeder, M. A. & Velozo, L. E. (2014). Reproducción vegetativa de *Cordia curassavica* (María negra). *Horticultura Argentina* 33 (81): 37–43.

[8] Téllez Valdés, O., Reyes, M., Dávila Aranda, P., Gutiérrez García, K., Téllez Poo, O., Álvarez Espino, R. X., González Romero, A. V., Rosas Ruiz, I., Ayala Razo, M., Murguía Romero, M. & Guzmán, U. (undated). *Guía Ecoturística. Las Plantas del Valle de Tehuacan-Cuicatlan*. Volkswagen, FESI-UNAM & Millennium Seed Bank Project, Royal Botanic Gardens, Kew.

[9] United States Department of Agriculture (USDA) (2013). *Weed Risk Assessment for* Cordia curassavica *(Jacq.) Roem. & Schult. (Boraginaceae) — Black Sage*. Animal and Plant Health Inspection Service (APHIS), USDA, Raleigh, North Carolina, USA.

Glossary

compiled by Steve Davis

Botanical, Horticultural, Mycological and Vegetation Terms

A

abaxial, the side of an organ that faces away from the axis that bears it; for example, the lower surface of a leaf.

abscission, (of leaves or leaflets, sometimes on flower or fruit stalks, rarely branches), detaching from the stems that bear them at a predetermined place, the abscission zone.

acaulous, without a stem, or without a visible stem.

achene, a small dry thin-walled fruit, not splitting when ripe, and containing a single seed.

acicular, needle-shaped.

aculeate, armed with prickles (as distinct from thorns).

acuminate, tapering to a long tip (usually of leaf tips).

acute, sharp, sharply pointed, the margins near the tip being almost straight and forming an angle of <90°.

adaxial, the side of an organ towards the axis on which it is inserted (e.g. the upper surface of a leaf).

Afromontane, discontinuous mountain regions in Africa usually supporting a rich assemblage of plant species with many endemics, the vegetation and flora usually very distinct from that of surrounding lowlands.

agglomerated, densely crowded, but not stuck together.

agrolite, a soilless growing medium (e.g. perlite, vermiculite).

Columnar cactus, *Pachycereus weberi* (J.M.Coult.) Backeb., in San Rafael, Coxcatlán (Puebla) with UPP seed collector. (Photo: T. Ulian, RBG Kew.)

air layering, (also known as marcotting) a horticultural method of vegetative propagation in which a shoot of a woody plant is rooted while attached to the parent plant (usually by wrapping an incision in the stem with a moisture-retaining material, such as moss, and then surrounding with a moisture barrier, such as plastic film); when sufficient roots have formed the stem is cut from the parent plant and planted; see also **layering**.

alternate, inserted at different levels of the main line of development; as distinct from opposite.

angular, with an angle, as where two planes meet.

annual, completing its life cycle within one year or one growing season.

anther, the part of the stamen containing the pollen.

anthesis, time of fertilisation of the flower; time of receptivity of stigma or distribution of pollen; the time when the flower opens.

apetalous, without petals.

apex (plural **apices**)**,** distal end, tip.

apical, of the apex.

apiculate, ending in an abrupt, short point.

appendage, attached secondary part.

appressed, lying close and flat (e.g. branches or hairs on a stem).

arborescent, becoming tree-like.

areole, 1. usually flat area on each side of some mimosoid legume seeds that is surrounded by the pleurogram; 2. in Cactaceae, the spine-bearing cushion; extremely reduced branches (axillary buds) that usually bear spines.

aril, 1. an appendage partially or completely enveloping the seed, sometimes resembling a third integument, and arising from the hilum, funicle or any other part of the seed coat; 2. any fleshy cup-like structure containing a seed.

aromatic, producing volatile oils with discernible odours.

ascospore, a fungal spore produced sexually inside an ascus.

ascus (plural **asci**)**,** a membranous sac in which the spores of certain fungi develop.

asexual, sexless, without gender.

attenuate, gradually narrowing over a long distance.

axil, the angle between the stem and the leaf.

axillary, arising in an axil, the point between the stem and the leaf or another organ that arises from the stem.

B

bark, outermost layer of stems and roots in woody plants.

berry, an indehiscent simple fruit with one to many seeds immersed in a fleshy pulp, supported by an endocarp <2 mm thick, the pericarp not differentiated internally by a hardened endocarp or airspace.

bi-, (prefix) with two...

binomial, (in nomenclature) the name consisting of the genus name and species name.

bipinnate, doubly pinnate; i.e. the rachis bearing first-order axes which bear the leaflets; when the primary divisions of a pinnate leaf are themselves pinnate.

bisexual, having both sexes in the same flower, or in the same inflorescence.

blade, expanded part of leaf or petal.

bole, in trees, the part of the trunk below the lowermost branches; the unbranched part of the trunk above the buttresses.

bordered, with the margin a different colour.

bract, a modified and specialised leaf in the inflorescence, standing below partial peduncles, pedicels or flowers.

bracteole, a secondary bract, usually smaller than the bracts and always borne above them; a small modified leaf borne just below the flower, or anywhere along a pedicel above the bract.

branch, a lateral division of the growth axis.

bristle, 1. a slender and stiff cylindric emergence, about the size of a hair; 2. slender stiff continuation of midrib in inflorescence bract.

bud, immature shoot or flower.

bulb, underground storage organ; the bud(s) enclosed by fleshy scale leaves and/or leaf bases.

bushland, an open stand of bushes 2–7 m tall with a canopy cover of 40% or more.

Bushveld, low altitude scrub vegetation (a South African term).

buttress, mechanical supporting system at the base of a tree, usually a woody fin.

C

caducous, falling off soon after formation, not persistent.

calyx (plural **calyces**)**,** the outermost whorl of floral organs, often divided into sepals.

canopy, uppermost layer of vegetation usually of woodland or forest.

capitulum (plural **capitula**)**,** a compact cluster of ± sessile flowers; the capitulum may be surrounded by specialised bracts.

capsule, a dry dehiscent fruit composed of two or more united carpels, opening by valves, slits or pores.

carpel, one of the cells or locules of the syncarpous ovary.

caryopsis, the fruit in Poaceae, a small dry thin-walled fruit, with the single seed fused to the pericarp; a type of achene.

cataphyll, a scale leaf or scale-like leaf.

caudex, an enlarged storage organ at soil level, composed of the swollen stem or root, or both.

caudiciform, formed like a caudex, enlarged or swollen.

cauline, arising from, or inserted on, the stem.

central spine, (in cacti and similar succulents) the spine in the middle of the areole or spine shield, often larger or with a different colour from the others, the radial spines.

cerebriform, resembling the external fissures and convulutions of a brain.

chalaza, part of an ovule where the body joins the envelope.

chartaceous, thin and stiff, like paper.

circinnate, coiled inwards upon itself.

circular, round (in two dimensions).

clawed, with a very narrow part near the base, more distally expanded.

climber, a plant that grows upwards by attaching itself to other structures which it uses as supports.

columnar, in the form of a column or pillar.

coma, a tuft of long hairs at one (or both) end(s) of a seed; seed appendage to aid wind dispersal.

combinational dormancy, (of seed) a type of dormancy determined by the presence of both physical and physiological dormancy.

commissure, the place of joining.

compound, composed of several parts (e.g. leaf with leaflets).

compressed, flattened.

concave, hollow, like the inside of a bowl.

congeneric, belonging to the same genus.

conical, cone-shaped.

contracted, narrowing, or (of inflorescence) narrow and dense.

convex, with a rounded surface, like the outside of a bowl.

coppice, to cut back to near ground level at regular intervals.

coppice shoot, new branches arising after coppicing.

cordate, 1. (of the base of a leaf) deeply notched so the whole base has a slight heart shape; 2. the shape of the whole leaf, which is ovate with a notched base and an acute apex.

coriaceous, leathery, tough.

corm, short underground swollen stem, a storage stem.

corolla, the second whorl of floral organs, inside or above the calyx and outside the stamens, consisting of free petals or of a joined tube and petal lobes.

corona, a series of appendages on the corolla or on the back of the stamens, or at the junction of the corolla tube and the corolla lobes; often united in a ring.

cortical, of the bark.

corymb, a ± flat-topped, racemose (indeterminate) inflorescence in which the branches or the pedicels start from different points but all reach to about the same level.

corymbiform, shaped like a corymb.

cotyledon, seed-leaf.

cover crop, plants grown to combat soil erosion.

crenate, of margins, notched with regular, rounded symmetrical teeth.

crested, with an elevated, irregular ridge.

Critically Endangered, IUCN Red List term for a plant that is on the brink of extinction; for precise definition see http://www.iucnredlist.org/technical-documents/categories-and-criteria.

crown, in trees, the cluster of branches and leaves borne at the top of the trunk, or the shape formed by the uppermost and outermost leaves.

cryoconservation, (of seeds) a process of cooling and storing viable tissues at very low temperatures, usually in liquid nitrogen, to maintain viability.

culm, stem of a grass or sedge.

cultivar, a cultivated variety of a species.

cultivated, grown by humans in a modified environment.

cuneate, tapering gradually, wedge-shaped.

cupule, cup-like structure at the base of fruits, formed by the dry, enlarged floral envelope.

cylindric, like a cylinder, i.e. long and narrow with a circular cross-section.

cyme, a sympodial inflorescence in which the central flower opens first, growth being continued by axillary buds arising below this central flower; sometimes used for a compound, ± flat-topped inflorescence.

cymose, with a cyme.

cypsela, a ripened ovary with attached floral parts (i.e. longitudinally oriented awns, bristles or similar structures).

D

dambo, seasonally waterlogged depressions covered with grassland (in Africa).

Data Deficient (DD), IUCN Red List term for a plant for which there is inadequate information to make an assessment of extinction risk; for precise definition see http://www.iucnredlist.org/technical-documents/categories-and-criteria.

deciduous, falling seasonally, losing all of its leaves for part of the year.

decussate, used of opposite organs (e.g. leaves), when alternate pairs are at right angles to each other.

dehiscent, opening spontaneously when ripe.

deltoid, shaped like an equal-sided triangle.

dentate, prominently toothed with acute symmetrical projections pointing outwards (e.g. of margins).

***diaspore*,** reproductive portion of a plant, such as a seed, fruit or fragment of fruit that is dispersed and may give rise to a new plant.

dichasium (plural **dichasia**)**,** a determinate or definite inflorescence with a terminal flower and two opposite lateral branches, these again ending in a terminal flower and each with two opposite lateral branches ending in flowers.

didymous, 1. in pairs; 2. divided into two lobes.

digitate, 1. like fingers; 2. (of a compound leaf) when the leaflets diverge from the same point; = palmate.

dioecious, with unisexual flowers, the male and female flowers on different plants; with male and female plants.

disc floret, small tubular flowers in the centre of a flower head of Compositae.

discolorous, with two different colours (e.g. the upper surface of a leaf dark green, the lower surface white).

disjunct, (plant geography) with widely separated distribution areas.

disk, a ± flat plate-shaped object.

distal, furthest from the place of attachment (e.g. the tip is the distal part of a leaf).

domatia, small cavities that are linked to the presence of ants or mites.

dormancy, a period of pause in growth and development, usually to overcome adverse environmental conditions (cold, drought); in seeds can be of various types (e.g. physical, physiological, morphological, morphophysiological or combinational).

dorsal, 1. the back of…; 2. the surface facing away from the axis, i.e. abaxial.

drupe, a stone fruit (e.g. plum, cherry), a fleshy indehiscent fruit with the seed(s) enclosed in a stony endocarp.

drupelet, the single drupes that together form the fruit.

E

ectomycorrhiza, a form of symbiotic relationship that occurs between a fungal symbiont and the roots of various plant species.

eglandular, without glands.

ellipsoid, a 3-dimensional shape that is elliptic in the vertical plane.

elliptic, broadest at the middle with two equal rounded ends.

emarginate, (of apex) with a distinct sharp notch.

embryo, the rudimentary plant contained in the seed, consisting of cotyledon(s), radicle and plumule.

emergent, coming out of, rising from; in a forest an emergent is taller than the surrounding canopy.

empty seed, seed that looks perfectly developed externally, but which lacks an embryo (e.g. determined by X-ray analysis or a cut test); a high percentage of empty seeds is a measure of low quality of a seed lot (opposite: **filled seed**).

Endangered (EN), IUCN Red List term for a plant in danger of extinction; for precise definition see http://www.iucnredlist.org/technical-documents/categories-and-criteria.

endemic, restricted to, unique to, not naturally found elsewhere.

endo-, (prefix) within.

endocarp, the innermost layer of a multi-layered fruit wall.

endosperm, the food storage tissue within a seed that commonly surrounds the embryo, absent from the seeds of some species if absorbed during development.

endozoochorous, dispersal of plants through the interior of animals, through ingestion and excretion of fruit or seed.

entire, 1. not divided; 2. (of margins) smooth, unbroken by serrations, teeth or other irregularities.

epicarp, the outermost layer of a multi-layered fruit wall.

epidermal, having to do with the outermost layer of cells.

epidermis, the outermost layer of cells.

epigeal, (of germination) above ground.

epiphyte, plant growing on and attached to another plant without deriving nourishment from it.

erect, upright.

evergreen, retaining its leaves throughout the year.

exfoliating, coming off in large, thin-layered flakes.

exotic, not native in a given area.

explant, tissue taken from a plant and grown in a culture medium.

ex situ, away from its (original) place.

extraction, (of seed) the process of separating seeds from whole plants or fruits or their parts.

F

fascicle, a cluster of similar organs (e.g. leaves or flowers) arising from ± the same point.

fasciculate, (of erect branches) in close bundles.

fibrillar, relating to a fibril (a slender thread or fibre).

filament, a stalk that bears an anther.

filiform, slender, thread-like.

filled seed, seed that has an embryo (e.g. determined by X-ray analysis or a cut test); a high percentage of filled seeds is a measure of high quality of a seed lot (opposite: **empty seed**).

floret, (in Compositae) a single flower.

foetid, stinking.

follicle, a pod arising from a single carpel, opening along the inner (adaxial) suture to which the seeds are attached.

forest, habitat dominated by trees; area >0.5 hectare with trees >5 m tall and a canopy of (almost) interlocking tree crowns, with more than one layer of woody vegetation.

frondose, resembling a pinnately compound fern frond.

fruit, the seed-bearing organ.

fulvous, yellow, tawny.

fungoid, resembling a fungus.

funicle, the stalk of the ovule or seed attaching it to the placenta; seed stalk.

funicular, deriving from the funicle.

fusiform, thick but tapering towards both ends.

G

gallery forest, forest restricted to the banks of a seasonal or permanent river or stream.

genus (plural **genera**)**,** Linnean group containing related species (usually of similar appearance) and bearing the same first name of the binomial.

geophyte, a plant whose growing point survives adverse seasons as a resting bud on an underground organ, such as a rhizome, bulb, tuber or root.

germination, (of seed) the process by which a seed develops into a seedling.

glabrescent, becoming glabrous or nearly so.

glabrous, smooth, without hairs, scales etc.

gland, a secretory area or mass on the surface, either embedded or ending a hair.

glandular, covered with glands or with a zone of secretion-producing tissue.

glaucous, covered with a waxy bluish grey or sea green bloom which rubs off easily.

gleba, fleshy spore-bearing inner mass of certain fungi.

globose, round, spherical.

graft, horticultural technique involving insertion of a scion (bud, shoot or twig) on to a rootstock (various forms, e.g. cleft graft, side graft).

granulose, composed of grains.

grassland, habitat dominated by grasses or herbs, with very few woody plants present.

gynoecium, the female element of a flower.

H

hair, an outgrowth of the epidermis consisting of one or more elongated cells.

Hartig net, (of certain fungi) hyphal network that extends into the root, penetrating between the epidermal and cortical cells of ectomycorrhizal plants; this network is the site of nutrient exchange between the fungus and host plant.

hemi-, half.

herb, plant without a persistent woody stem above ground.

heterostylous, with flowers of two or more types each having styles of different lengths.

hilum, the scar left on the seed from its attachment point to the placenta.

hirsute, with rather coarse, stiff hairs.

hispid, with long stiff hairs or bristles, more sharply bristly than hirsute.

hispidulous, minutely hispid.

hyaline, colourless, translucent substance (of spores and other microscopic structures in fungi).

hybrid, a cross between two species.

hypha (plural **hyphae**)**,** each of the branching filaments that make up the mycelium of a fungus.

hypogeous, under the surface of the soil.

I

imbibition, (of seed) absorption of water resulting in the activation of metabolic processes leading to germination.

imparipinnate, unevenly pinnate, i.e. pinnate with a single terminal leaflet.

indehiscent, (of fruits) not splitting open.

inflorescence, the part of the plant that bears the flowers, including all its bracts, branches and flowers, but excluding unmodified leaves.

infraspecific, (of taxa or variation) below the rank of species (e.g. subspecies, variety, form or race).

infructescence, the part of the plant that bears the fruits, including all its bracts, branches and fruits, but excluding unmodified leaves.

infundibuliform, funnel-shaped, i.e. abruptly widening from a narrow cylindric part to a wider distal part.

inselberg, isolated, rounded hill rising abruptly out of the plains with little vegetation and large areas of bare rock.

in situ, in its (original) place.

internode, the part of the stem between two nodes.

in vitro, in a test-tube or other laboratory environment.

in vivo, in the living organism.

involucre, a series of bracts, usually close together and appressed, below or around a compact head of flowers (as in Compositae).

IUCN Red List, a comprehensive inventory of the conservation status of species using precise criteria to assess extinction risk; see http://www.iucnredlist.org/.

K

keel, (in Leguminosae), the two often partially united lowest/anterior petals that conceal the sexual parts of the flower.

keeled, bearing a ridge along the middle (like the keel of a boat).

kernel, (the usually edible) seed inside a hard covering of a nut or fruit stone.

kraal, (compost) organic material consisting of plant residues digested by animals in a kraal (an enclosure for livestock).

L

lacrymiform, tear-shaped, i.e. ovoid with a narrowing apex.

lamina, expanded part or blade of leaves or petals.

lanate, with long dense curly interwoven matted woolly hairs.

lanceolate, narrowly ovate and tapering to a point at the apex.

lateral, on or at the side or margin.

latex, milky juice, often sticky.

lax, loose, open, distinct from each other.

layering, a method of vegetative propagation in which a shoot is fastened down to the ground and forms roots while still attached to the parent plant (can also occur naturally); see also **air layering**.

leaf, chlorophyll-bearing lateral outgrowth from stem.

leaflet, one part of a compound leaf.

Least Concern (LC), IUCN Red List term for a plant that is in no particular danger of extinction; for precise definition see http://www.iucnredlist.org/technical-documents/categories-and-criteria.

legume, 1. the fruit pod of Leguminosae, derived from a single carpel, usually (though with many exceptions) opening along a suture into two halves, usually dry; 2. any species in the family Leguminosae.

lenticels, corky eruptions on bark that allow gas exchange.

lenticular, a 3-dimensional body that is circular in section and convex on both sides.

lepidote, clothed on the surface with small scales.

liana, woody climber, supported by other vegetation.

ligulate, 1. strap-shaped, narrow and with parallel sides; 2. with a ligule; 3. (in Compositae inflorescences) denoting the presence of florets with a ligule.

ligule, 1. a distal projection of the leaf sheath; 2. (in Compositae), a 5-toothed strap-shaped floret.

linear, narrow and much longer than wide, with parallel margins.

lobe, 1. a division to about halfway of any organ; 2. a part of the calyx or corolla that is distinct from the lower, united/fused part.

lobulate, with small lobes.

locule, the cavity of the carpel in which the ovule or ovules are borne.

loculicidal, when a ripe capsule splits into the cells, i.e. splits not at the lines of junction between the locules (i.e. septa) but along the midrib or dorsal suture.

Lower Risk-Least Concern (LR-lc), previous IUCN Red List term; for precise definition of current Red List terms see http://www.iucnredlist.org/technical-documents/categories-and-criteria.

Lower Risk-Near Threatened (LR-nt), previous IUCN Red List term; for precise definition of current Red List terms see http://www.iucnredlist.org/technical-documents/categories-and-criteria.

lunate, half-moon-shaped.

M

mantle, (of ectomycorrhizal fungi) the hyphal sheath that covers the root tip of the host plant.

marsh, permanently or periodically inundated area, rich in nutrients, ± neutral, dominated by grasses or rushes, with few woody species.

mature, (of a fruit) when fully grown and ripe, ready to distribute seeds.

mericarp, seed-containing parts of a fruit that do not form a single unit and that each derive from a carpel, these parts usually dehisce independently from each other when ripe.

mesocarp, the middle layer of a multi-layered fruit wall, often distinguished as such when fleshy or succulent.

mesophytic, vegetation adapted to normal conditions, avoiding both very wet and arid conditions.

microvesiculate, with minute vesicles (cavities).

midrib, the main vascular supply and support structure of a simple leaf-blade or leaflet, a continuation of the petiole, running the full length of the leaf.

miombo, deciduous woodland (in Africa) dominated by woody Leguminosae species.

monocarpic, flowering (and possibly fruiting) only once, then dying.

monochasium, inflorescence with a terminal flower and one bracteole subtending a lateral flower (adjective – monochasial).

monoecious, with all flowers bisexual, or with male and female flowers on the same plant.

monoliform, resembling a string of beads.

montane, pertaining to mountainous regions.

mopane woodland, (in Africa) dominated by the mopane tree, *Colophospermum mopane* (Benth.) J.Léonard.

morphological dormancy, (of seed) a type of dormancy caused by the embryo being underdeveloped or undifferentiated, germination taking place when the embryo has fully differentiated.

morphophysiological dormancy, (of seed) a type of dormancy caused by the embryo being underdeveloped or undifferentiated (i.e. morphological dormancy) and whose development needs the application of a pre-treatment to break the physiological dormancy.

mucilage, slime or jelly-like excretion, chemically composed of high molecular weight carbohydrate.

mucronate, ending abruptly in a short stiff point.

muthi market, traditional medicine market (term especially applied in South Africa).

mycelium, the vegetative part of a fungus, consisting of a network of fine white filaments (hyphae).

mycorrhiza (plural **mycorrhizae**)**,** symbiotic fungus in or on the roots.

mycorrhizal, with symbiotic fungi in or on the roots.

N

native, undoubtedly indigenous, species occurring naturally in a given area.

naturalised, non-native, introduced species that has become established and reproduces freely.

Near Threatened (NT), IUCN Red List term for a plant at risk of becoming vulnerable to extinction in the near future; for precise definition see http://www.iucnredlist.org/technical-documents/categories-and-criteria.

nectar, sweet fluid extruded by glands as an attractant to pollinators.

nitrogen-fixation, the process by which bacteria (less often other organisms) convert atmospheric nitrogen into organic compounds that can be taken up by plants.

node, the area of a stem where a leaf is attached or used to be attached.

Not Threatened (nt), previous IUCN conservation category for a plant which is neither rare nor threatened; for precise definitions of current Red List terms see http://www.iucnredlist.org/technical-documents/categories-and-criteria.

nut, a one-seeded indehiscent fruit with a hard dry pericarp (the shell) that is derived from a one loculed ovary.

nutlet, diminutive of nut.

O

obconical, conical with the narrow part near the base and the wide part near the apex.

oblanceolate, narrowly obovate with an attenuate base and an acute apex.

oblique, 1. (in leaves) when the two sides of the leaf are unequal near the base; 2. (in an ovary) when the ovary is at an angle to the symmetric plane.

oblong, longer than broad, with the margins parallel for most of their length.

obovate, egg-shaped (2-dimensional) with the broadest part near the apex.

obovoid, egg-shaped (3-dimensional), with the broadest part towards the apex.

obpyriform, (of a 3-dimensional shape) like an inverted pear, i.e. with the broadest part proximal.

obtuse, (of an apex or base) not pointed, blunt, ending in an angle of between 90–180°.

offset, a lateral shoot at the base of a plant that can be removed for propagation.

operculum (plural **opercula**), lid or cover.

opposed, placed opposite to.

opposite, (of leaves and branches) when two are borne on the same node but on diametrically opposed sides of the stem.

orbicular, 1. (2-dimensional) flat with a circular outline; 2. (3-dimensional) globose, in the shape of a sphere.

orthodox, seed storage behaviour denoting seeds which can be dried, without damage, to low moisture contents, usually much lower than those they would normally achieve in nature; over a wide range of storage environments their longevity increases with reductions in both moisture content and temperature, in a quantifiable and predictable way.

ovary, the (usually enlarged) part of the pistil that contains the ovules and eventually becomes the fruit.

ovate, egg-shaped (2-dimensional), about 1.5 x as long as broad, with the wider part below the middle.

ovoid, egg-shaped (3-dimensional), with the broad part below the middle or nearest the base.

ovule, the immature seed in the ovary before fertilisation.

P

pachycaul, pachycaulous, thick-stemmed and sparsely branched; trunk may be bottle-shaped.

palmate, (in lobed or compound leaves) when all lobes or leaflets originate from one central point (as fingers originate from the palm of the hand).

panicle, an inflorescence in which the main axis has several lateral branches, each of which is branched; (more specifically) an inflorescence in which both the main axis and any lateral branches are indeterminate (i.e. racemose or monopodial).

paniculate, with the inflorescence a panicle.

papillose, bearing many small soft nipple-like projections.

pappus, a series of bristles, hairs or scales round the base of the corolla and later around the apex of the fruit (as in Compositae).

papyraceous, papery; with the thickness or consistency of paper.

paripinnate, evenly pinnate, terminated by a pair of opposite leaflets.

parthenocarpy, with fruit developing without fertilisation of the ovule, with seedless fruit.

pedicel, the stalk of an individual flower in an inflorescence.

peduncle, 1. (of an inflorescence) the lower unbranched part or stalk, as distinct from the rachis; 2. a flower stalk bearing either a solitary flower, a cluster or the common stalk of several flowers.

pedunculate, (of inflorescences) stalked.

pendulous, hanging.

pentamerous, (of a flower) with its constituent parts in multiples of five.

perennial, living for several to many years.

perianth, collective term for the calyx and corolla.

pericarp, 1. the wall of the ripened ovary, divisible into epicarp, mesocarp and endocarp when a distinction between the three can be made; 2. fruit wall, sometimes includes the seed.

peridium, the protective layer that encloses a mass of spores in fungi.

petal, a single, usually free, unit of a completely divided corolla or second floral whorl.

petiole, leaf stalk.

photoperiod, the interval in a 24-hour period during which a plant (or a seed) is exposed to light.

photoperiodism, the alternation of light and dark periods that affects the physiological activities of many plants (e.g. onset of flowering).

physical dormancy, (of seed) a type of dormancy caused by an impermeable seed coat.

physiological dormancy, (of seed) a type of dormancy caused by inhibiting chemicals in the seed which prevent embryo growth.

pilose, hairy with short thin hairs.

pinna, pinnae, leaflet of a pinnate leaf, or first division of a pinnate leaf where this division is itself divided into leaflets.

pinnate, divided into a central axis and several lateral ribs or leaflets (like a feather).

pioneer, a species colonising new environments (e.g. after clear-cutting or fire) and starting a plant succession.

pistil, the female organ of a flower, consisting when complete of ovary, style and stigma.

placenta, the part of the ovary to which the ovules or seeds are attached.

plano-convex, flat on one side, convex on the other.

pleurogram, u-shaped or elliptic fracture line on the lateral faces of some Leguminosae seeds, which surrounds the areole.

pod, 1. a dry dehiscent fruit with a firm outer layer enclosing a hollow centre with one or more seeds; 2. a legume formed of a single carpel.

poikilohydrous, with its water content determined by the surrounding atmosphere, becoming dormant in the dry season after losing most of its water, rehydrating when water becomes available.

pollen, powder-like fertilising agent carried by the anthers.

pollination, the transfer of pollen from anther to stigma.

porrect, pointing upwards at a slight angle from the vertical.

pre-treatment, (of seed) method employed to break seed dormancy (e.g. scarification).

pricking out, a technique of transplanting seedlings from where they have germinated to new locations to provide them with space to develop; usually seedlings are separated when large enough to handle (e.g. when the first set of leaves have emerged after the cotyledons) and planted individually into new locations in a seed tray or pot.

prickle, a sharp outgrowth, detachable without tearing the organ.

prostrate, lying flat.

proximal, nearest to the point of attachment, basal.

pruinose, covered with a waxy, frost-like powder or bloom.

pseudo-, 1. seemingly (but not really); 2. a prefix denoting a resemblance to another state or organ.

puberulent, minutely pubescent, the hairs hardly visible to the naked eye.

puberulous, with a rather dense covering of very short soft hairs.

pubescent, with dense fine, short, soft hairs; downy.

pulp, juicy or fleshy tissue of a fruit.

pyrene, (of a fruit) the stone, the seed plus a hard layer of endocarp surrounding the seed.

pyriform, pear-shaped.

R

raceme, a monopodial inflorescence in which the flowers are borne on pedicels along a central axis, with the terminal flowers being the youngest and last to open; there are many types of racemes.

racemose, in the form of a raceme, resembling a raceme.

rachis, 1. (in compound leaves) that part of the main axis distal to the petiole that bears the leaflets; 2. (in inflorescences) that part of the main axis distal to the peduncle that bears the flowers.

radial spine, (in cacti and similar succulents) the spines on the edge of the areole or spine shield, often smaller or with a colour different to the central spine.

radiate, spreading from, or arranged round, a common centre.

radical, (of leaves) arising so close to the base of the stem as to appear to come from the root; as opposed to cauline leaves, which grow from the stem.

radicle, the first root arising from the germinating seed.

rainforest, forest with continuous canopy cover, and a climate with no or few dry periods.

Rare (R), previous IUCN conservation category; for precise definitions of current Red List terms see http://www.iucnredlist.org/technical-documents/categories-and-criteria.

ray floret, the florets of the margin of a head (capitulum) of Compositae when different from those of the centre (or disc) florets.

recalcitrant, seed storage behaviour denoting seeds which do not survive drying to any large degree, and are thus not amenable to long-term storage at sub-zero temperatures, although the critical moisture level for survival varies among species.

receptacle, 1. the expanded part at the end of the flower stalk on which the organs of a flower (i.e. sepals, petals, stamens and carpels) are inserted; 2. (in species with compound heads) also used for the expanded part of the head-stalk that bears the collected flowers (e.g. in Compositae).

recurved, bent or curved downward or backward.

Red List, see **IUCN Red List**.

reniform, kidney-shaped.

resin, hardened exudate from wounded stem or leaves that is soluble in alcohol but not in water.

resinous, with the scent or consistency of resin.

reticulate, net-veined.

rhizome, underground stem, distinguished from root by its nodes, buds or scale-like leaves.

rhizomorph, a root-like aggregation of hyphae in certain fungi.

rhomboid, 1. (of leaf shape) rhombic-like, nearly square with the petiole at one of the acute angles; 2. (of 3-dimensional shape) 4-angular, with the angles obtuse.

riparian, habitat along a river or stream, influenced by the watercourse.

ripe, mature, complete for its function.

riverine forest, forest along a river or stream, influenced by the watercourse.

root, the branching lower portion of the axis of a (higher) plant, usually growing down into the soil and functioning in anchorage and absorption of water and nutrients.

root-crown, 1. the place where the root changes into the stem at ground level; 2. sometimes the hairy or bracteate apical part of the perennial rootstock where the annual shoots are burned or grazed off.

rootstock, 1. an established plant on to which a cutting or bud (scion) from another plant is grafted as a means of vegetative propagation (most common usage in this book); 2. underground stems and/or roots, often perennating; 3. rhizome, stem on or below ground sending out rootlets and distally leaves.

rosette, a circle of tightly packed leaves, a basal rosette is at ground level, spreading from a stem with short internodes at that point.

ruderal, growing in waste (or previously disturbed) places.

S

samara, a dry indehiscent fruit with a wing (longer than the seed-bearing part) developed to one side.

savanna, open deciduous woodland over a continuous grass/herb layer; also open grassland or wooded grassland.

scabrous, rough to the touch, with small pointed protrusions.

scandent, climbing.

scarification, (of seed) weakening, opening, or otherwise altering the seed coat to encourage germination (by various means, e.g. mechanical, thermal, chemical).

schizocarp, fruit splitting into its carpellary constituents or one-seeded portions (i.e. mericarps).

scion, a cutting (usually a young shoot) or bud which is grafted on to a rootstock as a means of vegetative propagation.

scrambler, plant growing upwards supporting itself on other vegetation or on objects but not twining or attaching itself.

scrambling, growing upwards through other vegetation or objects but not twining.

scrub, bushland or stunted forest.

secondary, 1. vegetation type following disturbance or destruction of original (primary) vegetation; 2. not primary, subordinate.

seed, the structure produced from a fertilised ovule by which all seed plants reproduce, consisting of an embryo and usually a seed coat, with endosperm; reproductive part of a fruit.

seed coat, the outer coat of the seed, usually comprising two layers: testa (outer coat) and tegmen (inner coat).

seedling, juvenile plant recently arisen from the seed.

semi-, (prefix) half.

sepal, a single part of the outermost whorl of floral organs, the calyx; usually green, protecting the corolla in bud.

septum (plural **septa**)**,** partition of fruit or ovary.

sericeous, silky, with closely appressed soft straight hairs and with a shiny silky sheen.

serrate, toothed like a saw, with regular acute and angled teeth pointing towards the apex.

sessile, without a stalk.

shrub, self-supporting woody plant branching at or near the ground or with several stems from the base.

shrubland, open or closed stand of shrubs <2 m tall.

silviculture, the growing and cultivation of trees; the practice of controlling the establishment, growth, composition and quality of forests to meet diverse needs and values.

simple, 1. (of leaves) not divided into leaflets; 2. (of inflorescences) with only one order of branching; 3. (of fruits) resulting from the ripening of a single ovary.

solitary, single, not in clusters.

spatheole, (in Poaceae) 1. the bladeless sheath subtending the inflorescence; 2. the modified leaf sheath encasing part of the inflorescence.

spatulate, shaped like a small spatula: oblong, with an extended basal part.

species, Linnaean unit of plant classification; group of populations of similar morphology and constant distinctive characters, thought to be capable of interbreeding and producing offspring.

spheroidal, shaped like a sphere.

spike, a racemose inflorescence with the flowers alternate and sessile along a common unbranched axis, flowers single or in short clusters.

spikelet, (in Poaceae) structure of two sterile bracts (the glumes) with a small axis and a number of florets.

spine, a sharp-pointed, hardened structure derived from a leaf, stipule, root or branch, but always originating from the vascular or woody part.

spinescent, ± spiny, ending in a sharp point.

spinulose, with small spines.

spore, a reproductive cell capable of developing into a new individual without fusion with another reproductive cell.

sporocarp, spore-producing organ.

squamula (plural **squamulae**)**,** small scale(s).

stamen, the male organ of a flower, consisting of a stalk (filament) bearing the connective and container(s) (anthers) that bear the pollen.

standard, the large upper/posterior petal (outside in the bud) of some Leguminosae corolla.

stem, the main axis of a plant, bearing roots, leaves and/or flowers.

sterile, (used of sexual parts, such as anthers) barren, not functional.

stigma, the pollen receptor on the gynoecium.

stipulate, with stipules.

stipule, leaf-like, spine-like or scale-like appendages of the leaf, usually in pairs at the base of the petiole.

striate, with parallel longitudinal grooves.

strigose, with sharp stiff hairs lying ± parallel to and close to the surface.

style, the part of the gynoecium between the ovary and the stigma, often slender and sometimes lacking when the stigma sits on the ovary.

sub-, (prefix) 1. nearly, almost; 2. below, under.

submontane, of or characteristic of the foothills of a mountain or mountain range.

subshrub, small shrub with partially herbaceous stems.

subspecies, (subsp.) subdivision of species, each subspecies being geographically or ecologically isolated from each other and with fewer distinguishing characters than demarcate a species; often used merely in a hierarchical sense of being between a species and a variety.

subulate, awl-shaped, like a stout needle tapering to a fine point.

succulent, 1. (adjective) juicy, pulpy; 2. (noun) a plant with thick, fleshy and swollen stems and/or leaves, adapted to dry environments (e.g. *Aloe*, Cactaceae).

sucker, a shoot arising below ground from the roots some distance from the main stem.

supra-, (prefix) above, over.

swamp, wetland with the water level permanently at or above the soil level and with a cover of trees, shrubs or reeds/sedges.

symbiosis, living together of dissimilar organisms, either to mutual advantage or without advantage.

sympodial, of a sympodium, without a single main stem.

syncarpous, (of a flower) with united carpels.

synonym, (in nomenclature) a surplus scientific name, belonging to a taxon which already has a valid name; where two or more names are applied to the same taxon they are called synonyms but only one of these can be correct — usually this is the oldest validly published name (principle of priority) but the correct name may be a conserved or non-rejected name.

T

taproot, the primary root, going straight down.

taxon, a named group of any rank (e.g. species, subspecies, variety).

taxonomy, classification, ordering into groups according to relationships.

teeth, small sharp protuberances.

tegument, a general term for the seed coat (without differentiating between the layers).

tendril, slender coiling structure derived from branch, leaf or inflorescence, used in climbing.

tepal, a division of the perianth, i.e. a sepal or petal, used especially when it is unclear which is which.

terete, circular in cross-section (usually of a cylindric structure lacking grooves or ridges).

terminal, 1. at apex of part under discussion; 2. (of inflorescences) ending the axis, as opposed to axillary.

ternate, arranged in a whorl or cluster of three.

testa, the outer coat of the seed.

thicket, a dense stand of trees, climbers or shrubs with a closed canopy.

thorn, short pointed woody structure derived from a reduced branch.

tomentose, densely covered in short soft hairs, somewhat matted.

translucent, letting some light through, not quite transparent.

tree, perennial woody plant with secondary thickening, with a clear main trunk; the distinction between tree and shrub is fluid, but generally accepted to be dependent on the single trunk, and on height, a tree being at least 2–3 m tall.

tri-, (prefix) with three...

trichome, hair, bristle, prickle or scale.

truncate, ending abruptly in a ± straight line, as if cut off.

truncheon, a thick cutting (sometimes a branch) used for vegetative propagation.

trunk, the main axis of a tree from the roots to where the crown branches.

tube, a hollow cylinder.

tuber, 1. a thickened branch of an underground stem, serving as storage organ, distinguished by bearing leaves/leaf scars and axillary buds; 2. a swollen root or branch of a root acting as a reserve store of nourishment or water.

tuberculate, covered with knobbly or wart-like protuberances.

tubular, cylindric and hollow.

turbinate, top-shaped, obconical and narrowed towards the point.

U

umbel, a (racemose or indefinite) inflorescence with branches arising from ± the same point on a common peduncle.

umbellate, with umbels.

umbelliform, in the shape of an umbel.

understorey, sub-canopy layer(s) of vegetation in forest or woodland; usually denoting shrub and small tree layer.

V

valvate, meeting exactly at the margins without overlapping.

variegated, (of leaves) irregularly coloured with two or more colours.

variety, (var.) infraspecific taxon below the rank of subspecies with one or several distinguishing characters, not geographically disjunct from other conspecific taxa.

vascular tissue, tissue consisting mostly of strands or vessels (water-conducting cells), as opposed to cellular tissue (i.e. a grouping of one or more types of cells that together carry out a specific function; a level of complexity between cells and organs).

vegetative propagation, asexual reproduction, reproduction not through seed and fruit but by bulbils, runners, plantlets, stolons etc.

vein, strand of vascular tissue in a flat organ, often visible on the surface.

verrucose, warty, with little bumps.

verticillate, (of leaves) in a whorl, i.e. several arising at the same node, arranged regularly around the stem.

viability, (of seed) the germination capacity.

viable, (of seed) capable of germination.

villose, with long soft weak hairs.

Vulnerable (VU), IUCN Red List term for a plant threatened in its survival; for precise definition see http://www.iucnredlist.org/technical-documents/categories-and-criteria.

W

whorl, a set of similar organs arranged in a circle around a central axis.

whorled, arranged in a circle around a central axis (e.g. leaves around a stem).

wild, spontaneous, not cultivated or introduced.

wildings, seedlings that are growing wild or escaped from cultivation (which can be transplanted and used for propagation).

wing, 1. lateral petal of some Leguminosae flowers; 2. a flattened extension to any organ, e.g. leaf rachis or fruit margin.

winged, (of a 3-dimensional body) with flattened to blade-like ridges on either side.

wood, the hard fibrous material that forms the main substance of the trunk or branches of a tree or shrub.

wooded grassland, a ± continuous grass/herb layer with woody plants covering 10–40%.

woodland, a stand of trees with an open canopy (40–80% cover) and generally with an understorey of narrow-leaved grasses, herbs and shrubs.

woody, made of wood or wood-like tissue.

X

xerophilous, xerophytic, adapted to growing and reproducing in areas with low water availability.

Z

zygomorphic, with bilateral symmetry, i.e. either side of an (imaginary) central line being a mirror image of the other.

Index of scientific names

Hyphaene petersiana palm tree.
(Photo: T. Ulian, RBG Kew.)

Index of vernacular/common names

Fruit cracking and seed extraction of *Melia volkensii* in Kenya.
(Photo: T. Ulian, RBG Kew.)

Planting activities in Mali.
(Photo: T. Ulian, RBG Kew.)

U.N.T.M
U.N.T.M

Annex – Summary of overall habit and main uses of the species

Species	Habit	Animal food	Environ-mental use	Food	Fuels	Materials	Medicine	Poisons	Social use
BOTSWANA									
Adansonia digitata L.	Tree	•		•		•	•		•
Bauhinia macrantha Oliv.	Shrub - small tree	•		•			•		
Cassia abbreviata Oliv.	Tree		•		•	•	•		
Citrullus lanatus (Thunb.) Matsum. & Nakai	Annual herb	•		•					
Colophospermum mopane (J.Kirk ex Benth.) J.Léonard	Tree	•			•	•	•		
Cucumis africanus L.f.	Annual herb			•			•		
Grewia flava DC.	Shrub	•		•		•	•		
Guibourtia coleosperma (Benth.) J.Léonard	Tree		•	•		•	•		
Harpagophytum procumbens (Burch.) DC. ex Meisn.	Perennial herb						•		
Hyphaene petersiana Klotzsch ex Mart.	Palm tree	•		•		•			
Kalaharituber pfeilii (Henn.) Trappe & Kagan-Zur	Hypogeous fungus			•			•		
Lippia javanica (Burm.f.) Spreng.	Shrub						•		
Myrothamnus flabellifolia Welw.	Shrub						•		
Schinziophyton rautanenii (Schinz) Radcl.-Sm.	Tree			•		•			•
Sclerocarya birrea subsp. *caffra* (Sond.) Kokwaro	Tree	•		•		•	•		•
Stomatostemma monteiroae (Oliv.) N.E.Br.	Perennial climber			•			•		
Strychnos cocculoides Baker	Shrub - small tree			•		•	•	•	
Strychnos pungens Soler.	Tree - shrub			•		•	•		
Tylosema esculentum (Burch.) A.Schreib.	Perennial herb			•			•		
Ximenia caffra Sond.	Shrub - small tree	•		•		•	•		
KENYA									
Albizia coriaria Welw. ex Oliv.	Tree	•	•		•	•	•		
Annona senegalensis Pers.	Shrub - small tree	•	•	•		•	•		
Balanites aegyptiaca (L.) Delile	Tree	•	•	•	•	•	•	•	
Berchemia discolor (Klotzsch) Hemsl.	Shrub - small tree	•	•	•		•	•		
Blighia unijugata Baker	Tree	•	•		•	•	•		
Carissa spinarum L.	Shrub - climber	•	•	•		•	•		
Centrapalus pauciflorus (Willd.) H.Rob.	Herb	•				•	•		
Combretum collinum Fresen.	Tree - shrub	•			•	•	•		

Species	Habit	Animal food	Environ-mental use	Food	Fuels	Materials	Medicine	Poisons	Social use
Croton macrostachyus Hochst. ex Delile	Tree - shrub	•	•		•		•		
Croton megalocarpus Hutch.	Tree	•	•		•	•	•		
Dovyalis macrocalyx (Oliv.) Warb.	Shrub - small tree		•	•		•	•		
Markhamia lutea (Benth.) K.Schum.	Tree	•	•		•	•	•		
Melia volkensii Gürke	Tree	•	•		•	•		•	
Ocimum gratissimum L.	Woody herb - shrub			•		•	•	•	
Olea europaea subsp. *cuspidata* (Wall. & G.Don) Cif.	Tree - shrub	•	•	•	•	•	•		•
Piliostigma thonningii (Schumach.) Milne-Redh.	Tree	•	•	•	•	•	•		
Searsia pyroides (Burch.) Moffett	Shrub - small tree	•	•	•	•	•	•		•
Senegalia brevispica (Harms) Seigler & Ebinger	Shrub - small tree	•	•		•		•		
Senegalia polyacantha subsp. *campylacantha* (Hochst. ex A.Rich.) Kyal. & Boatwr.	Tree			•		•			•
Senegalia senegal (L.) Britton	Shrub - tree	•	•	•	•	•	•		
Senna didymobotrya (Fresen.) H.S.Irwin & Barneby	Shrub - tree		•		•		•		
Sesbania sesban (L.) Merr.	Tree - shrub	•	•		•	•	•		
Stereospermum kunthianum Cham.	Tree - shrub	•	•		•	•	•		
Tetradenia riparia (Hochst.) Codd	Shrub - small tree		•	•			•		
Vachellia tortilis (Forssk.) Galasso & Banfi	Tree	•	•	•	•	•	•		
Vangueria madagascariensis J.F.Gmel.	Shrub - small tree	•		•		•	•		
Vernonia amygdalina Delile	Shrub - small tree		•	•	•		•		
Vitex doniana Sweet	Tree	•	•	•	•	•	•		
Vitex keniensis Turrill	Tree		•	•	•	•			
Zanthoxylum gilletii (De Wild.) P.G.Waterman	Tree		•			•	•		
MALI									
Alchornea cordifolia (Schumach. & Thonn.) Müll.Arg.	Tree - shrub	•	•			•	•		
Anthocleista djalonensis A.Chev.	Tree						•		
Bobgunnia madagascariensis (Desv.) J.H.Kirkbr. & Wiersema	Tree - shrub				•		•	•	
Carapa procera DC.	Tree		•			•	•	•	
Coix lacryma-jobi L.	Annual herb	•				•			
Cola cordifolia (Cav.) R.Br.	Tree		•		•	•	•		•
Combretum micranthum G.Don	Shrub - tree - liana	•			•	•	•		
Entada africana Guill. & Perr.	Tree - shrub	•				•	•	•	

Species	Habit	Animal food	Environ-mental use	Food	Fuels	Materials	Medicine	Poisons	Social use
Erythrina senegalensis DC.	Tree		•		•		•	•	
Evolvulus alsinoides (L.) L.	Herb						•		
Flemingia faginea (Guill. & Perr.) Baker	Shrub						•		
Gymnosporia senegalensis (Lam.) Loes.	Shrub - small tree				•	•	•		
Lawsonia inermis L.	Shrub - small tree		•		•	•	•		•
Lippia multiflora Moldenke	Shrub - small tree	•					•	•	
Nauclea latifolia Sm.	Shrub - small tree	•	•		•	•	•		
Securidaca longepedunculata Fresen.	Shrub - small tree	•			•	•	•	•	
Stylosanthes erecta P.Beauv.	Perennial herb - subshrub	•					•	•	•
Vachellia nilotica (L.) P.J.H.Hurter & Mabb.	Tree	•	•		•	•	•		
Vitex madiensis Oliv.	Shrub - small tree			•		•	•		
Zanthoxylum zanthoxyloides (Lam.) Zepern. & Timler	Tree - shrub				•	•	•		•
SOUTH AFRICA									
Adenium swazicum Stapf	Shrub		•				•	•	
Aloe arborescens Mill.	Shrub		•				•		
Argyrolobium tomentosum (Andrews) Druce	Shrub		•				•		•
Artemisia afra Jacq. ex Willd.	Shrub						•	•	
Berchemia zeyheri (Sond.) Grubov	Tree - shrub			•		•	•		
Coddia rudis (E.Mey. ex Harv.) Verdc.	Shrub - small tree		•	•			•		
Cordyla africana Lour.	Tree - shrub		•	•		•	•		
Dioscorea strydomiana Wilkin	Shrub-like caudex geophyte						•		
Drimia delagoensis (Baker) Jessop	Perennial herb					•	•		•
Eucomis autumnalis (Mill.) Chitt.	Perennial herb		•				•		
Gerbera ambigua (Cass.) Sch. Bip.	Perennial herb		•				•		•
Gomphocarpus physocarpus E.Mey.	Annual or perennial herb - shrub		•				•		•
Gunnera perpensa L.	Perennial herb						•		
Hypoxis hemerocallidea Fisch., C.A.Mey. & Avé-Lall.	Perennial herb		•				•		
Kigelia africana (Lam.) Benth.	Tree			•		•	•		•
Pachypodium saundersii N.E.Br.	Shrub		•				•		•
Prunus africana (Hook.f.) Kalkman	Tree		•			•	•		

Species	Habit	Animal food	Environ-mental use	Food	Fuels	Materials	Medicine	Poisons	Social use
Scadoxus puniceus (L.) Friis & Nordal	Perennial herb		•				•		
Trichilia emetica Vahl	Tree		•			•	•	•	
Warburgia salutaris (G.Bertol.) Chiov.	Tree - shrub		•				•		
MEXICO									
Agave kerchovei Lem.	Shrub	•	•	•		•			
Amphipterygium adstringens (Schltdl.) Standl.	Tree				•	•	•		
Bursera morelensis Ramírez	Tree	•			•	•	•		
Castela tortuosa Liebm.	Shrub						•		
Ceiba aesculifolia subsp. *parvifolia* (Rose) P.E.Gibbs & Semir	Tree			•		•	•		
Cyrtocarpa procera Kunth	Tree	•		•	•	•	•		
Escontria chiotilla (F.A.C.Weber ex K.Schum.) Rose	Cactus	•	•	•	•				
Gymnolaena oaxacana (Greenm.) Rydb.	Shrub	•					•		
Gymnosperma glutinosum (Spreng.) Less.	Perennial herb - subshrub						•		
Hylocereus undatus (Haw.) Britton & Rose	Epiphytic cactus		•	•			•		
Jatropha neopauciflora Pax	Shrub		•		•	•	•		
Lantana camara L.	Shrub		•				•		
Lippia origanoides Kunth	Shrub			•			•		
Myrtillocactus geometrizans (Mart. ex Pfeiff.) Console	Cactus		•	•					
Neobuxbaumia tetetzo (F.A.C.Weber ex K.Schum.) Backeb.	Cactus			•		•			
Prosopis laevigata (Humb. & Bonpl. ex Willd.) M.C.Johnst.	Tree - shrub	•	•	•	•	•	•		
Stenocereus pruinosus (Otto ex Pfeiff.) Buxb.	Cactus	•	•	•	•	•			
Stenocereus stellatus (Pfeiff.) Riccob.	Cactus		•	•	•				
Varronia bullata subsp. *humilis* (Jacq.) Feuillet	Shrub						•		
Varronia curassavica Jacq.	Shrub	•		•			•		

BACK COVER IMAGES

Cola cordifolia (Cav.) R.Br., Mali. (Photo: S. Sanogo, IER.)

Dioscorea strydomiana Wilkin, South Africa. (Photo: P. Gómez-Barreiro, RBG Kew.)

Tylosema esculentum (Burch.) A.Schreib., Botswana. (Photo: E. Mattana, RBG Kew.)

Melia volkensii Gürke, Kenya. (Photo: T. Ulian, RBG Kew.)

Neobuxbaumia tetetzo (F.A.C.Weber ex K.Schum.) Backeb., Mexico. (Photo: D. Franco-Estrada, FESI-UNAM.)

RBG Kew's Millennium Seed Bank, Wakehurst, Ardingly, West Sussex, UK. (Photo: W. Stuppy, RBG Kew.)